Advances in Precision Nutrition, Personalization and Healthy Aging

Alexander G. Haslberger
Editor

Advances in Precision Nutrition, Personalization and Healthy Aging

Springer

Editor
Alexander G. Haslberger
Department of Nutritional Sciences
University of Vienna
Vienna, Austria

ISBN 978-3-031-10152-6 ISBN 978-3-031-10153-3 (eBook)
https://doi.org/10.1007/978-3-031-10153-3

This Springer imprint is published by the registered company Springer Nature Switzerland AG
The registered company address is: Gewerbestrasse 11, 6330 Cham, Switzerland

Foreword

The fast advances in nutritional research and the translation of the progress into practical approaches for health promotion and disease prevention in the last decades were amazing. The way from the first dietary guidelines to the analysis of molecular pathways of nutritional regulation, such as the role of mutations, epigenetics, microbiota and metabolites, rapidly resulted in the possibilities for a precision nutrition.

Nutrition is a global science that originally was envisaged to biologically analyse and integrate the processes related to food transformation into energy and nutritious components for cell functions and homeostasis. It is now evident that such physical, chemical and metabolic reactions participate in human development to ensure life expectancy and well-being, with a growing and inseparable relevance for personal, population and planetary health. Nowadays, nutritional challenges and nutritionist's interests are being focused on health and wellness involving physical, emotional, intellectual, cognitive, spiritual, environmental and occupational facets. Moreover, according to the newer emerging health scenarios, food intake should be assessed in relationship with social, satisfaction, satiety, security, safety and sustainable dimensions. In this context, two apparently alternative approaches, one derived from a global public health perspective and another derived from a precision-personalised nutrition paradigm, should be harmonised and deliberated since they are complementary to each other, and as such, personalised, participative, preventive and predictive strategies are all needed in order to maintain a healthy status as well as to prevent and manage diseases (Martínez-González et al. 2021).

The present book covers many important recent aspects in this area, starting from an analysis of trends to the scientific background in genetics, epigenetics, microbiota, metabolomics, neurology and healthy ageing to methods of the analysis of big data derived data, clinical praxis, emerging new products between nutrition and nutraceuticals plus consumers aspects with omics technologies and bigdata/bioinformatic tools supported on machine learning approaches.

I know Alexander Haslberger now for many years from research cooperation and his scientific work. He is driven by a strong motivation for basic science but also for an ethical, social and environmentally responsible translation of scientific progress into praxis. His work with big international organisations such as WHO trained his view for international and global requirements.

I also know many other authors for their brilliant scientific work in this area.

I think the present book will give stimulating new views on developments in this area for readers from the relevant sciences but also for all citizens and consumers interested in the developments of personalised health prevention and nutrition with precision perspectives to compute all available phenotype- and genotype-related factors and determinants at all age stages.

Madrid, Spain
May 2022

Prof. J. Alfredo Martínez

Reference

Martínez-González, M. A., Kim, H. S., Prakash, V., Ramos-Lopez, O., Zotor, F., & Martinez, J. A. (2021). Personalised, population and planetary nutrition for precision health. BMJ Nutrition, Prevention & Health, 4(1), bmjnph-2021-000235. https://doi.org/10.1136/BMJNPH-2021-000235

Prof. J. Alfredo Martínez holds Ph.D. in Nutrition as well as Pharm.D. and M.D. degrees. He currently is Director of the Nutrition Precision Program at IMDEA Madrid.

He is or has been Prof. at UNAV, USC and UPV and enjoyed training or invited stays at MIT, Nottingham, Berkeley, Harvard, Oxford and King's College London. He is currently President of IUNS.

Contents

Trends in Personalised Precision Nutrition, Objectives

1

Petra Rust and Alexander G. Haslberger

Contents

ABSTRACT

The transition from undernutrition to overnutrition in many parts of the world as well as the integration of the fast developments of molecular biology has strongly impacted nutritional sciences. Especially the increasing understanding of interactions between genetics, epigenetics, microbiota, the immune system, and nervous system and consequences from lifestyle and the exposome has paved the way for the need to understand individually highly different metabolic responses to foods and nutritional needs.

Often nutrition is seen as a rather complex field between natural sciences and translational sciences, focusing on highly complex molecular pathways as well as challenging epidemiological surveys, considering difficult interactions between nutrition and various lifestyle factors.

P. Rust · A. G. Haslberger (✉)
Department for Nutrition, University Vienna, Vienna, Austria
e-mail: alexander.haslberger@univie.at

P. Rust
e-mail: Petra.rust@univie.ac.at

A. G. Haslberger (ed.), *Advances in Precision Nutrition, Personalization and Healthy Aging*, https://doi.org/10.1007/978-3-031-10153-3_1

"Although food and nutrition have been studied for centuries, modern nutritional science is surprisingly young. The first vitamin was isolated and chemically defined in 1926. Research on the effects on nutrition in non-communicable chronic diseases, such as cardiovascular disease, diabetes mellitus, obesity, and cancers, is even more recent, accelerating over the past two or three decades and especially after 2000" (Mozaffarian et al. 2018).

However, to understand some tensions in the development of modern nutrition research one needs to consider early nutritional concepts: Anaxagoras (500–428 BC), a Greek philosopher, mentioned that "foods we eat contain components that were needed for the growth of the body". He believed that "everything is in everything, at all times", and physical characteristics (hair, nails, flesh, etc.) were generated from foods that contained those same substances. Plato's (428/427–348/347 BC) idea of a healthy diet consisted of a balance of cereals, fruits, vegetables, dairy products, with a strong emphasis on moderate consumption of meat and wine. His belief was that excess food from one source would lead to future ailments (Medieval European Nutrition, Health Ahoy, n.d.; Mozaffarian et al. 2018; Pléh 2012).

But already vacillating between concepts of nature and nurture Hippocrates believed that "those who are constitutionally very overweight are more susceptible to die earlier than those who are thin" and he recognised that "when people ate mainly a fresh, plant-based diet, they developed fewer diseases". Therefore, he recommended that: "Let food be your medicine, and let medicine be your food" (Skiadas and Lascaratos 2001).

In the Middle Ages, e.g. Hildegard von Bingen chose herbal remedies for healing diseases (principle of subtility) so that "there would be no difference between remedies and foods". "Everything that is good for the body is a remedy". The organism as a whole is nourished and strengthened. St. Hildegard put it very simply: "Your food shall be your remedy". https://www.st-hildegard.com/en/, *Liber subtilitatum diversarum naturarum creaturarum, H. v. Bingen, 11 Jhd. She has* spelt as the best of all grains (Fig. 1.1).

One might suggest that already in the antique concepts of nutrition differentiated between foods for growth of the body and foods for health and disease prevention. These tensional views seem to remain until now.

By the mid-twentieth century all major vitamins had been isolated and synthesised. But war- and poverty-related severe energy and nutrient deficiency with millions of children dying drew the attention to dietary reference intakes to prevent malnutrition and diseases. The first recommended dietary allowances (RDAs) were a result of those concerns, when the League of Nations, British Medical Association, and the US government separately commissioned scientists to generate new minimum dietary requirements to be prepared. In 1941, these first RDAs were announced at the National Nutrition Conference on Defence, providing new guidelines for total calorie intake and needs of selected nutrients (Harper 2003; National Nutrition Conference for Defense 1941). Lately, WHO has now drawn

Fig. 1.1 Hildegard von Bingen, Physica. Liber subtilitatum diversarum naturarum creaturarum "Engane Lebenssmiddl soin engane Heilmiddl sei" 1310 https://de.m.wikipedia.org/wiki/Datei:Ildegarda_di_bingen,_physica,_renania,_1310_ca._%28ashburnham_1323%29.jpg

together the dietary evidence linked to both chronic diseases of adulthood and malnutrition (New concepts of a balanced diet, n.d., https://www.who.int/news-room/fact-sheets/detail/healthy-diet).

In Europe the European Food Safety Authority (EFSA) published the dietary reference values (DRVs) first in 1993 and updates these since then (https://www.efsa.europa.eu/en/topics/topic/dietary-reference-values). DRVs include the average requirement (AR), the population reference intake (PRI), the adequate intake (AI), and the reference intake range for macronutrients (RI). These reference values are not recommendations for individuals but support professionals on amounts of nutrients needed for health prevention and for establishing dietary guidelines. In Germany, Austria, and Switzerland, the D-A-CH reference values for nutrient intake are the basis for the practical implementation of a nutritious diet. They also specify quantities for the daily intake of energy and nutrients. Individualised reference value tables allow people in certain life situations, such as pregnancy and breastfeeding, to search for specific reference values and have them displayed (https://www.dge.de/wissenschaft/referenzwerte/tool/).

To be of help to consumers, reference values need to be translated into food-based dietary guidelines which have to consider cultural differences, dietary pattern, and evidence of the relationship between diet and health. Currently, more than 100 countries worldwide have developed food-based dietary guidelines (https://www.fao.org/nutrition/education/food-dietary-guidelines/en/).

First approaches of established general nutritional guidelines included personalised concepts. The USA had its first new food pyramid in 1992. Named "MyPyramid", it was designed in accordance with the US Dietary Guidelines for

Americans and jointly published in April 2005 by the US Department of Health and Human Services (HHS) and the US Department of Agriculture (USDA). These guidelines are "the cornerstone of federal nutrition policy and education" and are based on "what experts have determined to be the best scientific knowledge about diet, physical activity and other issues related to what [Americans] should eat and how much physical activity [they] need". Following an extensive public campaign USDA announced that, "MyPyramid is about the ability of Americans to personalize their approach when choosing a healthier lifestyle that balances nutrition and exercise". Central to the campaign was a web-based tool, now modified to MyPlate available at https://www.choosemyplate.gov/ when a person enters her/his age, sex, body weight and height, and activity level into the online form a "personalised" MyPlate Plan is created. This plan lists the recommended number of daily servings for five food categories—grains, vegetables, fruits, dairy products, and protein sources like seafood, meat, eggs, and pulses—along with recommendations for foods that should be consumed in limited amounts (added sugar, saturated fat, and sodium). It further delineated information on food groups as well as shopping assistance. Elderly get support like "What's On Your Plate? Smart Food Choices for Healthy Aging" (https://www.nia.nih.gov/health/healthy-eating).

With 11 million deaths, 255 million disability-adjusted life years (DALYs) attributable to poor dietary pattern, (GBD 2019; http://ghdx.healthdata.org/gbd-2019), it is important to have achievable dietary recommendations.

Currently, nutritional recommendations and traditional nutritional intervention approaches are being replaced by personalised strategies. However, there is a need for evidence to support efficiency and additional benefits of precision nutrition beyond traditional dietary interventions. Discussions notice that dietary guideline reflects the daily intake required to meet the nutritional needs of 97.5% of the healthy population; i.e. they are not geared to specific individual needs or to sick people (2021). The European Food Safety Authority (EFSA) recognises that "physiological requirements vary between individuals depending on genetic and epigenetic differences, age, sex [and] physiological status"; however, it is assumed that nutrient requirements follow a normal distribution (EFSA 2010).

1.1 The Rise of Molecular Nutrition

Understanding of nutritional requirements was heavily influenced by developments in molecular biology especially the Human Genome Project (1990–2003) showing the role of the multiple genes involved in metabolisms. But even more, projects like the 1000 Genome Project, launched 2008, provided a detailed catalogue of human genetic variation and boosted the search for candidate genes for personal metabolic specificities or diseases (Project Consortium et al. 2010).

Consequently, multiple genome wide association studies (GWAS) aimed to identify associations of genotypes with phenotypes by testing for differences in the allele frequency of genetic variants between individuals who are ancestrally similar (Uffelmann et al. 2021) but differ phenotypically. Data derived from these GWAS built the basis for genetic testing concepts for disease susceptibilities as well as for individualised nutritional advice. Unfortunately, many commercial companies adopted the technically easy and cheap methods for analysis of single nucleotide polymorphisms (SNPs) to provide direct to consumer analysis and nutritional counselling. In Figure 1.2 this is still under severe critiques because of methodological shortcomings, small meaningfulness because of low generally penetrance of SNPs, missing information about epigenetics, microbiota or metabolomics, and insufficient data interpretation and direct consumer contact. (Prasad et al. 2016; Hodge and Greenberg 2017).

Lately, the upcoming of findings from epigenetics demanded the need to understand gene environment interactions, especially the role of the exposome, for understanding our metabolisms. The epigenomic profile for a certain phenotype is often a result of the complex interplay between multiple genetic and environmental factors. This complex interaction poses an enormous challenge to visualise and interpret data. Furthermore, due to the dynamic nature of the epigenome, it is critical to determine causal relationships from the many correlated associations (Cazaly et al. 2019) (Fig. 1.3).

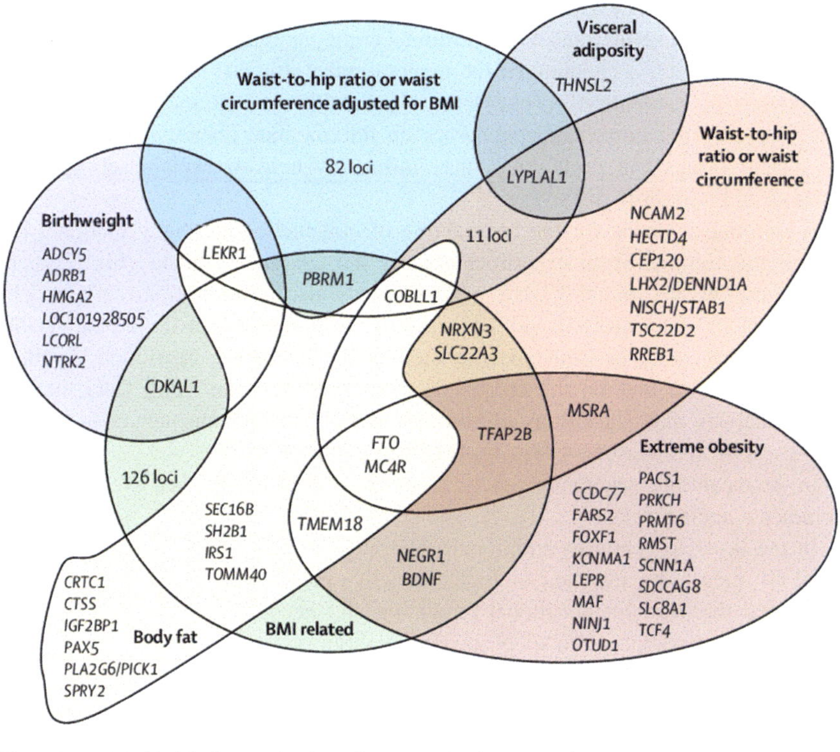

Fig. 1.2 Major GWAS discoveries for adiposity traits. Findings of genome wide association studies (GWAS) (Goodarzi 2018)

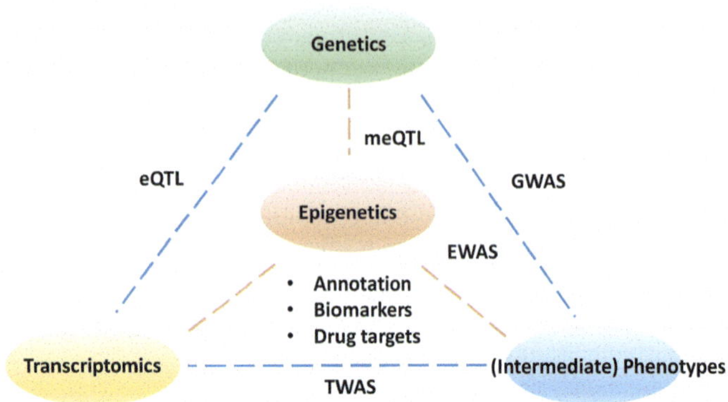

Fig. 1.3 Pillars to understanding the functional impact of epigenetics data. The epigenetic links need to be made with sequence variants in genetics as well as changes in transcriptomics. Wiederholung zum Text. GWAS—genome wide association studies; EWAS—epigenome wide association studies; meQTL—methylation quantitative trait loci; eQTL—expression quantitative trait loci; TWAS—transcriptome wide association study (Cazaly et al. 2019)

Especially the studies of Jirtle et al. showing the possibility to resolve toxin induced epigenetic mutations with nutritional component even trans-generationally (Dolinoy et al. 2007) encouraged the exploration of epigenetically active dietary compounds as prevention or intervention in health care (Remely et al. 2015). Understanding the impact of epigenetics on intermediate phenotypes, for example metabolomics and proteomics, may ultimately help to explain aetiology of diseases and help drug discovery.

Additional evidence for the central role of epigenetics in ageing modified by various diets comes from experiments using the epigenetic clocks (Fig. 1.4). It is well established that the DNA methylation landscape of normal cells undergoes a gradual modification with age, termed as "epigenetic drift" indicating the biological age. Effects from specific diets, such as caloric restriction, fasting, or Mediterranean diet on the epigenetic clock, can be seen using CpG methylation analysis and algorithms of the epigenetic clock (Fransquet et al. 2019; Orozco-Solis and Sassone-Corsi 2014; Quach et al. 2017).

In the meantime, several epigenetic clocks have been established even claiming to indicate ageing risks.

In the last two centuries additionally the huge impact of highly different personal GI microbiota and their biologically active metabolites on our metabolisms broadened the view on nutritional regulation and needs (Fig. 1.5). Tracing personal different responses to diets to the metabolic activities of microbiota even established concepts for nutritional advice (Zeevi 2015).

Around 2010 the first "1,000 days of life" theory developed as was boosted by the findings of a prenatal establishment of GI microbiota and their interaction with the prenatal immune system and epigenetics. This time spanning roughly between conception and one's second birthday has been established to be a unique period

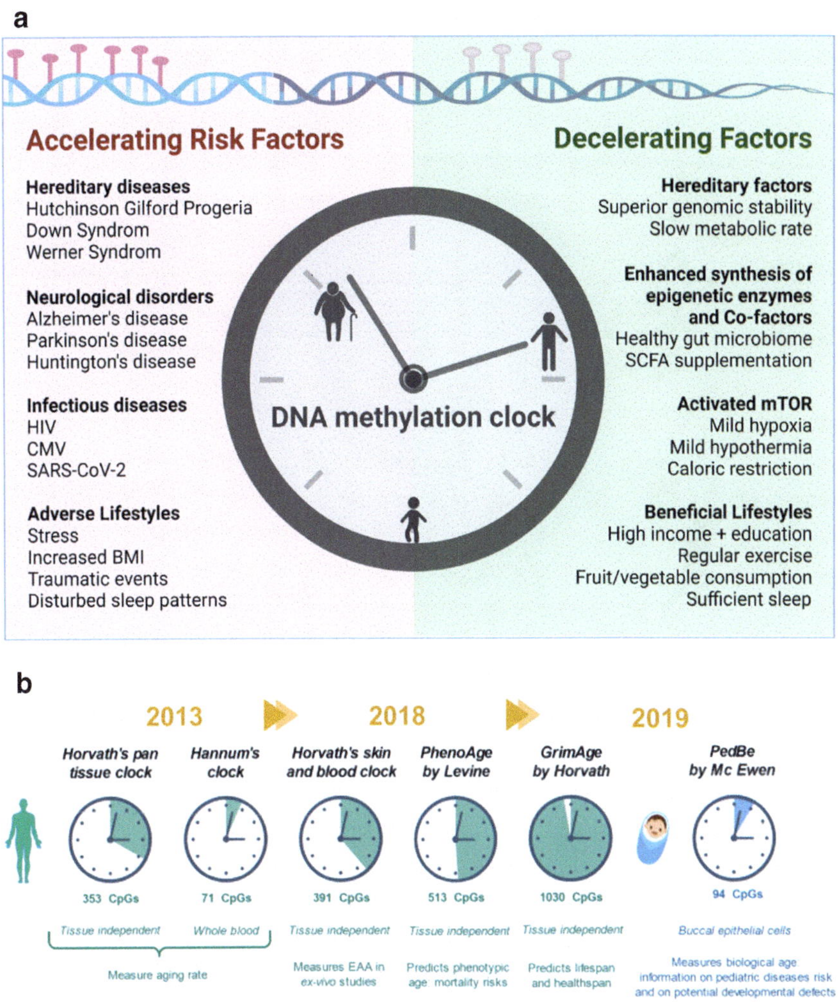

Fig. 1.4 **a** Factors that either negatively or positively affect epigenetic ageing. Epigenetic ageing as determined with DNA methylation clocks is accelerated in consequence of several diseases and further negatively affected by some concomitant conditions of certain lifestyles. On the other hand, there are some hereditary and environmental factors that are associated with slowed epigenetic ageing (Klutstein et al. 2022), **b** two first epigenetic clocks predicting chronological age were both reported in 2013 by Horvath et al. and Hannum et al. Hannum's clock was developed on 656 people aged from 19 to 101 years old using whole blood samples. The model requires input of gender and body mass index. Methylation was measured using Illumina Infinium HumanMethylation 450 BeadChip assay, a high-density DNA methylation array with single CpG site resolution. The test was devised on a group of 482 people. First tests were performed taking approx. 485,000 CpG markers into account. It was then fine-tuned to a set of 71 markers showing strong methylation-age relationship. Interestingly, these markers are located near genes involved in age-related diseases (Klutstein et al. 2022; Topart et al. 2020)

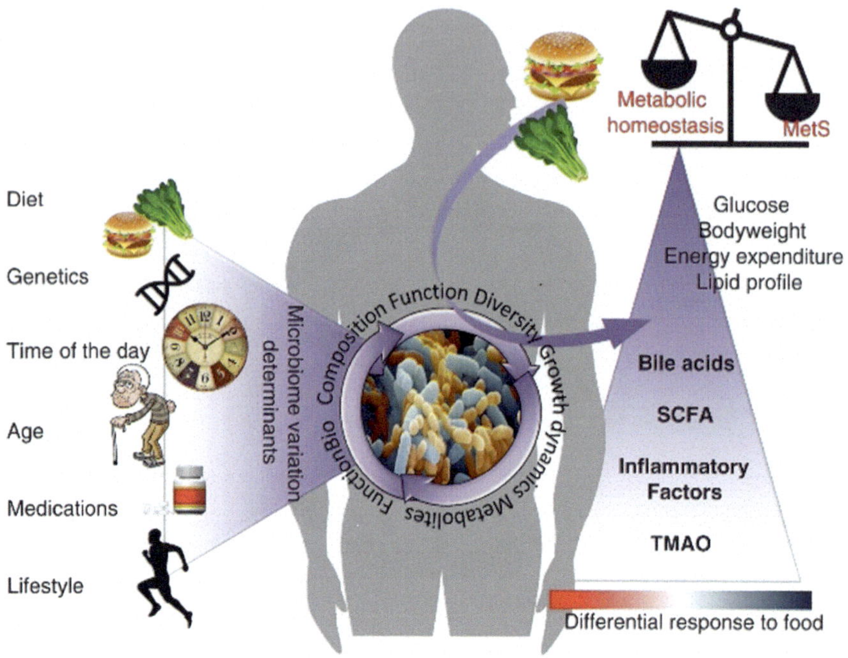

Fig. 1.5 Variations in the microbiome mediate differential effects of the environment on metabolic homeostasis. Multiple host and environmental factors contribute to interindividual variations in the microbiome. This, in turn, leads to a person-specific microbiome regulation of metabolic homeostasis. SCFA—short chain fatty acids; TMAO—trimethylamine-*N*-oxide; MetS—metabolic syndrome (Shapiro et al. 2017)

of opportunity when the foundations of optimum health, growth, and neurodevelopment across the lifespan are established. Imprinting of the immune system, the epigenetic system, and microbiota occurs in this time. Maternal nutrition and maternal microbiota have a central influence on these developments (Fuhler 2020).

The microbiota develops according to the needs of the host over the lifespan. A good balance of gut bacteria early in life influences our health, an imbalance—called dysbiosis—is associated with the development of various diseases (IBS, allergies, DM, obesity). Factors that influence the microbiota of the newborn include: maternal vaginal and intestinal microbiota, mode of delivery, infant diet, antibiotic use, gestational age, siblings, pets, and hygiene. There is a large potential of maternal nutrition during gestation and lactation on the maternal gut, breast milk, and the infant gut microbiota. Dietary fibre intake of the breastfeeding women influences the production of short-chain fatty acids by intestinal bacteria causing an increase in regulatory T-cells and protection against respiratory diseases. In addition, the mother's diet could influence the oligosaccharide profile of breast milk. Observational studies indicate an association between maternal nutrition and the milk microbiome (Fig. 1.6).

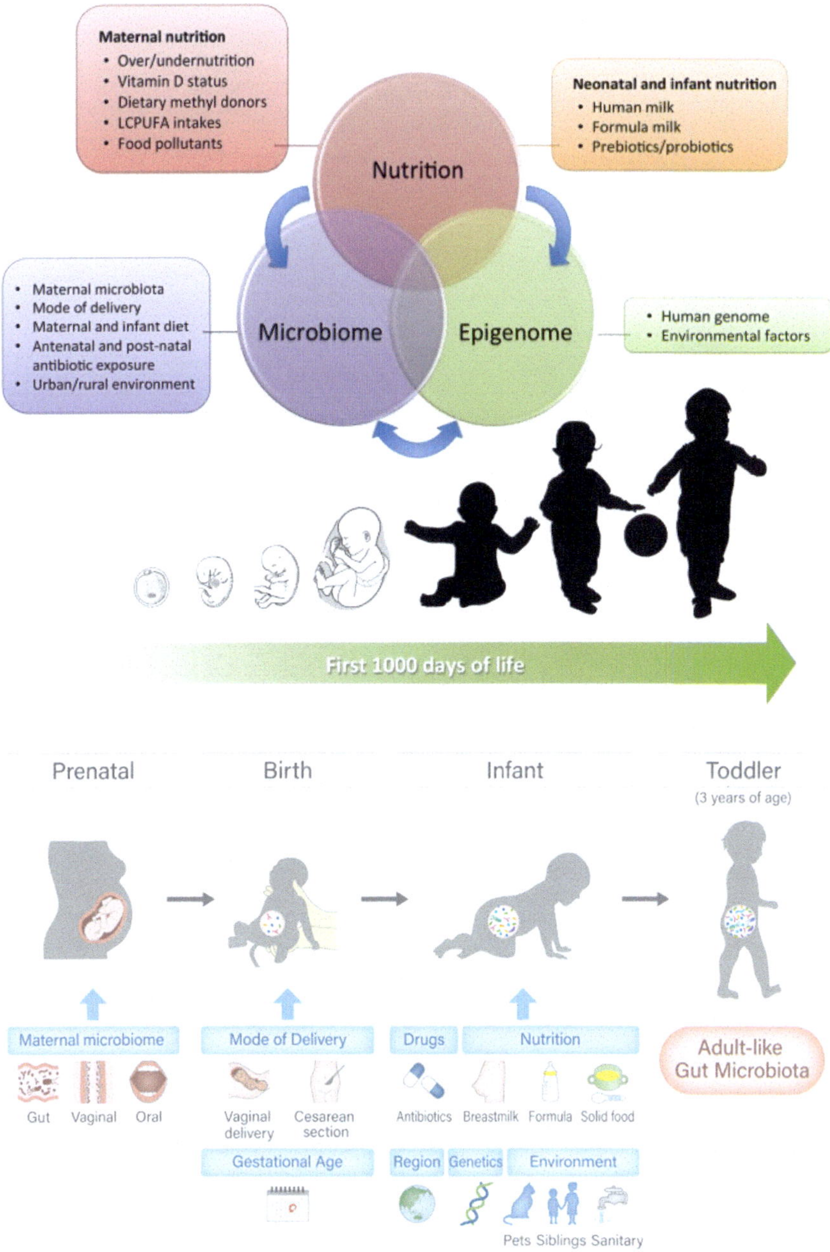

Fig. 1.6 Interrelation between maternal and neonatal nutrition, gut microbiota, and epigenetics during the first 1000 days of life (Indrio et al. 2017; Akagawa et al. 2021; Sindi et al. 2021; Kapourchali et al. 2020)

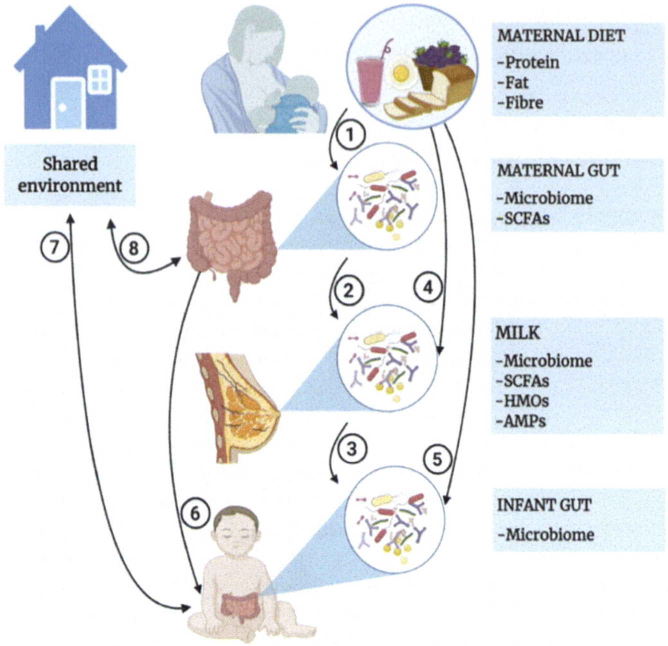

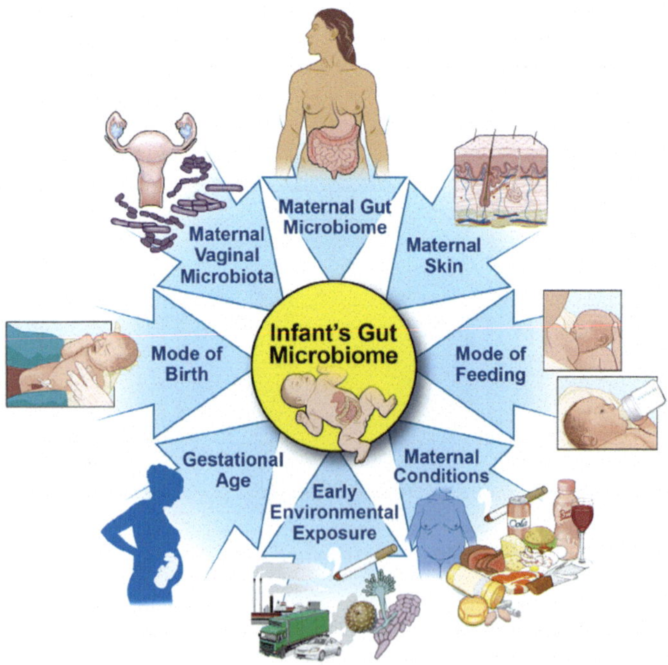

Fig. 1.6 (continued)

Intergenerational transmission of obesogenic microorganisms is also discussed.

Also, the definition of the hallmarks of ageing contributed to the understanding of ageing (Fig. 1.7). Presumably epigenetic consequences of nutrition and food ingredients on the regulation of crucial genes of many hallmarks such as DNA stability and repair, senescence, telomere attrition, or mitochondrial functions emphasise the importance of nutrition epigenetic interactions.

Moreover, the concept of the crucial importance of the benefit of anti-oxidative activities of food ingredients is being modified by the understanding of the need of small doses of reactive oxygen species (ROS) for triggering activation of health-promoting genes. The upcoming of concept of mitohormesis might deeply change aspects of nutritional requirements. Promoting health and lifespan by increased levels of reactive oxygen species (ROS) (Martinvalet and Walch 2022; Ristow and Schmeisser 2014) (Fig. 1.8).

Fig. 1.7 Hallmarks of ageing

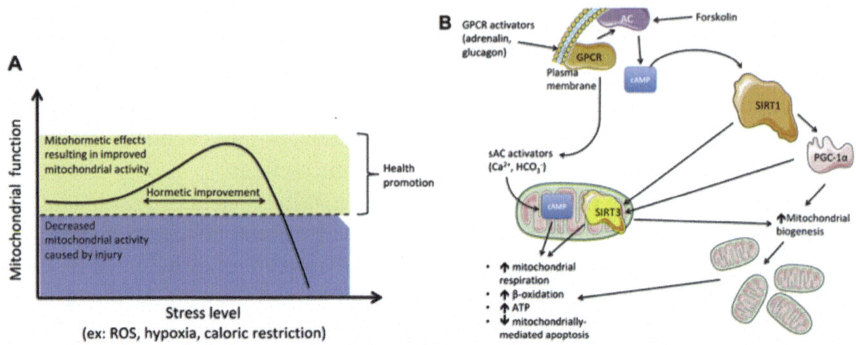

Fig. 1.8 Mitohormetic response. ROS and other mitochondrial toxicants are well known to cause the development of a mitohormetic response, when presented at low values. In fact, a small harmful effect will boost a response of overdrive; i.e. the cell will try to elevate mitochondrial activity to combat the injury, whether by increasing mitochondrial numbers and active respiratory components, to mitophagic (removing the more damaged units) and fission (increasing the number of mitochondria, while diminishing their overall surface area per unit, thus increasing $\Delta\Psi$ and ATP generation) events. However, this is a tough balancing act to pull through, for the tipping point where activity rapidly decreases can be easily traversed, resulting in the more commonly known toxic effects of ROS and other mitochondrial toxicants. B. The role of cAMP on mitochondrial metabolism and mitohormesis. cAMP signalling affects mitochondrial homeostasis and metabolism, for it can result in an increase in NAD^+ and the activation of sirtuin 1 (SirT1), which deacetylates (and thus hyperactivates) peroxisome proliferator-activated receptor gamma coactivator-1α (PGC-1α), the master regulator of mitochondrial biogenesis, leading to elevated mitochondrial numbers and thus increased overall cellular mitochondrial activity. Similarly, cAMP can lead to the activation of SirT3 within the mitochondrial matrix, leading to the deacetylation of several proteins, resulting in the elevation of mitochondrial activity (Palmeira et al. 2019)

Lifestyle-nutrition interactions with metabolism came at the centre of interest with the EPIC project. The European prospective investigation into cancer and nutrition (EPIC 1992–2000) study was one of the largest cohort studies in the world, with more than half a million (521,000) participants. EPIC was designed to investigate the relationships between diet, nutritional status, lifestyle and environmental factors, and the incidence of cancer and other chronic diseases. https://epic.iarc.fr/index.php.

An enormous amount of nutritional studies showed the interaction between nutrition and incidence of diseases such as cancer or metabolic diseases and concluded, e.g. that "cancer is a preventable disease that requires major lifestyle changes" (Anand et al. 2008).

The European NUGO project established 2004 (Baccini et al. 2008) focused on the promotion of molecular nutrition research in Europe and the development and promotion of mechanistic research in nutrition, food, and health by application of *omics* technologies also looking for translational aspects. H. Daniel pointed out that "health is considered as a key market driver. When taken into the food and nutrition sector, the key question is, how health-promotion can be achieved at the level of the individual and the foods consumed. What can be predicted is that a wide range of web-based health services will become available within the next years and those will also employ numerous electronic devices that allow assessment of food intake and measurements of a variety of lifestyle parameters (exercise, sleep, leisure time) and health indicators (blood pressure and glucose level, metabolite profiles, etc.)". https://www.ncl.ac.uk/media/www nclacuk/humannutritionresearchcentre/files/Newcastle2016HDaniel.pdf.

1.2 The Way to Personalisation

The work of a broad number of international organisations and studies paved the way to personalised medicine and personalised nutrition. However, the need for a preventive, personal health care including the use of markers which indicate the development of pathologies that allow intervention already before the development of symptoms has been seen since the seventies of the last century (Merchant 1978).

"The EU hosted international project Public health Genomics (PHGEN) aimed to European best practice guidelines for quality assurance, provision and use of genome-based information and technologies" to support the Member States (and other relevant stakeholders) to more efficiently and effectively work together at European level in addressing the challenges deriving from emerging genome-based information and technologies and to prepare for the paradigm shift of personalised health care in time (Brand, n.d.; Gutierrez-Ibarluzea 2013; The Public Health Genomics European Network (PHGEN), n.d.).

Starting in 2006, mainly aspect from hereditary genetic diseases was addressed, but at the end of the project 2012, PHGEN appreciated the importance to include gene environment interactions. PHG chairman Ron Zimmern summarised as a vision for public health: "In essence, personalised prevention refers to efforts to prevent disease at the level of the individual; this complements rather than replaces classic, population-based public health efforts, but recognises that the most accurate and effective predictive prevention for an individual is based on their unique biological, environmental and behavioural risk factors, and their personal situations, preferences and drivers. Biological risk can be assessed using genomic and other biomarkers" https://www.phgfoundation.org/blog/towards-personalised-prevention.

The special role of biomarkers for the prevention, assessment, and management of developing diseases is now broadly accepted (Chow et al. 2017; International Programme on Chemical Safety, n.d.; Mahaman et al. 2022; Sweeney et al. 2021).

The paradigm shift from reactive to predictive preventive and personalised medicine was discussed in medicine in the early twenty-first century: paradigm and anticipation—EPMA position paper 2016 (Golubnitschaja et al. 2016) (Fig. 1.9).

The establishment of reliable markers and the understanding of lifestyle, genetic, and epigenetic interactions forced the way of medicine into precision or personalised medicine. "According to the precision medicine initiative, precision medicine is an emerging approach for disease treatment and prevention that takes

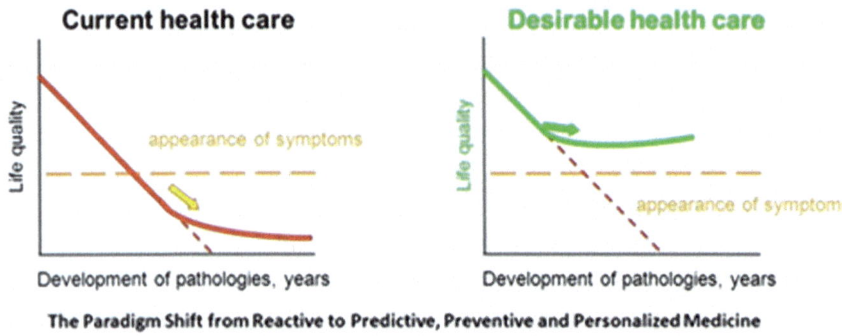

Fig. 1.9 Towards preventive personalised medicine (Golubnitschaja et al. 2016)

into account individual variability in genes, environment, and lifestyle for each person". "This approach will allow to predict more accurately which treatment and prevention strategies for a particular disease will work in different groups of people. It is in contrast to a one-size-fits-all approach, in which disease treatment and prevention strategies are developed for the average person, with less consideration for the differences between individuals" (What Is Precision Medicine?: MedlinePlus Genetics, n.d.).

Especially in the field of cancer the analysis of molecular markers of samples from liquid biopsies allows even the identification of subclones in tumours and allows personalised treatment beyond general tissue-oriented therapies (Amelio et al. n.d.) (Fig. 1.10). The benefit of combining genetic mutation-analysis epigenetic Mi RNA and CpG methylation analysed with self-learning algorithm's for early cancer detection was recently shown (Tomeva et al. 2022).

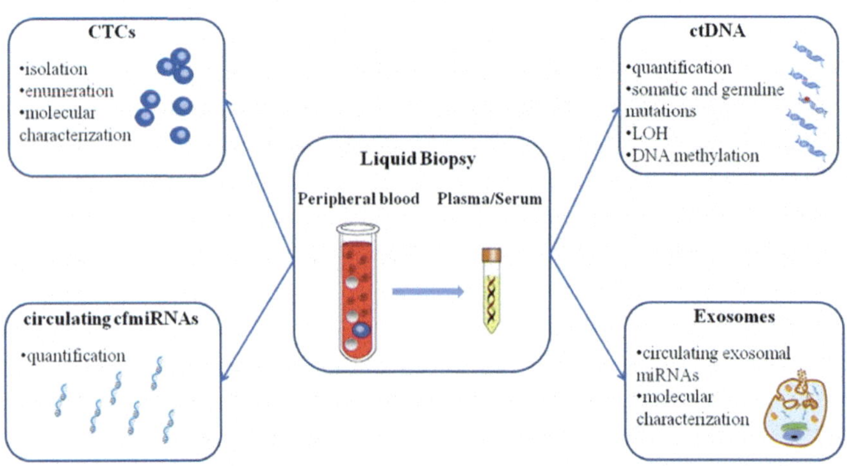

Fig. 1.10 Markers, liquid biopsy (Giannopoulou et al. 2019)

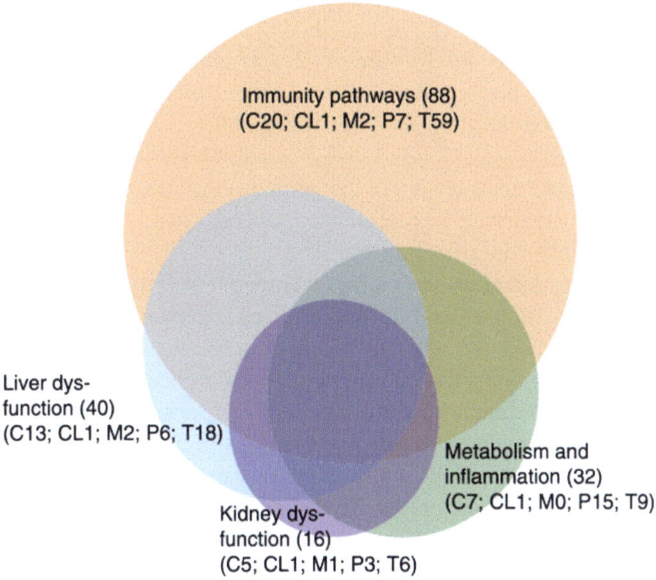

Fig. 1.11 Personalised ageing. Ageotypes (Ahadi et al. 2020)

Another evidence for the use of personalised approaches comes from the omics wide analysis of markers within the hallmarks of ageing and the identifications of personal specific mechanisms which drive accelerated ageing. Recently personal metabolism-specific ageing pattern has been identified and addressed as age types by deep longitudinal profiling (Ahadi et al. 2020) (Fig. 1.11).

In parallel to medicine also in nutritional sciences developed the concept of personalised nutrition based on biomarkers. Whereas there is some discussion to what extent the term precision medicine and personalised medicine are overlapping personalised nutrition and precision nutrition are yet at the centre of interest in nutritional research (de Toro-Martín et al. 2017; González-Muniesa and Alfredo Martínez 2019; Tuncay and Ergoren 2020) and discussions about methodological needs for nutritional advice, financial, ethical, and legal considerations for public health systems (Árnason 2012; de Toro-Martín et al. 2017; Food4Me: The Ethical and Legal Challenges of Personalised Nutrition|Eufic, n.d.; González-Muniesa and Alfredo Martínez 2019).

The EU-funded Food4Me project performed a multi-centre study to show that an internet delivered personalised nutrition advice could improve people's lifestyle. The project envisaged (Celis-Morales et al. 2017) advice on personalised nutrition at three levels: the person's diet only, the diet combined with knowledge of the person's phenotype (measurable traits, such as physical and biochemical measurements, e.g. height, weight, or cholesterol level), and the diet, phenotype and genotype.

For this Food4Me developed a novel, internet delivered, food frequency questionnaire for dietary analysis. The new method that included a digital photographic atlas was used to quantify food intake, took approximately 20 min to complete, and was validated in 2 peer-reviewed studies. https://www.eufic.org/en/healthy-living/article/personalised-nutrition-food4me-project/https://cordis.europa.eu/project/id/265494/reporting.

The results of these analyses imply that attitudes towards personalised nutrition appear to be primarily driven by perceptions of benefit and how achievable it is to access or adopt (Celis-Morales et al. 2017).

Personalised nutrition at the individual level requires costly and time-consuming collection of information, as well as models that are capable of accurately generating personalised advice for the individual. A more feasible approach may be to personalise diets at the group level. Recent studies indicated that individuals may be grouped according to unique metabolic responses to foods and dietary changes. Grouping individuals based on similarities in their metabolic phenotype, metabotypes, is a novel concept; however, different definitions and concepts for metabotypes are under development. The underlying idea behind metabotyping is to identify metabolic phenotypes based on factors such as diet, anthropometric measures, clinical parameters, metabolomics data, and the gut microbiota. "An optimal diet can then be tailored to fit each metabotype specifically" (Palmnäs et al. 2020) (Fig. 1.12).

There is a discussion whether people with a high risk for certain diseases such as cardiometabolic diseases have special metabotypes (Adams et al. n.d.; Grabowski 2020; Hillesheim and Brennan 2020; O'Donovan et al. 2015, 2017; Palmnäs et al. 2020; Riedl et al. 2017). The results of various studies suggest that an optimised metabotype approach is capable of delivering targeted nutritional

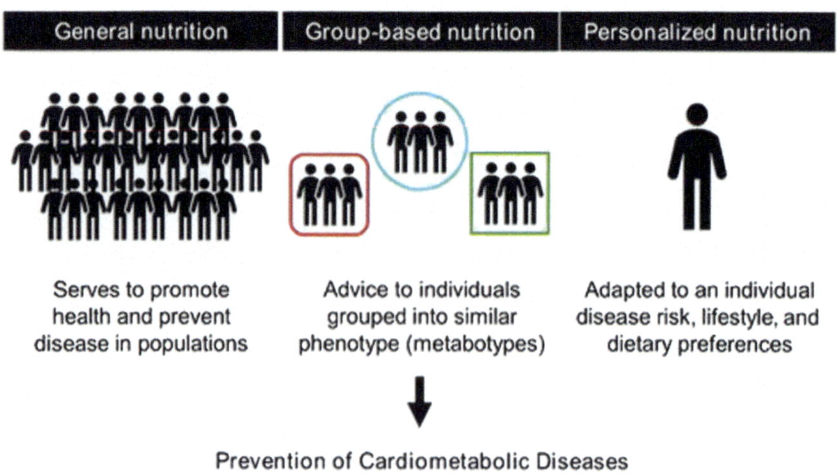

Fig. 1.12 Metabotyping and group-based nutrition in the context of the conventional population-based guidelines and personalised nutrition (Palmnäs et al. 2020)

counselling to healthy adults and is very comparable to individualised counselling. With this information an optimised metabotype approach could be effective in changing diet quality. Clearly metabotyping and group-based nutrition want to enter conventional population-based guidelines and personalised nutrition.

Caution in the claim of benefits of personalised nutrition comes from reviews that scientific evidence is mostly based on observational studies with a low level of reproducibility. However, personalisation is likely to enable sustained behavioural change (Ordovas et al. 2018).

1.3 Consequences of Personalisation

The change of lifestyle mostly needed for improved nutrition is still a central problem due to social and economic aspects (Ekpanyaskul and Padungtod 2021; Lichtenstein et al. 2006). A study in Middle Eastern immigrants reports that facilitators for lifestyle modifications (LSM) are connected to presence of family support. Identification of sociocultural barriers and facilitators for LSM is crucial for successful health promotion (Olaya-Contreras et al., n.d.) (Fig. 1.13).

Methods for improving long-term adherence to lifestyle changes have been studied in Mexican Americans (Foreyt et al., n.d.) and included self-monitoring, cognitive restructuring, stress management, and social support (Fig. 1.14).

The European Union-funded *PROTEIN* project (*PeRsOnalised nutriTion for hEalthy livINg*) consortium produces a novel adaptable mobile application based on sound nutrition and physical activity advice from experts in their field, accessible to all population groups, with differing health outcomes, whose behaviour can

Perceptions, experiences and further barriers for healthier lifestyles	Challenges with regular meals	▶ Loneliness at home affects willingness to eat regularly. ▶ Eating fastfood instead of regular meals. ▶ Not perceiving the need for regular mealtimes and unestablished mealtime schedules. ▶ Eating repeatedly snacks and sweets. ▶ Sometimes overeating or fasting.
	Food preparation, taste preferences and portion size	▶ Preferences for eating food with 'good smell and taste' over dishes prepared in a healthier way (boiled). ▶ Difficulties to introduce to the diet healthier food, healthier cooking and low-calorie products. ▶ Custom of eating large portions.
	Family traditions and expectations	▶ The children decide what to eat in the family hindering a healthy diet. ▶ Following family traditions when gathering to eat hindering the willingness to eat moderately. ▶ Family resistance to exercise. ▶ Challenges changing the role of the mother; limiting the mother's time and possibilities to exercise.
	Other barriers influencing motivation level	▶ Climate and related factors lessening outdoor physical activity and increasing overeating. ▶ Depressed mood isolation affecting willingness to participate in outdoor physical activities. ▶ Difficulty to establish an exercise routine after many years of physical inactivity that worsens with concomitant pain.

Fig. 1.13 Barriers to lifestyle changes

Fig. 1.14 Improving lifestyle changes

Table 1 Strategies for promoting changes in dietary and activity behaviours

Strategy	*Example(s)*
Self-monitoring	Food or exercise diary
Stimulus control	Reminder notes to exercise; no snack foods in the house
CBT techniques	
– Cognitive restructuring	Addressing discrepancies between desired weight loss and realistic weight loss; challenging patients' beliefs that self-worth is based on weight
– Stress management/ inoculation	Progressive muscle relaxation, diaphragmatic breathing, meditation
– Relapse prevention	Normalising slips and lapses; practice high-stress coping skills; visualisation
Social support	Participation in support group; exercising with friend or family member; taking classes at community college

CBT = Cognitive behavioural therapy.

be tracked with a variety of sensors and health hazard perception (Wilson-Barnes et al. 2021).

Another approach to personalised nutrition could be strategies to compose individualised mixtures of bioactive, foods components, according to the results of an analysis of markers and an identification of molecular pathways at risk, e.g. in the area of healthy ageing. In Figure 1.15 especially information about individual very different bioavailability, different fermentation of nutraceuticals to biologically active metabolites or epigenetic regulation is of central importance for personalised precise interventions. Personally different responses to fasting and fasting mimetics were shown, for example, using sirtuin-inducing plant compounds (Lilja et al. 2020, 2021).

Clearly there is a need to develop foods and additives for specific consumer groups at risk. Unfortunately, these developments are hampered by often unsubstantiated health claims of industry or difficult, often confusing and internationally diverse regulations for marketing and labelling. This can often only be dealt by big industries and so contributes to monopolisation and the loss of innovative approaches e.g. the combination of pharmaceutical compounds with food additives.

In summary nutrition, in parallel to medicine, is moving towards personalisation where still a number of obstacles and requirements need to be overcome such as educating health professionals to correctly interpret genetic and epigenetic data, creating ways to motivate positive behaviour change in patients, and correctly implementing personalised nutrition into practice. It is necessary to collect real world data on food intake, physical and social behaviour which the "digital environment" enables as never before. Personalised nutrition needs more comprehensive phenotyping and improved algorithms based on artificial intelligence to

Fig. 1.15 Personalised additives following analysis of markers of healthy ageing

predict the effect on an individual's diet on management, https://doi.org/10.1002/mnfr.202200077. Furthermore, ethical and legal guidelines, as well as standardised regulations for tests, need to be put in place to assure patients health is not being harmed (Ferguson et al. 2016) (Fig. 1.16).

The way to a preventive personalised health care also addresses the aspects of the upcoming understanding of **salutogenesis**, a health approach focusing on factors that support human health and well-being, rather than on factors that cause disease (pathogenesis). More specifically, the "salutogenetic model is concerned with the relationship between health, stress, and coping" (Lindström and Eriksson 2005) and strongly relates to nutrition, dealing with challenges to healthy eating in a health-promoting manner (Swan et al. 2015).

The objective of the present work is to deepen the understanding of progress and obstacles in the development of personalised precision nutrition. Especially the progress in different areas in the establishment of markers and their integration into a holistic concept of preventive personalised health care is discussed. Developments in genetics, epigenetics, microbiota, metabolomics, and their analysis in improved algorithms are shown. Challenges in the interpretation of markers are discussed. Consequences for possible personalised improvements of nutrition and lifestyle such as exercise and practical implementation conclude the book.

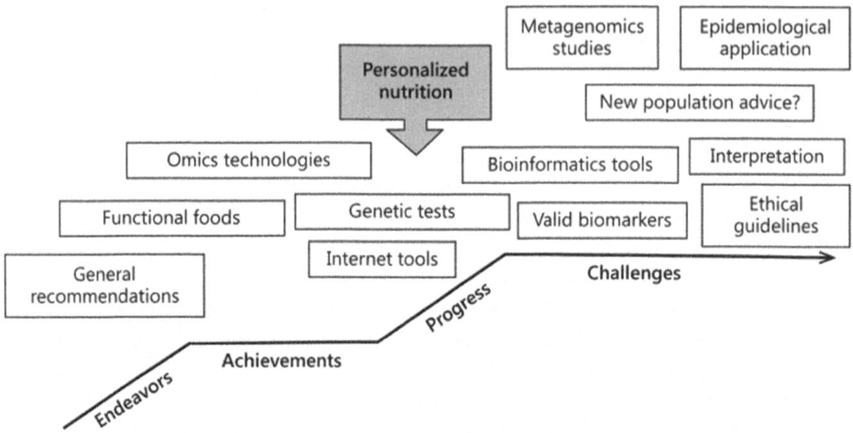

Fig. 1.16 Achievements already made and challenges faced by personalised nutrition (Prasad et al. 2016)

References

Ahadi S, Zhou W, Schüssler-Fiorenza Rose, SM, Sailani MR, Contrepois K, Avina, M., Ashland M, Brunet A, Snyder M (2020) Personal aging markers and ageotypes revealed by deep longitudinal profiling. Nat Med 26(1): 83. https://doi.org/10.1038/S41591-019-0719-5

Akagawa S, Akagawa Y, Yamanouchi S, Kimata T, Tsuji S, Kaneko K (2021) Development of the gut microbiota and dysbiosis in children. Biosci Microbiota Food Health 40(1):12–18. https://doi.org/10.12938/bmfh.2020-034. Epub 25 Aug 2020. PMID: 33520564; PMCID: PMC7817514

Amelio I, Bertolo R, Bove P, Buonomo OC, Candi E, Chiocchi M, Cipriani C, Daniele N di, Ganini C, Juhl H, Mauriello A, Marani C, Marshall J, Montanaro M, Palmieri G, Piacentini M, Sica G, Tesauro M, Rovella V, Melino G (n.d.) Liquid biopsies and cancer omics. https://doi.org/10.1038/s41420-020-00373-0

Anand P, Kunnumakara AB, Sundaram C, Harikumar KB, Tharakan ST, Lai OS, Sung B, Aggarwal BB (2008) Cancer is a Preventable Disease that Requires Major Lifestyle Changes. Pharm Res 25(9):2097. https://doi.org/10.1007/S11095-008-9661-9

Árnason V (2012) The personal is political: Ethics and personalized medicine. Ethical Perspect 19(1):103–122. https://doi.org/10.2143/EP.19.1.2152681

Baccini M, Bachmaier EM, Biggeri A, Boekschoten MV, Bouwman FG, Brennan L, Caesar R, Cinti S, Coort SL, Crosley K, Daniel H, Drevon CA, Duthie S, Eijssen L, Elliott RM, van Erk M, Evelo C, Gibney M, Heim C, de Wit N (2008) The NuGO proof of principle study package: a collaborative research effort of the European Nutrigenomics Organisation. Genes Nutr 3(3–4): 147. https://doi.org/10.1007/S12263-008-0102-5

Brand A (n.d.) Public Health Genomics-public health goes personalized? https://doi.org/10.1093/eurpub/ckq197

Cazaly E, Saad J, Wang W, Heckman C, Ollikainen M, Tang J (2019) Making sense of the epigenome using data integration approaches. Front Pharmacol 10: 126. https://doi.org/10.3389/FPHAR.2019.00126/BIBTEX

Celis-Morales C, Livingstone KM, Marsaux CFM, Forster H, O'Donovan CB, Woolhead C, Macready AL, Fallaize R, Navas-Carretero S, San-Cristobal R, Kolossa S, Hartwig K, Tsirigoti

L, Lambrinou CP, Moschonis G, Godlewska M, Surwiłło A, Grimaldi K, Bouwman J, Mathers JC (2015) Design and baseline characteristics of the Food4Me study: a web-based randomised controlled trial of personalised nutrition in seven European countries. Genes Nutr 10(1): 1–13. https://doi.org/10.1007/S12263-014-0450-2/TABLES/3

Celis-Morales C, Livingstone KM, Marsaux CFM, Macready AL, Fallaize R, O'Donovan CB, Woolhead C, Forster H, Walsh MC, Navas-Carretero S, San-Cristobal R, Tsirigoti L, Lambrinou CP, Mavrogianni C, Moschonis G, Kolossa S, Hallmann J Godlewska M, Surwiłło A, Mathers JC (2017)Effect of personalized nutrition on health-related behaviour change: evidence from the Food4Me European randomized controlled trial. Int J Epidemiol 46(2): 578–588. https://doi.org/10.1093/IJE/DYW186

Chow SL, Maisel AS, Anand I, Bozkurt B, de Boer RA, Felker GM, Fonarow GC, Greenberg B, Januzzi JL, Kiernan MS, Liu PP, Wang TJ, Yancy CW, Zile MR (2017) Role of Biomarkers for the Prevention, Assessment, and Management of Heart Failure: A Scientific Statement From the American Heart Association. Circulation 135(22):e1054–e1091. https://doi.org/10.1161/CIR.0000000000000490

de Toro-Martín J, Arsenault BJ, Després JP, Vohl MC (2017) Precision Nutrition: A Review of Personalized Nutritional Approaches for the Prevention and Management of Metabolic Syndrome. Nutrients 9(8):913. https://doi.org/10.3390/NU9080913

Dolinoy DC, Huang D, Jirtle RL (2007) Maternal nutrient supplementation counteracts bisphenol A-induced DNA hypomethylation in early development. Proc Natl Acad Sci USA 104(32):13056–13061. https://doi.org/10.1073/PNAS.0703739104

Ekpanyaskul C, Padungtod C (2021) Occupational Health Problems and Lifestyle Changes Among Novice Working-From-Home Workers Amid the COVID-19 Pandemic. Saf Health Work 12(3):384–389. https://doi.org/10.1016/J.SHAW.2021.01.010

Encyclopedia Deutsche Gesellschaft für Ernährung, Österreichische Gesellschaft für Ernährung, Schweizerische Gesellschaft für Ernährungsforschung, Schweizerische Vereinigung für Ernährung (2021) (Hrsg.): Referenzwerte für die Nährstoffzufuhr. Bonn, 2. Auflage, 7. aktualisierte Ausgabe

Ferguson LR, De Caterina R, Görman U, Allayee H, Kohlmeier M, Prasad C, Choi MS, Curi R, De Luis DA, Gil Á, Kang JX, Martin RL, Milagro FI, Nicoletti CF, Nonino CB, Ordovas JM, Parslow VR, Portillo MP, Santos JL, Martinez JA (2016) Guide and position of the International Society of Nutrigenetics/Nutrigenomics on personalised nutrition: part 1-fields of precision nutrition. J Nutrigenet Nutrigenomics 9(1):12–27. https://doi.org/10.1159/000445350

Food4Me: The ethical and legal challenges of personalised nutrition | Eufic. (n.d.). Retrieved 3 Feb 2022, https://www.eufic.org/en/collaboration/article/food4me-the-ethical-and-legal-challenges-of-personalised-nutrition

Foreyt JP, Carlos WS, Ii P (n.d.) The challenge of diet, exercise and lifestyle modi®cation in the management of the obese diabetic patient. Retrieved 3 Feb 2022, http://www.stockton-press.co.uk/ijo

Fransquet PD, Wrigglesworth J, Woods RL, Ernst ME, Ryan J (2019) The epigenetic clock as a predictor of disease and mortality risk: A systematic review and meta-analysis. In Clinical Epigenetics 11(1): 62. BioMed Central Ltd. https://doi.org/10.1186/s13148-019-0656-7

Fuhler GM (2020) The immune system and microbiome in pregnancy. Best Pract Res Clin Gastroenterol 44–45:101671. https://doi.org/10.1016/J.BPG.2020.101671

Giannopoulou L, Zavridou M, Kasimir-Bauer S, Lianidou ES (2019) Liquid biopsy in ovarian cancer: the potential of circulating miRNAs and exosomes Transl Res 205: 77–91. https://doi.org/10.1016/J.TRSL.2018.10.003

Golubnitschaja O, Baban B, Boniolo G, Wang W, Bubnov R, Kapalla M, Krapfenbauer K, Mozaffari MS, Costigliola V (2016) Medicine in the early twenty-first century: paradigm and anticipation-EPMA position paper 2016. EPMA J 7(1). https://doi.org/10.1186/S13167-016-0072-4

Gonçalo RA, David A, Adam A, Lisa DB, Richard MD, Richard AG, Matt EH, Gil AM (2010) A map of human genome variation from population-scale sequencing The 1000 Genomes Project Consortium Nature 467. https://doi.org/10.1038/nature09534

González-Muniesa P, Alfredo Martínez J (2019) Precision Nutrition and Metabolic Syndrome Management. Nutrients, 11(10). https://doi.org/10.3390/NU11102411

Goodarzi MO (2018) Genetics of obesity: what genetic association studies have taught us about the biology of obesity and its complications. Lancet Diabetes Endocrinol 6(3):223–236. https://doi.org/10.1016/S2213-8587(17)30200-0

Grabowski (2020) https://doi.org/10.18388/pb.2020_329

Gutierrez-Ibarluzea I (2013) The Public Health Genomics European Network (PHGEN). https://www.academia.edu/33414188/The_Public_Health_Genomics_European_Network_PHGEN_

Harper AE (2003) Evolution of Recommended Dietary Allowances—New Directions? 7(1); 509–537. https://doi.org/10.1146/ANNUREV.NU.07.070187.002453

Hillesheim, Brennan (2020) https://nutritionandmetabolism.biomedcentral.com/articles/10.1186/s12986-020-00499-z

Hodge SE, Greenberg DA (2017) How can we explain very low odds ratios in GWAS? I. Polygenic Models. Human Heredity 81(4):173–180. https://doi.org/10.1159/000454804

Indrio F, Martini S, Francavilla R, Corvaglia L, Cristofori F, Mastrolia SA, Neu J, Rautava S, Spena GR, Raimondi F, Loverro G (2017) Epigenetic matters: the link between early nutrition, microbiome, and long-term health development. Front Pediatr 5:178–178. https://doi.org/10.3389/FPED.2017.00178

INTERNATIONAL PROGRAMME ON CHEMICAL SAFETY (n.d.).

Kapourchali FR, Cresci GAM (2020) Early-life gut microbiome—The importance of maternal and infant factors in its establishment. Nutr Clin Pract 35(3):386–405. https://doi.org/10.1002/ncp.10490. Epub 24 Apr 2020. PMID: 32329544

Klutstein M, Reizel Y, Galow AM., Peleg S (2022) How to slow down the ticking clock: age-associated epigenetic alterations and related interventions to extend life span. Cells 11(3): 468. https://doi.org/10.3390/CELLS11030468

Lichtenstein AH, Appel LJ, Brands M, Carnethon M, Daniels S, Franch HA, Franklin B, Kris-Etherton P, Harris WS, Howard B, Karanja N, Lefevre M, Rudel L, Sacks F, van Horn L, Winston M, Wylie-Rosett J (2006) Diet and lifestyle recommendations revision 2006: a scientific statement from the American Heart Association Nutrition Committee Circulation 114(1): 82–96. https://doi.org/10.1161/CIRCULATIONAHA.106.176158

Lindström B, Eriksson M (2005) Salutogenesis. In J Epidemiol Community Health 59(6): 440–442. BMJ Publishing Group Ltd. https://doi.org/10.1136/jech.2005.034777

Mahaman YAR, Embaye KS, Huang F, Li L, Zhu F, Wang JZ, Liu R, Feng J, Wang X (2022) Biomarkers used in Alzheimer's disease diagnosis, treatment, and prevention. Ageing Res Rev 74. https://doi.org/10.1016/J.ARR.2021.101544

Martinvalet D, Walch M (2022) Editorial: the role of reactive oxygen species in protective immunity. Front Immunol 5844. https://doi.org/10.3389/FIMMU.2021.832946

Merchant HW (1978) Personal preventive medicine: a need and a responsibility. J Am Den Ass (1939) 97(3): 450–452. https://doi.org/10.14219/jada.archive.1978.0299

Mozaffarian D, Rosenberg I, Uauy R (2018) History of modern nutrition science—implications for current research, dietary guidelines, and food policy. BMJ 361. https://doi.org/10.1136/BMJ.K2392

National nutrition conference for defense (1941) J Am Med Ass 116(23): 2598–2599. https://doi.org/10.1001/JAMA.1941.02820230042013

New concepts of a balanced diet (n.d.).

Medieval European Nutrition | Health Ahoy. (n.d.). Retrieved 1 Feb 2022, https://healthahoy.com/ancient-nutrition/medieval-european-nutrition/

Olaya-Contreras P, Balcker-Lundgren K, Siddiqui F, Bennet L (n.d.) Open access Perceptions, experiences and barriers to lifestyle modifications in first-generation Middle Eastern immigrants to Sweden: a qualitative study. https://doi.org/10.1136/bmjopen-2018-028076

Ordovas JM, Ferguson LR, Tai ES, Mathers JC (2018) Personalised nutrition and health. BMJ (Clinical Research Ed.) 361. https://doi.org/10.1136/BMJ.K2173

Orozco-Solis R, Sassone-Corsi P (2014) Circadian clock: Linking epigenetics to aging. In Current Opinion in Genetics and Development 26: 66–72 Elsevier Ltd. https://doi.org/10.1016/j.gde.2014.06.003

Palmeira CM, Teodoro JS, Amorim JA, Steegborn C, Sinclair DA, Rolo AP (2019) Mitohormesis and metabolic health: The interplay between ROS, cAMP and sirtuins. Free Radical Biol Med 141:483–491. https://doi.org/10.1016/J.FREERADBIOMED.2019.07.017

Palmnäs M, Brunius C, Shi L, Rostgaard-Hansen A, Torres NE, González-Domínguez R, Zamora-Ros R, Ye YL, Halkjær J, Tjønneland A, Riccardi G, Giacco R, Costabile G, Vetrani C, Nielsen J, Andres-Lacueva C, Landberg R (2020) Perspective: Metabotyping—a potential personalized nutrition strategy for precision prevention of cardiometabolic disease. Adv Nutr 11(3): 524−532. https://doi.org/10.1093/ADVANCES/NMZ121

Pléh C (2012) The history of the nature/nurture issue. Behavioral and Brain Sciences 35(5):376–377. https://doi.org/10.1017/S0140525X12001094

Prasad C, Velázquez-Arellano A, Serhan CN, Portillo MP, Simopoulos AP, Kohlmeier M, Ferguson LR, Curi R, Zulet MA, Allayee H, Nicoletti CF, Martinez JA, Görman U, De Caterina R, Martin RL, Nonino CB, Milagro FI, Parslow VR, Ordovas JM, de Luis DA (2016) Guide and position of the International Society of Nutrigenetics/Nutrigenomics on personalised nutrition: part 1 - fields of precision nutrition. J Nutrigenet Nutrigenomics 9(1):12–27. https://doi.org/10.1159/000445350

Quach A, Levine ME, Tanaka T, Lu AT, Chen BH, Ferrucci L, Ritz B, Bandinelli S, Neuhouser ML, Beasley JM, Snetselaar L, Wallace RB, Tsao PS, Absher D, Assimes TL, Stewart JD, Li Y, Hou L, Baccarelli AA, Horvath S (2017) Epigenetic clock analysis of diet, exercise, education, and lifestyle factors. Aging (Albany NY), 9(2): 419. https://doi.org/10.18632/AGING.101168

Remely M, Stefanska B, Lovrecic L, Magnet U, Haslberger AG (2015) Nutriepigenomics: The role of nutrition in epigenetic control of human diseases. Curr Opin Clin Nutr Metab Care 18(4):328–333. https://doi.org/10.1097/MCO.0000000000000180

Riedl (2017) https://doi.org/10.1007/s00394-019-01988-5

Ristow M, Schmeisser K (2014) Mitohormesis: Promoting Health and Lifespan by Increased Levels of Reactive Oxygen Species (ROS). Dose-Response 12(2):288. https://doi.org/10.2203/DOSE-RESPONSE.13-035.RISTOW

Shapiro H, Suez J, Elinav E (2017) Personalized microbiome-based approaches to metabolic syndrome management and prevention. J Diabetes 9(3):226–236. https://doi.org/10.1111/1753-0407.12501

Skiadas PK, Lascaratos JG (2001) Dietetics in ancient Greek philosophy: Plato's concepts of healthy diet. Eur J Clin Nutr 55(7):532–537. https://doi.org/10.1038/SJ.EJCN.1601179

Swan E, Bouwman L, Hiddink GJ, Aarts N, Koelen M (2015) Applying the salutogenic framework to nutrition research and practice. Am J Health Promot 30(2):71–73. https://doi.org/10.4278/ajhp.140127-CIT-46

Sweeney T, Quispe R, Das T, Juraschek SP, Martin SS, Michos ED (2021) The use of blood biomarkers in precision medicine for the primary prevention of atherosclerotic cardiovascular disease: a review. Expert Review of Precision Medicine and Drug Development 6(4):247–258. https://doi.org/10.1080/23808993.2021.1930531

(PDF) The Public Health Genomics European Network (PHGEN). (n.d.). Retrieved 4 Feb 2022, https://www.researchgate.net/publication/236880479_The_Public_Health_Genomics_European_Network_PHGEN

Tomeva E, Switzeny OJ, Heitzinger C, Hippe B, Haslberger AG (2022) Comprehensive Approach to Distinguish Patients with Solid Tumors from Healthy Controls by Combining Androgen Receptor Mutation p.H875Y with Cell-Free DNA Methylation and Circulating miRNAs. Cancers 14(2): 462. https://doi.org/10.3390/CANCERS14020462

Topart C, Werner E, Arimondo PB (2020) Wandering along the epigenetic timeline. Clin Epigenetics 12(1):1–13. https://doi.org/10.1186/S13148-020-00893-7/TABLES/1

Tuncay C, Ergoren MC (2020) A systematic review of precision nutrition and Mediterranean Diet: A personalized nutrition approaches for prevention and management of obesity related disorders. Clinical Nutrition ESPEN 38:61–64. https://doi.org/10.1016/J.CLNESP.2020.04.005

Uffelmann E, Huang QQ, Munung NS, de Vries J, Okada Y, Martin AR, Martin HC, Lappalainen T, Posthuma D (2021) Genome-wide association studies. Nature Reviews Methods Primers 1(1):1–21. https://doi.org/10.1038/s43586-021-00056-9

What is precision medicine? MedlinePlus Genetics. (n.d.). Retrieved 3 Feb, 2022, https://medlineplus.gov/genetics/understanding/precisionmedicine/definition/

Wilson-Barnes S, Gymnopoulos LP, Dimitropoulos K, Solachidis V, Rouskas K, Russell D, Oikonomidis Y, Hadjidimitriou S, Botana JM, Brkic B, Mantovani E, Gravina S, Telo G, Lalama E, Buys R, Hassapidou M, Dias SB, Batista A, Perone L, Hart K (2021) PeRsOnalised nutriTion for hEalthy livINg: The PROTEIN project. Nutr Bull 46(1):77–87. https://doi.org/10.1111/NBU.12482

Zeevi D (2015) Personalized nutrition by prediction of glycemic responses. Cell 163:1079–1094

Sindi AS, Geddes DT, Wlodek ME, Muhlhausler BS, Payne MS, Stinson LF. Can we modulate the breastfed infant gut microbiota through maternal diet? FEMS Microbiol Rev 11:fuab011. https://doi.org/10.1093/femsre/fuab011. Epub ahead of print. PMID: 33571360

Individualization, Precision Nutrition Developments for the 21st Century

2

María-Carmen López de Las Hazas and Alberto Dávalos

Contents

Abstract

Among the main challenges in nutrition research are development of strategies for providing dietary solutions that help people adjust their dietary needs and behavior at every stage of their life. An appropriate diet will maintain the body in good health and therefore prevent chronic diseases associated with

M.-C. L. de Las Hazas · A. Dávalos (✉)
Laboratory of Epigenetics of Lipid Metabolism, Madrid Institute for Advanced Studies (IMDEA)-Food, CEI UAM + CSIC, 28049 Madrid, Spain
e-mail: alberto.davalos@imdea.org

M.-C. L. de Las Hazas
e-mail: mcarmen.lopez@imdea.org

A. G. Haslberger (ed.), *Advances in Precision Nutrition, Personalization and Healthy Aging*, https://doi.org/10.1007/978-3-031-10153-3_2

dietary excess. Personalized nutrition is a novel approach that recommends food choices and eating patterns that meet individual needs and follow personal preferences. Over the last century, nutrition research has progressively incorporated small bodies of knowledge into the puzzle of personalization, including considering diet as a treatment for different diseases, biochemical markers, anthropometric markers, food frequency questionnaires, nutrigenetic and nutrigenomic information, and incipient nutritional genetic risk scores. Other factors will also need consideration, such as food sustainability, environmental protection, food security, cultural variations, allergies and intolerances, among others. This greatly complicates the matter of promoting personalized nutrition. Recent research aimed at predicting individual response to a nutrient includes use of deep phenotyping (i.e., through continuous postprandial monitoring), microbiota, and epigenetic data that will shape future precision nutrition approaches. Despite advances in personalized nutrition, many obstacles and challenges remain before its full benefits can transition from bench side to bedside. For instance, it requires specialized healthcare professionals, competitive costing, and potential customers ready to understand and accept new nutritional approaches. This chapter is an overview of how individualization has been shaping approaches to personalized nutrition including its social impact, business and value creation, social concerns, ethical and legal concerns, communication, and consumer attitudes toward personalized nutrition. Overall, developing precision nutrition must integrate biology, environment, and lifestyle. Although biology may remain fairly constant throughout life, both environment and lifestyle change constantly through epigenetic mechanisms. Moreover, integrating these data for every period of life will require new resources for large-scale data analysis, such as artificial intelligence and machine learning algorithms.

2.1 Introduction

Nutrition impacts public health through the prevention and treatment of different disorders and the promotion of well-being. Recent increases in many chronic diseases (i.e., CVD, cancer, neurodegenerative diseases, etc.) have been linked to unbalanced diets and inadequate physical exercise. The ancient saying "you are you eat" remains valid today since diet is now known to impact the expression of genes that regulate critical metabolic pathways.

Nutrition research integrates biological approaches ranging from the molecular to epidemiological to explore the relationship between eating patterns, nutritional status, and development of chronic disease. The biggest challenge in nutrition, however, is how to apply healthy eating and lifestyle patterns to improve health in any given individual. Traditionally, nutritional condition management has been associated with the application of dietary recommendations based on population-based, prospective cohort, and intervention studies (Becerra-Tomás et al. 2021; López-González et al. 2021). Dietary recommendations are normally developed generically for a population and consist of advice to follow a certain type of diet (i.e., low fat diet, Mediterranean Diet, etc.). Communicating this advice to the general population can take many forms although it is often done with an image such as the food pyramid. Approaches of this kind often have a limited impact on the frequency of individuals in a population developing healthy dietary habits. Another factor to consider is that not all dietary patterns are indicated for all people (Corella et al. 2007). Ideally, dietary approaches need to be designed for a specific individual based on their personal situation, considering factors such as weight, age, gender, and phenotypic information such as anthropometry, biochemical and metabolic analysis, and physical activity levels (Gibney and Walsh 2013). Taking these into account can increase diet compliance (Celis-Morales et al. 2017a).

Technological advances and sequencing of the human genome in 2001 opened the possibility of analyzing immense amounts of biological data and identifying significant genetic associations that can facilitate disease treatment (Cordero and Ashley 2012). Whole-genome sequencing analyses found that just 1% of the genome contains variations such as single nucleotide polymorphism (SNPs), copy number variations (CNVs), and other structural variants (Auton et al. 2015). This discovery has allowed researchers to begin understanding individual responses to dietary interventions. Identification of how diet, genotype, and their interaction influences individual response, and predispose people to developing disorders, such as obesity, inflammation, dyslipidemia, and oxidative stress, has opened myriad possibilities (Hesketh 2012) for diet personalization. Among these are nutrigenetics and nutrigenomics. Both are promising multidisciplinary fields that focus on studying the interactions between nutritional factors, genetic factors, and health outcomes (Ordovas 2004). Nutrigenetics focuses on how genotype influences the body metabolic response to nutrients and on the risk of nutrition-related diseases. Nutrigenetics also assess the variations in complex metabolic responses to specific individual nutrients as assessed by genome-wide association studies (GWASs). Nutrigenomics addresses the effect of nutrition (including macro-, micro- and anti-nutrients) on gene expression, the proteome and the metabolome (Ferguson et al. 2016) (Fig. 2.1). In this sense, functional genomics aims to determine how the individual components produce a particular phenotype.

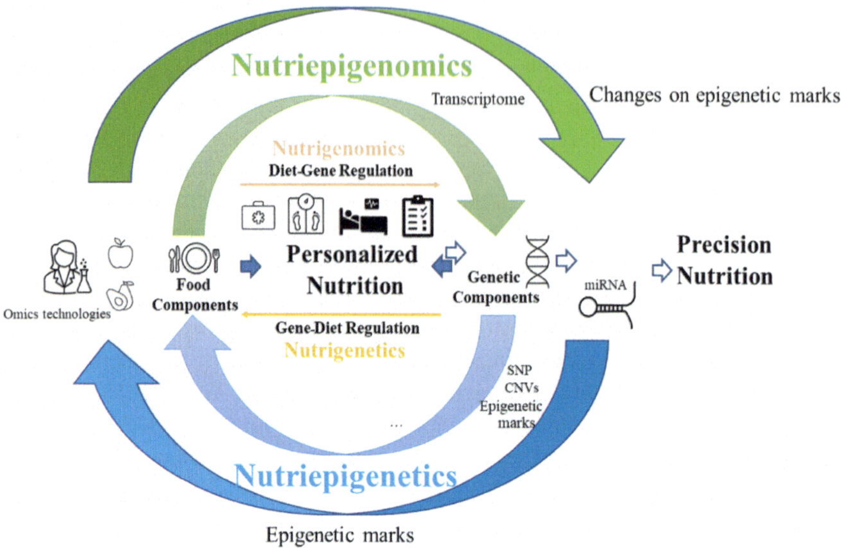

Fig. 2.1 Interplay of nutrigenetics and nutrigenomics and their integrative nutritional biomarkers in developing personalized nutrition

Integrating the knowledge generated in different omic areas (e.g., genomics, transcriptomics, proteomics, metabolomics and even metagenomics) requires bioinformatics to manage and interpret the functional genomics. New therapeutic tools are also needed to integrate functional genomics and design personalized approaches to manage the prevention and treatment of chronic diseases. Use of genetic information to predict an individual's response to a certain diet or product has transformed the traditional concept of nutrition. Specific dietary recommendations can be generated for individuals based on their response to dietary components, an approach called "personalized nutrition." It is an important element of personalized medicine and tries to establish nutritional recommendation guidelines for specific subgroups based on individual parameters (Ferguson et al. 2016). Understanding an individual's response (or lack thereof) to a dietary intervention is supported by quantifying their response to a nutrient(s) based on the interactions between metabolic, genetic, environmental, and social factors together with biological and cultural variations, including food preferences, allergies, and intolerances (San-Cristobal et al. 2020).

Nutrigenomics data can be applied to explore new mechanisms of action and biomarkers, as well as identify early symptoms and pivotal points of disease progression as parts of an individual's response to lifestyle intervention. Different genes and polymorphisms have been identified as critical regulators of metabolism. Therefore, when interpreted together with an individual's family clinical history, phenotypic data and social and personal habits, implementation of genetic risk scores can help healthcare providers to develop novel approaches to personalized/precision nutrition.

The inclusion of different epigenetic data (i.e., DNA methylation, miRNAs) will contribute to identify those epigenetic factors that influence the response to a dietary intervention and is affected by the interindividual variability. For instance, the modulation of circulating miRNAs after a dietary treatment (Mantilla-Escalante et al. 2021, 2019) or as response of a bioactive compound intake (Nuñez-Sánchez et al. 2015; Tomé-Carneiro et al. 2016) may serve as novel dietary tools to modulate epigenetic markers (Dávalos and Fernández-Hernando 2013). Indeed, certain miRNAs from diet may resist the gastrointestinal (GI) tract playing a role in the regulation of GI tract physiology and impact on host gene expression (Zhang et al. 2012; Pozo-Acebo et al. 2021) where they may produce a biological function (Pozo-Acebo et al. 2021; Dávalos et al. 2020).

In this scenario, a personalized dietary intervention based on those factors that affect epigenetic parameters has been identified. For instance, Casas-Agustench identified that the type of lipid consumed during pregnancy affects on offspring development and their susceptibility to metabolic disorders (Casas-Agustench et al. 2015).

One of the best examples to illustrate the existence of personalization in the diet is maternal milk. Maternal milk is the natural precision nutrition as it varies in its composition between (i) individuals; (ii) the time of the day; (iii) the stage of lactation; and in response to maternal nutrition and interindividual variation. It also changes its composition during feeding; for instance, the foremilk contains higher amounts of carbohydrates and lower content of fats; meanwhile, the hindmilk resembles cream, and it has higher fat content and is lower in carbohydrates. Moreover, recent evidences show that they also provides extracellular RNAs (i.e., miRNAs) that influence important biological processes in neonatal (Tomé-Carneiro et al. 2018).

The International Society of Nutrigenetics/Nutrigenomics states that personalized nutrition can be applied at three levels (Fig. 2.2). It begins with dietary recommendations based on eating pattern (i.e., eat more vegetables) through data collected by dietary recall, food diaries, and/or food frequency questionnaires. An individual's phenotypic data are then quantified using anthropometric measurements (i.e., body mass index—BMI) and blood biomarkers such as cholesterol and fatty acids to provide more detailed recommendations. Finally, inclusion of genetic information contributes to identifying those genetic variants that may impact an individual's response (Ferguson et al. 2016). One challenge in this approach is to study genotype–phenotype associations to better understand interindividual variability and thus more accurately tailor dietary approaches to individuals (Tracy 2008). Randomized clinical trials have shown that personalized nutrition is more effective than general nutritional advice in terms of changing eating habits and treating obesity (Celis-Morales et al. 2017a; Nielsen and El-Sohemy 2014). It is also more effective at changing perceptions and behaviors because specific recommendations are based on personalization (Roke et al. 2017; Nielsen et al. 2014).

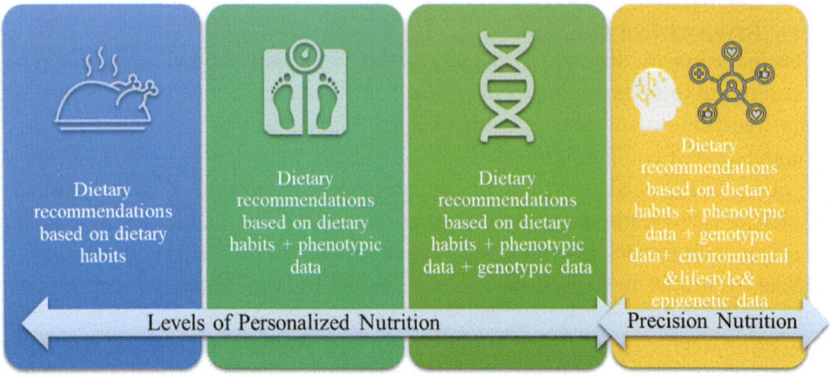

Fig. 2.2 Levels of advice in the development of a personalized nutrition approach

Precision nutrition approaches also seek to tailor treatments to an individual by including their lifestyle and environment. To date, the terms "personalized nutrition" and "precision nutrition" have been used randomly and almost interchangeably, but there is active debate as to what each should refer to. Precision nutrition is often understood to imply the use of quantitative genetics data and in-depth health profiling.

Personalized nutrition uses dietary data alongside biochemical, physiological, metabolomic, and genetic markers to accurately predict an individual's response to diet, calculate their specific nutritional requirements, and determine their risk of diet-related disease. Collecting much of this data requires new technologies, and innovations will probably continue in the field. This approach also influences consumer's knowledge, perceptions, and attitudes, inducing different behaviors and consequently adding value to specific products and opening different business models (Rodgers and Collins 2020). In addition to the above elements, the precision nutrition approach factors in lifestyle (i.e., physical activity), environmental factors (i.e., exposome and others), continuous metabolite monitoring (i.e., continuous glucose monitoring), and continuous health-parameter monitoring (i.e., hearth rate, blood pressure, etc.). All these data are incorporated (i.e., via machine learning algorithms and/or artificial intelligence) to accurately propose the precise diet most adequate for an individual at a given time in their life. This is done based on their nutritional needs and the results of large-scale dataset analysis and is aimed at providing them the most appropriate possible diet and consequently reducing their risk of diet-related diseases.

This chapter is an evaluation of the evolution of personalized nutrition and a summary of the challenges and possibilities of its use in the general population that includes a discussion of recent shifts in nutrition research and practice that have led toward personalization.

2.2 Evolution of Nutrition Science in the 20th Century Toward Personalization

Use of dietary therapy to treat different diseases began in the early 1900s when physiologists developed multiple approaches that became the foundations of current nutritional knowledge (Fig. 2.3). For instance, in 1908, Garrod considered alkaptonuria as a "hereditary error of metabolism" (Piro et al. 2010), and Addis designed a treatment for glomerulonephritis (or Bright's disease) using an individualized low-calorie, low-salt, and low-protein content diet (Bland 2019). Building on this base, in 1940s, Pauling described the biochemical importance of nutrition and the genetic uniqueness of specific nutrients (Bland 2019).

Modern nutrition science is considered to have begun when Christiaan Eijman won the 1929 Nobel Prize in Physiology or Medicine for his isolation of thiamine (vitamin B1) in 1926. His discovery was built on previous research such as the report that the hulk of unprocessed rice protected chickens against a beriberi-like condition (Mozaffarian et al. 2018). With the advent that vitamins are essential to the organism to avoid serious deficiency diseases, the first half of the twentieth century saw research was focused on the discovery, isolation, and synthesis of essential micronutrients to reduce the prevalence of many deficiency diseases. Working in the laboratory of Williams, Pauling discovered pantothenic acid, folic acid, and pyridoxine (vitamin B_6). In later years, Pauling's work supported the role of vitamin C in decreasing the duration and severity of cold symptoms (Pauling 1970). The pharmaceutical market and government agencies responded by focusing on single nutrients and specific disease states.

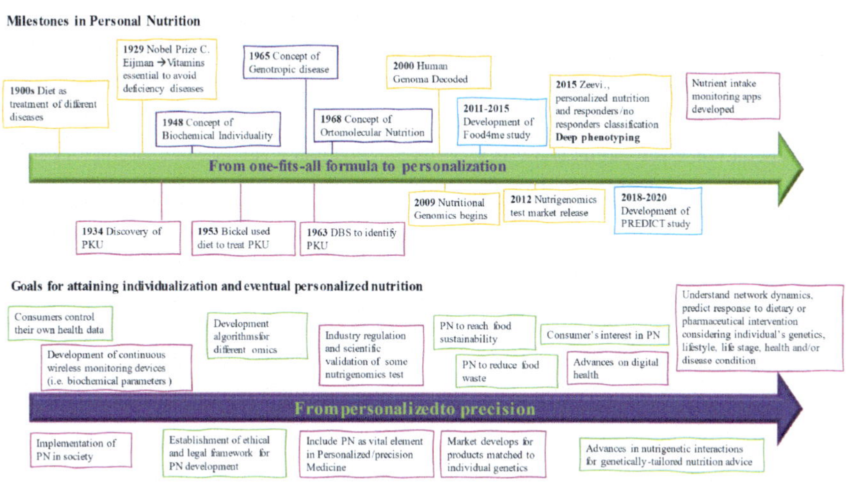

Fig. 2.3 Timeline of milestones in the development of personalized nutrition. The initial description and major advances in the understanding of the role diet play in development and treatment of different pathologies at an individual level. PN: personalized nutrition

The concept of personalization in nutrition began to appear in the work of Williams, who in 1948 introduced the concept of "biochemical individuality" and in 1965 the concept of "genetotrophic disease;" this is defined as a disease that occurs in an individual if the diet fails to provide a sufficient supply of one or more nutrients that are required at higher levels due to unique genetic needs (Williams 1950, Williams and Pelton 1965). In 1968, Pauling established the term "orthomolecular nutrition," defined as the science of improving and maintaining human health through the use of natural substances nutritious for the human organism. In this conception, if each cell of the body receives the optimal nutrients it requires to function properly, the organism's internal environment will be able to control certain diseases. This makes it is possible to prevent and treat many forms of disease via administration of optimal amounts of nutritious substances (Pauling 1968).

Another factor in development of nutrition science has been the discovery of diseases associated with genetic variants. A prime example is phenylketonuria (PKU), an autosomal recessive disorder of phenylalanine metabolism in which phenylalanine accumulation produces brain dysfunction (Van Spronsen et al. 2021), which was described in 1934 by Følling (1934). Discovery of PKU helped to better understand how nutrients and/or diet affect human health. Dietary treatment of PKU was developed in 1953 by Bickel et al. (Bickel et al. 1953), and screening to identify the disease in the population-based newborns, using dried blood spot (DBS) cards, was developed in 1963 by Guthrie and Susi (1963).

As nutrition science was burgeoning in the mid-20th century, so was the incidence of obesity. The increased prevalence of obesity began during the Nutritional Transition of the 1950s and derived from economic and social changes. The combination of reduced physical activity with modulation of eating patterns produced greater body mass index and consequently increased the prevalence of many chronic diseases (i.e., CVD) (Popkin 2015). At the time, it was believed that the principal cause of obesity was intake of fats and later of refined sugars (Keys et al. 1984; Kearns et al. 2016; Johns and Oppenheimer 2018). The food industry responded by developing new dietary products with reduced fat, saturated fats, cholesterol, and sugar contents, as well as others fortified with micronutrients. Technologies were developed that focused on reducing saturated fats such as partial hydrogenation of vegetable oils (Mozaffarian et al. 2018). Large cohort studies and intervention and randomized clinical trials have generated evidence of the impact of following certain dietary regimes and of specific nutrients on noncommunicable diseases (Mahmood et al. 2014; Martínez-González et al. 2014; Jacobs and Tapsell 2013). This knowledge helped in developing improved dietary guidelines and recommendations to follow certain healthier dietary patterns, such as traditional Mediterranean or vegetarian diets. These did not appear to affect the obesity pandemic, which continues unabated, and more recent research has identified overnutrition as the main contributor to this condition (Mozaffarian 2017).

2.2.1 Nutrition in the Post-genomic Era

Decoding of the human genome and consequent development of new technologies gave rise to nutritional genomics in 2009. It is an approach intended to facilitate the mechanisms involved in an individual's response to diet and to evaluate the variables affecting their organism (Ozdemir et al. 2009). The advent of nutrigenetics helped to augment the kind of recommendations that could be made based on genomic and biometric parameters. Indeed, 2012 saw the market expand with the introduction of a nutrigenetic test (Home—Genetic Testing Registry). These tests have tended to focus on genotype or genomics sequence-based analyses in search of biomarkers to incorporate into personalized diets. However, identification of variations at the proteome, metabolome, or epigenome levels will be important to developing more robust personalized diets and will allow monitoring of the organism's functioning at different life stages (Fig. 2.1).

Future progress in the science of PN will need to focus on the role of nutrition throughout the lifespan and address diet-related conditions through multifaceted interventions that go beyond just choosing healthy foods (Rodgers and Collins 2020). The fact that SNPs cannot explain different diseases such as obesity drove development of research aimed at quantifying the contribution of genetic variations to a wide range of phenotypes (i.e., fat mass and obesity-associated FTO gene) (JC, 2017). Consequently, there is currently a strong focus on deciphering the roles of certain SNPs, the interactions between gene variants or epistasis (Gratten and Visscher 2016), and interactions between the environment (Huang and Hu 2015), gut microbiota, gene expression modulation, the proteome, and the metabolome. In recent years, the effects of epigenetic modifications (i.e., histone acetylation, DNA methylation, miRNA expression) in health are gaining increasing importance. An excellent example is the case people who suffered through the Dutch Hunger Winter between 1944 and 1945; in early life, they manifested epigenetic dysregulation of the IFG2 gene which persisted throughout their lives (Heijmans et al. 2008).

Development of algorithms and application of statistical methods (i.e., multivariate techniques) allows integration of immense quantities of data and their translation into phenotypic and genotypic data and thus assessment of potential associations between eating patterns and metabolic outcomes that can then be used to generate recommendations for nutrient intake (Özdemir and Kolker 2016). These algorithms are a series of decision trees that lead from a specific phenotypic and/or genotypic characteristic to a concise recommendation for altering nutrient intake. This profound characterization of interindividual body response to a dietary stimulus will help in estimating dietary effects. For instance, Zeevi et al. developed an algorithm to predict individual postprandial glycemic response based on dietary habits, physical activity, anthropometrics, blood parameters, and gut microbiota (Zeevi et al. 2015). The PREDICT 1 study also assessed postprandial metabolic responses in a clinical setting and at home and found that gut microbiome had a greater influence than did meal macronutrients for postprandial lipemia but not for postprandial glycemia (Berry et al. 2020a), while genetic variants had a modest impact on glucose or postprandial predictions. In agreement with Zeevi study, the PREDICT 1 study also suggest that microbiome composition was predictive for a large panel of cardiometabolic blood markers (Asnicar et al. 2021).

New techniques will still need to be developed to assess health parameters using non-invasive and cost-effective samples. The multidisciplinary approaches of modern nutrition science will help to decipher the role of whole food, individual nutrients, social and sociocultural patterns, and genome/microbiome variation between individuals (Berry et al. 2020b; Toro-Martín et al. 2017). In the future, personalized nutrition practices may benefit from seamless planning of life sciences funding, research, and practice agendas that move from "farm to clinic to supermarket to society," and from "genome to proteome to metabolome" (Özdemir and Kolker 2016).

2.2.2 New Horizons in Personalized Nutrition

The National Institutes of Health (NIH) has launched a 2020–2030 Strategic Plan for NIH Nutrition Research that aims to increase biomedical research focused on the role of diet in diseases. Within this framework, the NIH provided an initial investment of $150 million for a large cohort study designed to answer key questions regarding what, when and how people should eat to promote health across the lifespan and how food may act as medicine (Rodgers and Collins 2020). The NIH Strategic Plan will contribute to deciphering the multidimensional approach of personalized nutrition and precision nutrition. For example, this will permit to identify different nutritional scores that can screen deeply and assess the influence of genetic variations. However, for effective personalized/precision nutrition to function, the research community will also need to focus on how nutrition advice is delivered in specific populations, preferably including all three approaches of personalization (Fig. 2.2). In this context, the use of nutrigenetic tests that target a limited number of SNP is a reality in the market today but will probably dramatically increase with other novel SNPs and the inclusion of valid nutritional GRS. They will be probably incorporated into all the features necessarily to translate precision nutrition to the market.

2.3 Individualization and Food Choices Based on Personalized/Precision Nutrition and Involvement of Diet in Chronic Diseases

The use of biomarkers to follow compliance with a given dietary regime will improve characterization of individual's diet. Metabolomics can help to identify different compliance biomarkers useful in exploring their health effects and identifying how microbial metabolites derived from food intake also mediate significant gut microbial metabolic activity (Ulaszewska et al. 2020). The food metabolome is that portion of the human metabolome derived directly from digestion and biotransformation of foods and their constituents. It represents a considerable and still largely unexploited source of novel dietary biomarkers that can be used to generate highly detailed and accurate measurements of dietary exposure. In depth, evaluation of the results will contribute to discovering different molecules and dietary factors associated with diseases (Scalbert et al. 2014). The interactions between the host and its gut microbiota are highly dynamic and complex but still largely unknown; they constitute a crucial framework for researching the gut-microbe-metabolic axis (Lamichhane et al. 2018). For instance, a study by Liu et al. found that the host can modulate the microbiome thorough miRNA secretion, which may provide an opportunity for modulation of the gut microbiome by miRNAs (Liu et al. 2016). Indeed, diet shapes gut microbiota composition and specific dietary patterns (i.e., Mediterranean Diet) are associated with beneficial microbiome-related metabolic profiles (Filippis 2016). In contrast, consumption of red meat is associated with alterations in gut microbiota that contribute to the development of CVD (Mei et al. 2021). Profiling gut microbiota is receiving more attention in nutritional intervention studies, and the composition and diversity of gut microbiota have been identified as potential risk factors for development of various chronic diseases (Ridaura et al. 2013).

Metabolic phenotyping—grouping people based on their metabolic characteristics—is a relatively new research field which may have great value in personalized nutrition. In longitudinal studies, metabotyping has shown that metabotypes may be associated with cardiometabolic risk factors and diet-related diseases, while its application in interventions can identify metabotypes with differential responses (Hillesheim and Brennan 2020).

The Food4Me study addressed the delivery of phenotype-based personalized nutrition. Participants were randomly distributed into four groups: (1) Non-personalized diet; (2) PN-based individual dietary intake; (3) PN including individual intake and phenotypic data (glucose, total cholesterol, carotenoids, w-3 index, characterization of 32 fatty acids, and vitamin D; and (4) group 3, plus genotypic data (MTHFR, FTO, TCF7L2, APOE E4, FADS1 genes). Dietary data were collected using a validated FFQ (Forster et al. 2014). Based on twenty-seven fasting metabolic markers measured by DBS, including cholesterol, individual fatty acids, and carotenoids, three metabotypes were aggregated from 180 randomly selected participants. Dietary advice encompassed characteristics of the metabotype and the decision trees, which include dietary factors not captured by the metabolites (total cholesterol, triacylglycerol, high-density lipoprotein cholesterol and glucose), as well as self-measured anthropometric characteristics (O'Donovan et al. 2015). When the appropriateness of the targeted dietary advice was compared with the individualized dietary advice, the resulting match was excellent (average match of 82%) and provided the same dietary message in both groups (O'Donovan et al. 2017). Further, research is still clearly needed to understand the biological mechanisms underlying the individual responses, especially detailed studies examining the underlying biology responsible for different metabotypes and deciphering the role of genetics and the microbiome (Hillesheim and Brennan 2020).

The use of deep phenotyping—comprehensive analysis of phenotypic abnormalities in which the individual components of the phenotype are monitored—to monitor different metabolic parameters will contribute to characterizing individual response, variability of symptom manifestations, etc. (Tracy 2008). An advantage of stratifying by metabolic profile is that it facilitates design of better nutrition advice focused on the individual. Very few studies have implemented extensive phenotyping (Zeevi et al. 2015; Berry et al. 2020a, 2020b). One example is the Maastricht study, an epidemiological study aimed at identifying pathophysiology and associated metabolic disturbance by monitoring 1000 individuals using traditional and advanced phenotyping (Schram et al. 2014). Wider phenotyping of humans and the possibility of monitoring health status through identification of individual profile metabotypes still needs to be explored via nutritional intervention studies if responses to nutrients are to be predicted.

Genetic risk scores (GRSs) are used to evaluate dietary interactions and may facilitate selection of more individualized and effective nutritional therapy by developing personalized, genotype-based approaches. In one study, a GRS was developed to predict obesity through analysis of a set of sixteen genetic variants associated with obesity and lipid metabolism (Table 2.1) (Goni et al. 2015). Other studies have used even larger sets of variants (i.e., 63 obesity-associated variants) to produce dietary recommendations aimed at reducing BMI in the United States (Casas-Agustench et al. 2014). Another approach has focused on intake of sugar-sweetened beverages, fried foods, and saturated fatty acids to evaluate the obesity risk score (Casas-Agustench et al. 2014; Olsen et al. 2016; Qi 2014; Qi et al. 2012; Brunkwall et al. 2016). These are just a few examples, but the general idea behind widespread GRS usage is to personalize treatment by incorporating different parameters into an analysis, for example, physical activity and circadian rhythms (i.e., analysis of *CRY1, PLIN1, CLOCK*) (Toro-Martín et al. 2017) or gut microbiomics (Bassaganya-Riera et al. 2021).

Chronic diseases generally manifest in highly variable ways between individuals and are associated with complex multifactorial variables, some as yet unknown. For instance, obesity is a complex multifactorial phenotype; interindividual variation in such phenotypes is thought to result from the action of multiple genes and environmental factors. Different strategies are therefore required to understand these diseases, such as identifying the role of specific genes in energy balance and adipose tissue biology. Other required techniques are genome-wide linkage scans to identify regions of interest and evaluation of the tissue-specific gene expression profile to compare lean versus obese individuals (Loos and Bouchard 2003). The epigenetic marks (DNA methylation, circulating miRNAs) involved in disease progression will also need to be developed (Mantilla-Escalante et al. 2021; Mantilla-Escalante 2019; Samblas et al. 2019; Kornfeld et al. 2013; Carmen Martínez-Jiménez et al. 2018; Aganzo et al. 2018). This can be done using a "dietary signature" created with genomic and epigenomic tools (i.e., transcriptomics, proteomics, metabolomics, miRNomics) and will help to identify the specific genes and proteins influenced by specific nutrients. In this context, it would be particularly useful to develop tools capable of incorporating the entire biological system

Table 2.1 Genotype of the 16 SNPs included in the GRS. Adapted from Goni et al. (2015)

Gene	SNP	Major/minor alleles
FTO	rs9939609	T/A
MC4R	rs17782313	T/C
MTHFR	rs1801133	C/T
PPARA	rs1800206	C/G
PPARG	rs1801282	C/G
APOA5	rs662799	T/C
APOE	rs429358	T/C
APOE	rs7412	C/T
LIPC	rs1800588	C/T
PLIN1	rs894160	G/A
NOS3	rs1799983	G/T
GCKR	rs1260326	C/T
LPL	rs328	C/G
CELSR2	rs12740374	G/T
CETP	rs1800777	G/A
LIPG	rs4939883	C/T

and monitoring all potential alterations in homeostasis. Deep phenotyping will contribute to better characterization of certain diseases and consequently improve evaluation of intervention outcomes. A major limitation of this technique is its high cost, need for trained professionals, and use of expensive omics tools. Improvement in different omics overtime will allow creation of customized diets based on the capacity of omics to identify and stratify the metabolites that contribute to modulating a certain health condition. Indeed, incorporation of different genetic outcomes will make for better precision in personalization even though many factors (Fig. 2.2) must be considered when designing individualized dietary regimes (Toro-Martín et al. 2017).

Advances in genomic sciences have clearly permitted a better understanding of the role of genetic variants and epigenetic signatures, as well as gene expression patterns, in the development of diverse chronic conditions (Ramos-Lopez et al. 2017). Various issues still need to be resolved before PN can come to fruition. For example, it is vital to confirm that PN interventions really do produce behavioral changes beyond those experienced in response to conventional dietary recommendations (Jinnette et al. 2021). Data from PN interventions indicate that phenotypic data, and the psychological, social, economic, and cultural factors that influence eating patterns have positive effects in designing PN recommendations (Jinnette et al. 2021), and should thus be considered when applying a widespread PN program in the general population. Another factor is that even though inclusion of genetic testing with dietary recommendations does promote positive changes in dietary habits (Horne et al. 2018; Celis-Morales et al. 2017b), this response is not equal in males and females. In fact, previous studies, including Food4me, indicate that females are more receptive to participation in nutrition studies than males, which can be explained by females' greater interest in health and nutrition (Jinnette et al. 2021; Livingstone et al. 2020, 2016). Socioeconomic status can also impact changes in eating and physical activity, meaning PN should focus on a wide range of social and behavioral factors in addition to genotypic analysis.

The data available to date suggest that improving PN dietary recommendations and ensuring that it is used in real life (not just in controlled trials) will require analysis of the added value associated with its implementation and the way in which nutrition professionals adopt and manage this new approach. Before scientific research on PN can be translated into daily clinical care, studies are needed identifying the strengths and limitations affecting consumers, exploring policies to protect personal information, and studying legal and ethical aspects.

2.4 Translating Personalized Nutrition for Society

The obesity pandemic has highlighted the acute need to develop novel approaches to reduce the incidence of different disorders and their treatment. Present dietary habits clearly need to be healthier.

Various studies have found that customized dietary advice is more effective than generic advice, leading to improvements in dietary behaviors and other associated parameters (Celis-Morales et al. 2017a; Aganzo et al. 2018). In a sub-study of the SMART trial, designed as 24-month randomized clinical trial of behavioral treatment for weight loss, participants receiving personalized advice attained 1.83 kg lower weights than those receiving "one-size-fits-all" information (Ambeba et al. 2015). In another study, inclusion of genotypic information did not motivate individuals to change their dietary behavior (Marteau et al. 2010). The Food4Me study found that even though tests were done for the fat mass and obesity-associated gene (FTO), transcription factor 7-like 2 gene (TCF7L2), apolipoprotein E4 genotype (ApoE4), desaturase decoding gene (FADS1), and methylenetetrahydrofolate reductase gene (MTHFR), participating individuals, were still unmotivated to change their behavior (Celis-Morales et al. 2017a). Why this occurs is still unknown; apparently, lifestyle habits are difficult to change.

In personalized nutrition, dietary recommendations will be prescribed at an individual level, but they need to be easy to follow, to translate into society, and to communicate to the public. For instance, a simple recommendation is to eat five fruits a day even though recommendations will be managed in response to a specific phenotype or genome (i.e., caffeine intake, PUFAs, systolic blood pressure, etc.). In theory, once a person is aware of their genotype, they will be more aware of what they eat and when they eat it. Perhaps, dietary intake will not change in the long term, but at least, the perception and behaviors related to specific recommendations can change (Roke et al. 2017).

Nutrigenomics and nutrigenetics will drive the food manufacturing and service industries to develop different products that consider personal taste, preferences, cultural habits, and the need to fulfill personalized dietary recommendations (Kussmann and Fay 2008). Many online tools already exist for collecting dietary intake data and producing dietary recommendations, but whether or not these contribute to improving eating patterns still requires validation (Ryan et al. 2015).

Few dietary interventions have focused on studying the effect of PN advice based on different phenotypic and genotypic characteristics. Among them, the Food4Me study focused on evaluating whether different dietary recommendations based on dietary pattern, phenotypical values, or genotype produced better compliance compared to traditional dietary advice (Celis-Morales et al. 2017a).

In the context of Food4Me study, results generally indicated that after 6 months, nutritional recommendations via the Internet induced greater compliance with healthier diets in the PN and lifestyle approach than conventional dietary recommendations. Effectiveness almost doubled when recommendations was further personalized by including individual dietary behavior. In concrete terms, participants reduced their consumption of red meat, saturated fat and salt, and increased their folate intake, producing significant improvements in dietary habits (Celis-Morales et al. 2017a).

Adding different biomarkers and/or genetic data to the recommendations produced no added value for diet quality but did act as a predictor of dietary intervention response (Celis-Morales et al. 2017a). Indeed, identification of phenotypes which are responsive to dietary recommendations may help to understand why this occurs, and then, this knowledge can be used to further enhance effectiveness of the dietary strategy. For instance, basal levels of circulating cholesterol contribute to classifying participants as responders or non-responders to dietary recommendations based on changes in cholesterol levels from baseline to month 6 (Kirwan et al. 2016). It was also found that the FTO genotype risk had a greater effect on a reduction of body weight and waist circumference in risk carriers than in non-risk carriers across different levels of personalized nutrition (Celis-Morales et al. 2017b). In summary, the potential use of metabolic and genetic profiles in identifying a person's response to an intervention could be crucial in developing precision nutrition.

2.4.1 Social Impact Regarding PN

Based on the complex and diverse individual characteristics influencing the factors affecting dietary interventions (i.e., dietary behavior), different nutrition science strategies have been developed to produce personalized dietary interventions more probable to impact societal dietary habits. One example is the introduction of new concepts that contribute to modifying the perceptions and impact of diet and its role in society (i.e., sustainable diet) (Clark et al. 2019). Integration of PN into society needs to incorporate (i) technical support using new technologies, development of new science and innovative tools, identification of obstacles and fears among the population; (ii) political motivation to include nutrition and health

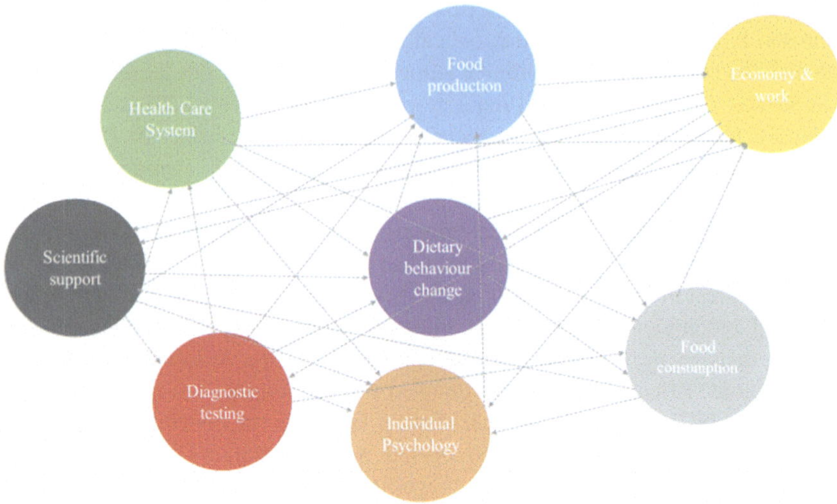

Fig. 2.4 Network of the personalized nutrition system map. Adapted from: eu: Food4Me (2015)

issues in the public healthcare system; and (iii) commercial efforts to familiarize consumers with PN. A transition process will occur in which modification of dietary eating patterns is supported by economic incentives, development of new tools and methods that highlight the interaction between nutrition and gene molecular data, and strong social pressure (Fig. 2.4). The most effective approach will involve collaboration between the public and private sectors (Bassaganya-Riera et al. 2021). The research community has advanced in PN development, but its widespread implementation in society is only feasible if public and private agents function as an integrated network.

2.4.1.1 New Data Collecting and Sampling Tools

A primary question in nutrition science is how to overcome the biases surrounding to nutrient intake. Anthropometric, dietary habits, and food intake data are important to collect for many reasons, including (i) development of nutritional recommendations; (ii) identifying nutrigenomic associations, and (iii) finding compliance biomarkers. They are also required for individualization and creating automatic diagnostic and monitoring tools for healthcare systems.

Although at the experimental stage, the concept of a Web-based study was introduced by Hercberg et al. (2010) as a prospective study (ongoing) and is expected to recruit $\approx$500,000 study participants in France. This study focuses on the relationship between nutrition and health and dietary patterns over the long term. All the data is to be collected remotely, will be digitized, and compared to a nationally representative survey (Andreeva et al. 2016). Similar intake was observed in both cohorts (electronic and survey) for carbohydrates, total lipids, protein, and total energy, but intake of fruits and vegetables, fiber, certain vitamins, and minerals was higher in the e-cohort. Alcoholic and non-alcoholic beverage intake was lower in the e-cohort. Real differences in intake, mode effects, and volunteer bias may each contribute to explaining the findings. These results suggest that new tools are needed to collect data and sample while avoiding volunteer bias. The Food4Me study used a similar Web-based model approach, as described above. These two studies are examples of the feasibility of new data collecting and sampling tools. Because they are relevant to bringing nutritional personalization to the general public, better tools are still needed.

Recent years have seen the introduction of different electronic data collecting tools for extensive phenotyping which may be useful in personalized nutrition. Mobile applications (apps) are in development to measure body parameters such as blood pressure, blood glucose concentration, sleep quality, heart rate, energy expenditure, and exercise intensity, among others (Lieffers and Hanning 2012).

Sample collection also poses a problem, and new tools will be needed before PN can become common. A common and cost-effective tool are DBS cards. They provide many scientific advantages are low cost and facilitate logistics (Deep et al. 2012). Analyses using DBS can provide data on health markers and nutritional status and can be used as compliance biomarkers for food intake data, i.e., vitamin E (Hoeller et al. 2016), carotenoids (Rubió et al. 2020), dietary phenols (Las Hazas et al. 2016), and even genomic and epigenomic biomarkers, etc. Sampling via DBS is currently limited due to the challenge of validating measurements and verifying comparability with traditional venous blood samples. In the future, DBS may also be employed to identify different phenotypic and epigenetic biomarkers.

2.4.2 PN-Associated Business and Value Creation Models

The high prevalence of obesity and chronic diseases in our societies represents an immense potential market for personalized nutrition services. These can range widely from the analytical sphere (e.g., tools, diagnostics, interpretation algorithms) to direct attention (shops, restaurants, wellness and sport centers, medical services, leisure, education, etc.) and other commercial services. To foment personalized nutrition services, new value creation models for PN need to be designed to augment consumer's perception of the role of food in health and the environment. These models should integrate different key points such as personal coaching principles, new technological tools that permit self-sampling diagnostics, and monitoring of lifestyle and food intake. The Food4Me project explored different business models and value creation concepts to merge insights from this process with emerging insights on scientific, technical, consumer, legal and ethical aspects.

One example of these potential markets is the recent expansion in the genetic testing market. This flood of new information may trigger potentially rapid behavioral changes intended to control adverse factors (Guasch-Ferré et al. 2018) although the scientific evidence supporting response to clinical recommendations is open to debate, and the extent of behavioral modifications after genetic information disclosure is unclear.

2.4.3 Social Concerns and Their Impact on PN Development

As personalized nutrition evolves, so will people's concerns about it. These include (i) the strength of using individual genetic information for personalizing dietary advice; (ii) extensive diagnostic testing may constitute a cost barrier to PN compliance; (iii) the perceived improbability of creating truly individualized nutritional recommendations (the most realistic approach is to identify large groups of individuals [nutritypes] with a similar profile for which a set of nutritional recommendations may function); (iv) developing personalized food products is economically unfeasible; (v) changes in product and dietary recommendations make the market skeptical of developing the required products; (vi) facilitating PN implementation in society could provide some benefits; (vii) consumers will have varying abilities to understand PN which could generate confusion and/or fear, and (viii) the perception of PN as a dietary health opportunity is not self-evident. All these aspects will need to be clarified for consumers before PN's benefits can become more generalized.

2.4.3.1 Ethical and Legal Concerns

Personalized nutrition services require personal health data, which raises new ethical, legal, and social issues. Personal data management should handle different issues of consumer or patient protection, but the patient or consumer will still have to assess the potential benefits of using the service and the risks of providing personal health or lifestyle data.

The increasing availability and affordability of genetic testing has prompted many consumers to purchase direct-to-consumer genetic tests and then bring their reports in for interpretation (Horne et al. 2021). Most of these kinds of genetic tests have minimal analytical or scientific validity and limited clinical utility. However, they also have different ethical, legal, and social implications (Horne et al. 2021). Despite their limited validity, genetic tests are increasingly popular, and, in the United States, for instance, the market is growing steadily (San-Cristobal et al. 2013).

Unlike traditional nutritional recommendations, PN has different associated legal issues involving individual rights. Various guidelines and regulatory frameworks affect PN services. Identifying the relevant legal norms applicable to PN, and the resulting barriers and requirements, has been approached by evaluating typical PN business models considering current legal requirements adopted, or soon to be adopted, by the EU or its member states, and/or international legal instruments, where they exist.

2.4.4 Consumer Attitudes Toward Personalized Nutrition

Another critical issue is consumer acceptance of personalized nutrition, especially since the data generated by PN may involve certain risks. The use and storage of an individual's genetic information as part of nutrigenomics generates different societal concerns. This information is highly personal, and it is therefore vital to know how these data are managed and how they could impact on individual's privacy. Consumer acceptance of new technologies affects perceptions of disease risk and responses to the application of personalized nutrition. Indeed, application of personalized nutrition phenotypic data, even when not involving genetic differences, raises privacy issues. Characterizing a phenotype requires sensitive personal information such as family and personal history, psychological well-being, and environment, social, and lifestyle practices.

Genomics has great potential in healthcare systems, but individuals' response to this treatment poses a challenge. Despite the extensive scientific research supporting it, individual motivation remains the principal challenge to successful implementation of personalized diets (Fallaize et al. 2013). In this sense, attitudes toward personalized nutrition may influence behavioral intention regarding its adoption and the potential success of marketing a personalized nutrition service. Furthermore, consumer's perceptions of its benefits can also influence how much they will be willing to pay for personalized nutrition recommendations. This is why careful communication about personalized nutrition is crucial to attracting individuals who could benefit from it but do not perceive those potential benefits (Stewart-Knox et al. 2013).

There is also debate about the willingness of consumers to pay more for the required additional analyses needed to generate Personalized Nutrition recommendations. They will need to consider their perceived risks, benefits, and self-efficacy, as well as their attitude about personalized nutrition and their behavioral intention to adopt it (Fig. 2.5).

Among the main results of the Food4Me is that consumers will pay for PN nutritional recommendations. Indeed, they are willing to pay up to 50% extra compared to regular services for analyses such as blood collection, DNA collection, and dietary advice (Poínhos et al. 2014). The main predictor of a positive attitude toward complying with PN is the motives for food choice. However, consumers' readiness to acquire greater knowledge about their health may not result in an intention to modulate their eating patterns. Negative attitudes toward PN compliance may be even stronger, particularly in terms of higher food price, sensory appeal, and familiarity (Nielsen et al. 2014; Stewart-Knox et al. 2008).

Personalized nutrition may be able to motivate changes in a population, but it is clear that intense feedback for individual patients is what positively effects nutrition self-efficacy scores (Stewart-Knox et al. 2008). When comparing between the

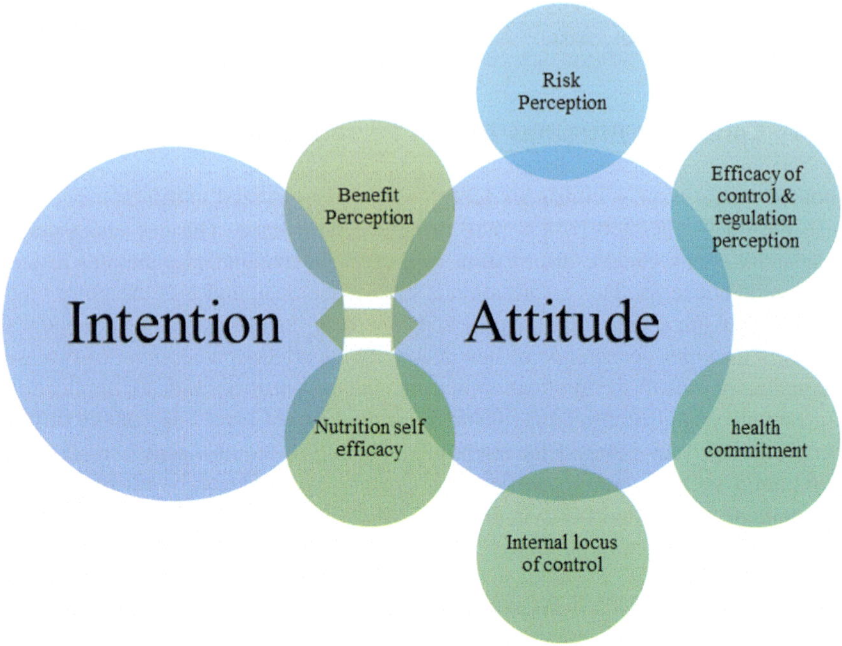

Fig. 2.5 Psychological determinants of personalized nutrition in European countries. Adapted from: Poínhos et al. (2014)

different approaches to personalized nutrition services in Food4Me participants, those based on an individual's current diet seem to be sufficient to elicit adequate changes in eating habits. Moreover, as consumers are more exposed to individualized/personalized medicine, they will probably become more positive toward the potential benefits of personalized nutrition (Celis-Morales et al. 2017a).

Communication will be an important aspect of introducing PN to the general public. Once ethical and legal concerns are resolved, and/or new legislation comes into effect regulating this specific aspect of human health regulation, promotion of PN can facilitate its widespread implementation. Given the multiple ways, people now access information and communicate; a key concern for PN will be to determine the best way to coach (traditional face-to-face interviews, Web-based interaction, chatbot interaction, or others) consumers into improving their eating patterns and attaining a level of personalized nutrition without fear of sharing personal genetic data, and other biomedical data, with public/private institutions. Social media applications may impact health behavior and so need to be considered when communicating for PN purposes. For instance, a study of Twitter found it influenced food-related behavior and attitudes, suggesting its potential impact in social health (Vydiswaran et al. 2020).

2.5 Future Outlook

The long-term effects caused by changes in food intake and lifestyle following administration of personalized advice are still unclear, but these approaches will need long-term support for participants/customers to generate sustainable changes. Models and studies to test how this can be achieved need to be developed. These may include, for example, novel reward systems to maintain interest and motivate participation by health insurers or other healthcare providers in the business. Again, collaboration between the public and private sectors will be vital to its functioning.

Non-invasive or minimally invasive monitoring devices (wireless continuous monitoring systems) will become important tools in future nutrition research and in developing precision nutrition approaches. These devices will also allow implementation of novel study designs and enable large-scale studies without the need to bring volunteers into study centers. They will also provide immediate feedback to study participants on changes in body functions, which could increase compliance and foster motivation, especially in scenarios of lifestyle interventions. Nutritional genetic risk scores will facilitate diet personalization, but stronger GRS needs to be created that can combine both nutrigenetic data and whole-genome sequencing data from precision medicine initiatives that truly predict the risk of common human diseases associated with diet. Algorithms to manage large-scale datasets that combine genetic, genomic, epigenetic, environmental, and other data types will be needed to integrate these data and individualize nutrition.

2.6 Concluding Remarks

Personalized nutrition will need to incorporate nutrigenomic knowledge as well as other lifestyle parameters such as physical activity, gut microbiomics, metabolomics and exposome habits at every stage of life. Myriad individual characteristics that may influence dietary eating patterns or behavior must also be incorporated, including cultural variations, allergies, and intolerances, many of which may have as yet unknown multigenic origins. For personalization to be effective, consumers will need to adopt positive attitudes and receptiveness toward usage of widespread genomics data in nutrition, accompanied by a positive social impact. New business and value creation models will also be needed, social, ethical and legal concerns addressed, communication implemented and consumer attitudes toward personalized nutrition refined. Personalized nutrition is aimed at modifying behavioral changes at every life stage much more effectively than one-size-fits-all diet strategies.

2.7 Financial Support

This research was funded by grants from the Spanish "Agencia Estatal de Investigación" and European FEDER Funds to AD (PID2019-109369RB-I00 and AGL2017-90623-REDT), to MCLH (RTI2018-093873-A-I00 and IJC2020-044353-I). Authors also acknowledge to Spanish Foundation of Arteriosclerosis ("Manuel de Oya" Nutrition grant 2021).

Conflict of Interest No conflict of interest is declared by authors.

References

Aganzo M et al (2018) Customized dietary intervention avoids unintentional weight loss and modulates circulating miRNAs footprint in Huntington's disease. Mol Nutr Food Res 62:1800619
Ambeba EJ et al (2015) The use of mHealth to deliver tailored messages reduces reported energy and fat intake. J Cardiovasc Nurs 30:35
Andreeva VA et al (2016) Comparison of dietary intakes between a large online cohort study (Etude NutriNet-Santé) and a nationally representative cross-sectional study (Etude Nationale Nutrition Santé) in France: addressing the issue of generalizability in e-epidemiology. Am J Epidemiol 184:660–669
Asnicar F et al (2021) Microbiome connections with host metabolism and habitual diet from 1098 deeply phenotyped individuals. Nat Med 27:321
Auton A et al (2015) A global reference for human genetic variation. Nature 526:68–74
Bassaganya-Riera J et al (2021) Goals in nutrition science 2020–2025. Front Nutr 318
Becerra-Tomás N et al (2021) Fruit consumption and cardiometabolic risk in the PREDIMED-plus study: a cross-sectional analysis. Nutr Metab Cardiovasc Dis 31:1702–1713
Berry SE et al (2020a) Human postprandial responses to food and potential for precision nutrition. Nat Med 26:964
Berry SE et al (2020b) Human postprandial responses to food and potential for precision nutrition. Nat Med 26:964–973
Bickel H, Gerrard J, Hickmans EM (1953) Influence of phenylalanine intake on phenylketonuria. Lancet (london, England) 265:812–813
Bland JS (2019) The evolution of personalized nutrition—from Addis, Pauling, and RJ Williams to the future. Integr Med A Clin J 18:10
Brunkwall L et al (2016) Sugar-sweetened beverage consumption and genetic predisposition to obesity in 2 Swedish cohorts. Am J Clin Nutr 104:809–815
Casas-Agustench P et al (2014) Saturated fat intake modulates the association between an obesity genetic risk score and body mass index in two US populations. J Acad Nutr Diet 114:1954–1966
Casas-Agustench P et al (2015) Consumption of distinct dietary lipids during early pregnancy differentially modulates the expression of microRNAs in mothers and offspring. PLoS ONE 10:e0117858
Celis-Morales C et al (2017a) Effect of personalized nutrition on health-related behaviour change: evidence from the Food4Me European randomized controlled trial. Int J Epidemiol 46:578–588
Celis-Morales C et al (2017b) Can genetic-based advice help you lose weight? Findings from the Food4Me European randomized controlled trial. Am J Clin Nutr 105:1204–1213
Clark MA, Springmann M, Hill J, Tilman D (2019) Multiple health and environmental impacts of foods. Proc Natl Acad Sci 116:23357–23362

Cordero P, Ashley EA (2012) Whole-genome sequencing in personalized therapeutics. Clin Pharmacol Ther 91:1001–1009
Corella D et al (2007) APOA5 gene variation modulates the effects of dietary fat intake on body mass index and obesity risk in the Framingham Heart Study. J Mol Med (berl) 85:119–128
Dávalos A, Fernández-Hernando C (2013) From evolution to revolution: miRNAs as pharmacological targets for modulating cholesterol efflux and reverse cholesterol transport. Pharmacol Res 75:60–72
Dávalos A et al (2020) Dietary microRNAs and cancer: a new therapeutic approach? Semin Cancer Biol. https://doi.org/10.1016/j.semcancer.2020.10.006
Deep A, Kumar P, Kumar A, Thakkar A (2012) Dry blood spot technique: a review. Int J Pharm Sci Rev Res 15:90–94
De Filippis F et al (2016) High-level adherence to a Mediterranean diet beneficially impacts the gut microbiota and associated metabolome. Gut 65
De Las Hazas MCLMCL, Motilva MJMJ, Piñol C, Macià A (2016) Application of dried blood spot cards to determine olive oil phenols (hydroxytyrosol metabolites) in human blood. Talanta 159:189–193
de Toro-Martín J, Arsenault BJ, Després JP, Vohl MC (2017) Precision nutrition: a review of personalized nutritional approaches for the prevention and management of metabolic syndrome. Nutrients 9
del Carmen Martínez-Jiménez V, Méndez-Mancilla A, Patricia Portales-Pérez D (2018) miRNAs in nutrition, obesity, and cancer: the biology of miRNAs in metabolic disorders and its relationship with cancer development. Mol Nutr Food Res 62
eu: Food4Me, project white paper and presentation—Eurohealth. https://eurohealth.ie/2015/05/08/eu-food4me-project-white-paper-and-presentation/.
Fallaize R, Macready AL, Butler LT, Ellis JA, Lovegrove JA (2013) An insight into the public acceptance of nutrigenomic-based personalised nutrition. Nutr Res Rev 26:39–48
Ferguson LR et al (2016) Guide and position of the international society of nutrigenetics/nutrigenomics on personalised nutrition: part 1—fields of precision nutrition. J Nutrigenet Nutrigenomics 9:12–27
Fölling A (1934) Über Ausscheidung von Phenylbrenztraubensäure in den Harn als Stoffwechselanomalie in Verbindung mit Imbezillität. Hoppe Seylers Z Physiol Chem 227:169
Forster H et al (2014) Online dietary intake estimation: the Food4Me food frequency questionnaire. J Med Internet Res 16
Gibney MJ, Walsh MC (2013) The future direction of personalised nutrition: my diet, my phenotype, my genes. Proc Nutr Soc 72:219–225
Goni L, Cuervo M, Milagro FI, Martínez JA (2015) A genetic risk tool for obesity predisposition assessment and personalized nutrition implementation based on macronutrient intake. Genes Nutr 10:1–10
Gratten J, Visscher PM (2016) Genetic pleiotropy in complex traits and diseases: implications for genomic medicine. Genome Med 8:1–3
Guasch-Ferré M, Dashti HS, Merino J (2018) Nutritional genomics and direct-to-consumer genetic testing: an overview. Adv Nutr 9:128–135
Guthrie R, Susi A (1963) A simple phenylalanine method for detecting phenylketonuria in large populations of newborn infants. Pediatrics 32
Heijmans BT et al (2008) Persistent epigenetic differences associated with prenatal exposure to famine in humans. Proc Natl Acad Sci USA 105:17046
Hercberg S et al (2010) The Nutrinet-Santé study: a web-based prospective study on the relationship between nutrition and health and determinants of dietary patterns and nutritional status. BMC Public Health 10
Hesketh J (2012) Personalised nutrition: how far has nutrigenomics progressed? Eur J Clin Nutr 67:430–435
Hillesheim E, Brennan L (2020) Metabotyping and its role in nutrition research. Nutr Res Rev 33:33–42

Hoeller U et al (2016) Application of dried blood spots to determine vitamin D status in a large nutritional study with unsupervised sampling: the Food4Me project. Br J Nutr 115:202–211
Home—Genetic Testing Registry (GTR)—NCBI. https://www.ncbi.nlm.nih.gov/gtr/
Horne J, Madill J, O'Connor C, Shelley J, Gilliland J (2018) A systematic review of genetic testing and lifestyle behaviour change: are we using high-quality genetic interventions and considering behaviour change theory? Lifestyle Genomics 11:49–63
Horne JR et al (2021) Guiding global best practice in personalized nutrition based on genetics: the development of a nutrigenomics care map. J Acad Nutr Diet. https://doi.org/10.1016/J.JAND.2021.02.008
Huang T, Hu FB (2015) Gene-environment interactions and obesity: recent developments and future directions. BMC Med Genomics 8:S2
Jacobs DR, Tapsell LC (2013) Food synergy: the key to a healthy diet. Proc Nutr Soc 72:200–206
Jinnette R et al (2021) Does personalized nutrition advice improve dietary intake in healthy adults? A systematic review of randomized controlled trials. Adv Nutr 12:657–669
Johns DM, Oppenheimer GM (2018) Was there ever really a "sugar conspiracy"? Twists and turns in science and policy are not necessarily products of malevolence. Science 359:747–750
Kearns CE, Schmidt LA, Glantz SA (2016) Sugar industry and coronary heart disease research: a historical analysis of internal industry documents. JAMA Intern Med 176:1680–1685
Keys A et al (1984) The seven countries study: 2289 deaths in 15 years. Prev Med (baltim) 13:141–154
Kirwan L et al (2016) Phenotypic factors influencing the variation in response of circulating cholesterol level to personalised dietary advice in the Food4Me study. Br J Nutr 116:2011–2019
Kornfeld J-W et al (2013) Obesity-induced overexpression of miR-802 impairs glucose metabolism through silencing of Hnf1b. Nature 494:111–115
Kussmann M, Fay LB (2008) Nutrigenomics and personalized nutrition: science and concept. Per Med 5:447–455
Lamichhane S, Sen P, Dickens AM, Orešič M, Bertram HC (2018) Gut metabolome meets microbiome: a methodological perspective to understand the relationship between host and microbe. Methods 149:3–12
Lieffers JRL, Hanning RM (2012) Dietary assessment and self-monitoring: with nutrition applications for mobile devices. 73 https://doi.org/10.3148/73.3.2012.e253
Liu S et al (2016) The host shapes the gut microbiota via fecal MicroRNA. Cell Host Microbe 19:32–43
Livingstone KM et al (2016) Profile of European adults interested in internet-based personalised nutrition: the Food4Me study. Eur J Nutr 55:759–769
Livingstone KM et al (2020) Characteristics of participants who benefit most from personalised nutrition: findings from the pan-European Food4Me randomised controlled trial. Br J Nutr 123:1396–1405
Loos RJF, Bouchard C (2003) Obesity–is it a genetic disorder? J Intern Med 254:401–425
López-González L et al (2021) Variety in fruits and vegetables, diet quality and lifestyle in an older adult mediterranean population. Clin Nutr 40:1510–1518
Mahmood SS, Levy D, Vasan RS, Wang TJ (2014) The Framingham Heart Study and the epidemiology of cardiovascular disease: a historical perspective. Lancet (london, England) 383:999–1008
Mantilla-Escalante DC et al (2019) Postprandial circulating miRNAs in response to a dietary fat challenge. Nutrients 11
Mantilla-Escalante DC et al (2021) Mediterranean diet enriched in extra-virgin olive oil or nuts modulates circulating exosomal non-coding RNAs. Eur J Nutr 60:1–15. https://doi.org/10.1007/S00394-021-02594-0
Marteau TM et al (2010) Effects of communicating DNA-based disease risk estimates on risk-reducing behaviours. Cochrane Database Syst Rev. https://doi.org/10.1002/14651858.CD007275.PUB2

Martínez-González MA, Dominguez LJ, Delgado-Rodríguez M (2014) Olive oil consumption and risk of CHD and/or stroke: a meta-analysis of case-control, cohort and intervention studies. Br J Nutr 112:248–259

Mei Z et al (2021) Dietary factors, gut microbiota, and serum trimethylamine-N-oxide associated with cardiovascular disease in the Hispanic Community Health Study/Study of Latinos. Am J Clin Nutr 113:1503–1514

Mozaffarian D (2017) Foods, obesity, and diabetes-are all calories created equal? Nutr Rev 75:19–31

Mozaffarian D, Rosenberg I, Uauy R (2018) History of modern nutrition science—implications for current research, dietary guidelines, and food policy. BMJ 361

Nielsen DE, El-Sohemy A (2014) Disclosure of genetic information and change in dietary intake: a randomized controlled trial. PLoS One 9

Nielsen DE, Shih S, El-Sohemy A (2014) Perceptions of genetic testing for personalized nutrition: a randomized trial of DNA-based dietary advice. J Nutrigenet Nutrigenomics 7:94–104

Nuñez-Sánchez MA et al (2015) MicroRNAs expression in normal and malignant colon tissues as biomarkers of colorectal cancer and in response to pomegranate extracts consumption: critical issues to discern between modulatory effects and potential artefacts. Mol Nutr Food Res 59:1973–1986

O'Donovan CB et al (2017) Metabotyping for the development of tailored dietary advice solutions in a European population: the Food4Me study. Br J Nutr 118:561–569

O'Donovan CB et al (2015) Use of metabotyping for the delivery of personalised nutrition. Mol Nutr Food Res 59:377–385

Olsen NJ et al (2016) Interactions between genetic variants associated with adiposity traits and soft drinks in relation to longitudinal changes in body weight and waist circumference. Am J Clin Nutr 104:816–826

Ordovas JM (2004) The quest for cardiovascular health in the genomic era: nutrigenetics and plasma lipoproteins. Proc Nutr Soc 63:145–152

Özdemir V, Kolker E (2016) Precision nutrition 4.0: a big data and ethics foresight analysis—convergence of agrigenomics, nutrigenomics, nutriproteomics, and nutrimetabolomics. OMICS 20:69–75

Ozdemir V et al (2009) Risk assessment and communication tools for genotype associations with multifactorial phenotypes: the concept of 'edge effect' and cultivating an ethical bridge between omics innovations and society. OMICS 13:43–61

Pauling L (1968) Orthomolecular psychiatry. Science 160:265–271

Pauling L (1970) Vitamin C and the common cold. Freeman, W.H

Piro A, Tagarelli G, Lagonia P, Quattrone A, Tagarelli A (2010) Archibald Edward Garrod and alcaptonuria: "Inborn errors of metabolism" revisited. Genet Med 128(12):475–476

Poínhos R et al (2014) Psychological determinants of consumer acceptance of personalised nutrition in 9 European countries. PLoS One 9

Popkin BM (2015) Nutrition transition and the global diabetes epidemic. Curr Diab Rep 15

Pozo-Acebo L, López de las Hazas M, Margollés A, Dávalos A, García-Ruiz A (2021) Eating microRNAs: pharmacological opportunities for cross-kingdom regulation and implications in host gene and gut microbiota modulation. Br J Pharmacol (2021). https://doi.org/10.1111/bph.15421

Qi Q et al (2012) Sugar-sweetened beverages and genetic risk of obesity. N Engl J Med 367:1387–1396

Qi Q et al (2014) Fried food consumption, genetic risk, and body mass index: gene-diet interaction analysis in three US cohort studies. BMJ 348

Ramos-Lopez O et al (2017) Guide for current nutrigenetic, nutrigenomic, and nutriepigenetic approaches for precision nutrition involving the prevention and management of chronic diseases associated with obesity. J Nutrigenet Nutrigenomics 10:43–62

Ridaura VK et al (2013) Cultured gut microbiota from twins discordant for obesity modulate adiposity and metabolic phenotypes in mice. Science 341

Rodgers GP, Collins FS (2020) Precision nutrition—the answer to "What to Eat to Stay Healthy." JAMA 324:735–736

Roke K et al (2017) Evaluating changes in omega-3 fatty acid intake after receiving personal FADS1 genetic information: a randomized nutrigenetic intervention. Nutrients 9

Rubió L et al (2020) Application of dried blood spot cards combined with liquid chromatography-tandem mass spectrometry to determine eight fat-soluble micronutrients in human blood. J.Chromatogr B Analyt Technol Biomed Life Sci 1152

Ryan NM, O'Donovan CB, Forster H, Woolhead C, Walsh MC (2015) New tools for personalised nutrition: the Food4Me project. Nutr Bull 40:134–139

Samblas M, Milagro FI, Martínez A (2019) DNA methylation markers in obesity, metabolic syndrome, and weight loss. Epigenetics 14:421–444

San-Cristobal R, Milagro FI, Martínez JA (2013) Future challenges and present ethical considerations in the use of personalized nutrition based on genetic advice. J Acad Nutr Diet 113:1447–1454

San-Cristobal R, Navas-Carretero S, Kohlmeier M, Alfredo Martínez J (2020) Precision nutrition interventions based on personalized genetic advice. Princ Nutr Nutr Fundam Individ Nutr 499–508. https://doi.org/10.1016/B978-0-12-804572-5.00065-3

Scalbert A et al (2014) The food metabolome: a window over dietary exposure. Am J Clin Nutr 99:1286–1308

Schram MT et al (2014) The Maastricht study: an extensive phenotyping study on determinants of type 2 diabetes, its complications and its comorbidities. Eur J Epidemiol 29:439–451

Stewart-Knox BJ et al (2008) Attitudes toward genetic testing and personalised nutrition in a representative sample of European consumers. Br J Nutr 101:982–989

Stewart-Knox B et al (2013) Factors influencing European consumer uptake of personalised nutrition. Results of a qualitative analysis. Appetite 66:67–74

Tomé-Carneiro J et al (2016) Hydroxytyrosol supplementation modulates the expression of miRNAs in rodents and in humans. J Nutr Biochem 34:146–155

Tomé-Carneiro J et al (2018) Breast milk microRNAs harsh journey towards potential effects in infant development and maturation. Lipid encapsulation can help. Pharmacol Res 132:21–32

Tracy RP (2008) 'Deep phenotyping': characterizing populations in the era of genomics and systems biology. Curr Opin Lipidol 19:151–157

Ulaszewska MM et al (2020) Two apples a day modulate human: microbiome co-metabolic processing of polyphenols, tyrosine and tryptophan. Eur J Nutr 59:3691–3714

van Spronsen FJ et al (2021) Phenylketonuria. Nat Rev Dis Prim 7:1–19

Vydiswaran VGV et al (2020) Uncovering the relationship between food-related discussion on Twitter and neighborhood characteristics. J Am Med Inform Assoc 27:254–264

Williams RJ (1950) Concept of genetotrophic disease. Nutr Rev 8:257–260

Williams RJ, Pelton RB (1965) Individuality in nutrition: the genetotrophic principle. Science 148:669

Zeevi D et al (2015) Personalized nutrition by prediction of glycemic responses. Cell 163:1079–1094

Zhang L et al (2012) Exogenous plant MIR168a specifically targets mammalian LDLRAP1: evidence of cross-kingdom regulation by microRNA. Cell Res 22:107–126

Precision Nutrition from the View of Genetics and Epigenetics

3

Lucia Migliore and Fabio Coppedè

Contents

Abstract

Precision nutrition takes advantages of omics technologies allowing investigating genome-wide genetic and epigenetic variants, and the global levels of messenger RNAs, proteins and metabolites, with the final goal to personalize the diet according to the individual's biological needs, in order to promote healthy aging and reduce the risk of age-related diseases. Nutritional genomics is a broad discipline encompassing nutrigenetics that aims to clarify how individuals respond to nutrients according to their genetic background; nutrigenomics that investigates changes in gene expression levels and the resulting levels of proteins and metabolites induced by dietary factors, and nutriepigenomics that studies the epigenetic changes induced by nutrients. These disciplines

L. Migliore (✉) · F. Coppedè
Department of Translational Research and of New Surgical and Medical Technologies, Medical School, University of Pisa, Via Roma 55, 56126 Pisa, Italy
e-mail: lucia.migliore@med.unipi.it

A. G. Haslberger (ed.), *Advances in Precision Nutrition, Personalization and Healthy Aging*, https://doi.org/10.1007/978-3-031-10153-3_3

provide complementary results to better personalize the diet according to the individual's integrated metabolism. After describing several examples of nutrigenetics, we will focus on the increasing evidence of interactions between dietary/nutritional factors and the epigenome that starting from the intrauterine life, can regulate gene expression levels and metabolic demands, potentially resulting in disease development later in life.

3.1 Introduction

Recently, research in nutrition is strongly oriented, as well as medicine in general, toward personalized nutrition that is a form of precision nutrition that takes into account inter- and intra-individual differences, in the field of genomics, metabolomics, proteomics and microbiota, in order to tailor the diet according to the individual's biology (Mullins et al. 2020). Differences exist in the genetic profile between individuals and specific ethnic groups, and these affect nutrient requirements, metabolism, and response to nutritional and dietary interventions (Ferguson et al. 2016). Evolutionary studies have revealed that humans genetically adapted to their ancestral diets and local environments, resulting in population-based differences in allele frequencies of common single nucleotide polymorphisms (SNPs) of genes required for the metabolism of some of the most common nutrients (Mullins et al. 2020). Nutrigenetics was conceived as the discipline pertaining to the interaction of nutritional and genetic factors that may play a role in disease etiology (Brennan and Mulligan 1975). Indeed, the main goal of nutrigenetics is to investigate how the inter-individual genotypic variability, particularly SNPs, influences an individual's response to dietary intake (Marcum 2020). The advent of high-throughput omics technologies has rapidly allowed to obtain genome-wide data in an affordable manner, leading to the discovery of many genetic variants associated with nutrient absorption and utilization, lipid metabolism, and fat accumulation that in turn can lead to gene–diet interactions of relevance for human diseases (Mullins et al. 2020). Omics means "global", and the omics technologies of genomics, epigenomics, transcriptomics, proteomics and metabolomics, coupled to the study of the individual's microbiome, have led to the development of nutrigenomics. Albeit many definitions of nutrigenomics exist in the literature, the broadest views it as a branch of science that uses high-throughput omics technologies to investigate the impact of diet and nutrition on gene expression levels (Brennan and de Roos 2021). Epigenetic mechanisms, including DNA methylation and histone tail modifications, do not alter the underlying DNA sequence, but regulate the chromatin structure and gene expression levels. The increasing evidence that the expression levels of several genes depends on epigenetic mechanisms, and that most of them are regulated by dietary/nutritional factors, has led to the development of the fields of nutriepigenetics and nutriepigenomics. Nutriepigenetics investigates mechanisms through which nutrients and dietary patterns may lead to epigenetic modifications resulting in changes in the expression levels of specific genes. Nutriepigenomics uses whole genome approaches for the same purpose (Ferguson et al. 2016). Nutritional genomics is the broad term encompassing nutrigenetics (the study of the different effects of nutrients according to our genetic constitution), nutrigenomics (the study of how nutrients may affect gene expression and the resulting levels of proteins and metabolites), and nutriepigenomics (the study of how nutrients change the chromatin structure and gene expression levels, without changing the DNA sequence) (Camp and Trujillo 2014). However, nutrigenetics, nutrigenomics, and nutriepigenomics complement each other providing a comprehensive picture of an individual's integrated metabolism, with the final goal of optimizing individual's health through a personalized nutrition, and the terms nutritional genomics and nutrigenomics are often used synonymously (Camp and Trujillo 2014; Marcum 2020).

3.2 Nutrigenetics and Nutrigenomics

Nutritional genetics or nutrigenetics investigates the impact of genetic variation on an individual's response to dietary intake, especially in terms of how genetic variation influences the metabolic state and risk for disease (Marcum 2020). Despite that two genomes can differ in SNPs, insertions and deletions of short DNA fragments (INDELs), copy number variants (CNVs), and larger rearrangements denoted structural variants (SVs), most of the nutrigenetics studies have focused on SNPs (Mullins et al. 2020).

Interesting examples in nutrigenetics are genetic polymorphisms in alcohol dehydrogenase 1B (*ADH1B*) and cytochrome P450 1A2 (*CYP1A2*) genes that influence alcohol and caffeine metabolism, respectively. Particularly, the mutant alleles rs1229984 and rs2066702 of *ADH1B* are associated with increased rates of ethanol metabolism and reduced risk of alcohol dependence, while the rs762551SNP of *CYP1A2* results in three genotypes (C/C, C/A and A/A) leading to slow (C/C and C/A) and rapid (A/A) caffeine metabolizers (Mullins et al. 2020).

Methylenetetrahydrofolate reductase (MTHFR) is one of the most important enzymes in folate metabolism and converts 5,10-methylenetetrahydrofolate to 5-methyltetrahydrofolate, the methyl donor for the remethylation of homocysteine to methionine, this latter required for the production of *S*-adenosylmethione (SAM), the universal intracellular methyl donor compound. The most well-known *SNP* of *MTHFR*, namely, rs1801133, results from a transition mutation from C to T at nucleotide 677 (c.677C > T), and the homozygous genotype (T/T) is responsible for a reduction of up to 70% in MTHFR enzymatic activity (Ueland et al. 2001). An important factor influencing MTHFR protein activity in homozygous T/T carriers is the availability of dietary folates. MTHFR works as a dimer protein and physiological levels of folate stabilize the dimer, but the *MTHFR* 677 T allele renders the enzyme thermolabile, particularly in homozygous (T/T) individuals that are prone to dimer destabilization under conditions of reduced folate bioavailability (Martínez-Frías 2008). The *MTHFR* 677TT genotype has been often linked to hyperhomocysteinemia and impaired DNA methylation, and associated with increased risk of various human conditions, including pregnancy complications (Coppedè 2021a), congenital disorders such as neural tube defects (Tabatabaei et al. 2020), Down syndrome (James et al. 1999; Coppedè et al. 2010) and congenital heart disease (Liu et al. 2020), autism spectrum disorders (Wei et al. 2021), and age-related diseases including cancer (Wang et al. 2021), cardiovascular diseases (Chita et al. 2020), and Alzheimer's disease (AD) (Coppedè et al. 2012; Yi et al. 2019).

Genes involved in lipid metabolism have often been linked to an increased risk for cardiovascular and neurological complications (Ordovas 2009; Liu et al. 2013). One of best examples is the *APOE* gene coding apolipoprotein E (ApoE). Apolipoproteins are proteins associated with lipid particles, which mainly function in lipid transport from one tissue or cell type to another. In peripheral tissues, ApoE is primarily produced by the liver and macrophages, and mediates cholesterol metabolism in an isoform-dependent manner. In the central nervous system (CNS), ApoE is mainly produced by astrocytes, and transports cholesterol to neurons via ApoE receptors, which are members of the low-density lipoprotein receptor (LDLR) family (Liu et al. 2013). Two SNPs (rs429358 and rs7412) within the *APOE* gene generate three common alleles of the *APOE* gene, namely $\varepsilon 2$, $\varepsilon 3$, and $\varepsilon 4$, with reported worldwide frequencies of about 8.4%, 77.9%, and 13.7%, respectively (Liu et al. 2013). The *APOE* $\varepsilon 4$ allele represents the major genetic risk factor for AD. Indeed, one copy of $\varepsilon 4$ increases AD risk by ~threefold and two copies by ~12-fold. The common $\varepsilon 3$ allele is neutral, while the $\varepsilon 2$ allele seems to confer protection against AD. The *APOE* $\varepsilon 4$ allele contributes to AD pathogenesis by impairing microglial responsiveness, lipid transport, synaptic integrity and plasticity, glucose metabolism, and cerebrovascular integrity and function. It also promotes amyloid-β (Aβ) aggregation, and influences tau pathology and tau-mediated neurodegeneration (Yamazaki et al. 2019). In addition, the *APOE* $\varepsilon 4$ allele is associated with an increase in LDL-cholesterol levels and with increased risk of cardiovascular diseases (Khalil et al. 2021).

Several variants in obesity-related genes can affect weight gain or loss in genetically predisposed subjects (Martinez et al. 2008; Vitolo et al. 2017). One of the best examples is represented by rs9939609 in the fat mass and obesity-associated (*FTO*) gene that predisposes carriers of the minor allele to weight gain and increased risk of obesity (Da Silva et al. 2018).

These are only some of several examples of nutrigenetics pertaining to SNPs affecting the individual's metabolic activities and risk of human diseases, and many more can be found in the literature (Ferguson et al. 2016; Mullins et al. 2020). Haplotypes are groups of gene variants, often SNPs, which are physically close to each other on the same DNA strand, and can be more reliable in predicting protein function and activity than a single SNP. For example, rs1801133 (c.677C > T) and rs1801131 (c.1298A > C) of the *MTHFR* gene are in strong linkage disequilibrium, resulting in common and less frequent haplotypes. Particularly, the rare 677 T/1298C haplotype seriously impairs protein stability and activity, being often prenatally lethal in homozygosis (Ulvik et al. 2007).

Genetic variants alone do not explain all the complexity of gene-nutrients interactions and their relevance for healthy aging or disease, and nutrigenomics is the discipline that uses omics approaches to understand how dietary factors influence gene expression levels and the resulting levels of proteins and metabolites, combining transcriptomics, proteomics, metabolomics, and other omics technologies (Ferguson et al. 2016; Marcum 2020). Within this context, increasing evidence reveals that investigators take advantage of nutriepigenetics/nutriepigenomics technologies to complement nutrigenetics/nutrigenomics approaches, combining data on gene variants, epigenetic modifications, and gene expression levels to better characterize the inter-individual variability in the response to dietary factors. For example a recent genome-wide methylation study in 1250 individuals, showed that alterations of the one-carbon metabolism, resulting from interaction between plasma homocysteine and the *MTHFR* c.677C > T polymorphism, are associated with site-specific changes in DNA methylation affecting gene expression levels (Nash et al. 2019). Similarly, associations between vitamin B6 deficiency, *MTHFR* genotypes, and global DNA methylation levels, were observed in pregnant women (La Merrill et al. 2012), and riboflavin supplementation altered global and gene-specific DNA methylation levels in adults with the *MTHFR* 677TT genotype (Amenyah et al. 2020a). The next sections of this chapter further explain the complex interplay between dietary factors and the epigenome.

3.3 Epigenetic Mechanisms

Epigenetic mechanisms include DNA methylation and hydroxymethylation that regulate gene expression levels; numerous post-translational modifications of histone tail aminoacids, including acetylation, methylation, phosphorylation, ubiquitylation and sumoylation, overall regulating the chromatin structure, and the regulation of gene expression levels mediated by short and long non-coding RNA molecules (ncRNAs). Among these, DNA methylation is one of the most extensively studied and consists of the addition of a methyl group to the DNA mediated by DNA methyltransferases (DNMTs), resulting in gene silencing when occurring in the promoter region. SAM, which is generated within the methionine-cycle of one-carbon metabolism, is the methyl donor compound for DNA methylation. Therefore, DNA methylation is sensitive to the bioavailability of dietary folates and related B-group vitamins and nutrients working as methyl donor compounds or cofactors in the one-carbon metabolic pathway (Coppedè 2021b). However, as further discussed in the next paragraphs, the different epigenetic mechanisms work in concert to adapt gene expression levels in response to the cellular metabolic demands. Indeed, most of the enzymes that add or remove the epigenetic marks are in turn regulated by intracellular metabolites, and several metabolic genes, such as those involved in lipid, glucose, one-carbon, and redox metabolism are regulated by epigenetic mechanisms (Coppedè 2021b) Fig. 3.1.

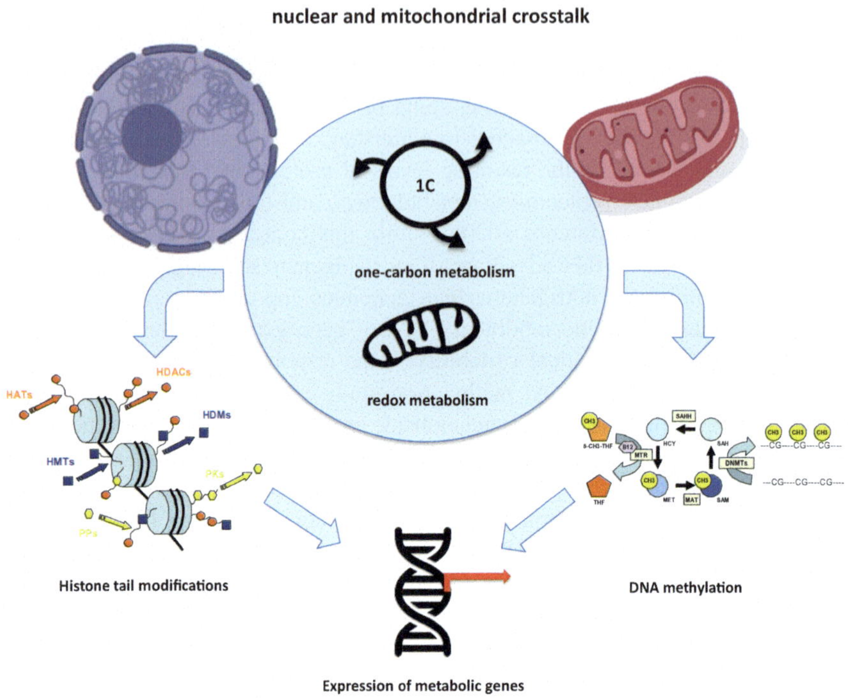

Fig. 3.1 A bidirectional crosstalk between the nuclear and mitochondrial genomes allows a coordinated regulation of gene expression levels according to metabolic demands and is mediated by one-carbon metabolites and redox cofactors that regulate the activity of several epigenetic enzymes

3.4 The DOHaD Theory: The Importance of the Maternal Diet in Animal and Human Models

The epigenome is an important target for changes induced by environmental factors such as nutrition, chemical pollutants, early traumatic experiences, temperature changes, and exercise (Feil and Fraga 2012). It is important to underline that the effect of the environment on the epigenome does not concern only the period after birth, but is able to influence in an incisive way also the development in utero. It is believed that the mother's lifestyle and environmental conditions during pregnancy can have long-term effects on the health of the offspring. The molecular mechanisms with which maternal influence would act is likely mediated by epigenetic modifications, able to interfere with the fetal programming of genes linked to diseases with adult onset (cancer, degenerative diseases, autoimmune disorders…), in accordance with the theory of embryo-fetal origins of diseases (Developmental Origins of Health and Disease, DOHaD), proposed by David Barker in the early 2000s (Barker et al. 2002; Barker 2007). Interesting results have been obtained with different animal and human models regarding the influence of the maternal diet on the offspring. In the Agouti mouse model maternal supplementation during pregnancy with donors of methyl groups can change the phenotype of the progeny by interfering with the expression of a gene responsible for coat color (and other phenotypic traits) (Waterland and Jirtle 2003). In fact, given the direct relationship between the folate metabolic pathway and DNA methylation mechanisms, alterations in nutritional status can directly influence DNA methylation patterns. In particular, methyl group donor nutrients such as methionine, folic acid, betaine, and choline have been implicated in the alterations of methylation patterns, as these nutrients are directly correlated with the biosynthesis of SAM, the precursor necessary for DNA methylation. In general, supplementation with methyl group donors appears to increase global methylation levels, while deficiency is associated with global hypomethylation. Among the genes that are affected by folate levels, there are many correlated with carcinogenesis, with inflammatory processes, with fetal growth and development. For example, deprivation of the essential amino acid methionine and folate deficiency are associated with liver and colon cancer in animals and humans. A recent randomized trial has shown that dietary fat composition also influences DNA methylation in adipocytes. Many studies have shown that metabolic syndrome and related disorders are linked to epigenetic changes detected in peripheral blood DNA (Feinberg 2018). Another model to investigate the effects of maternal diet on the epigenome regards nonhuman primates. A high fat diet (35% fat) was established that produced obesity in pregnant monkeys (macaques). In comparison with animals with a control diet (13% fat), the offspring of the obese monkeys were obese. Moreover epigenetic changes were identified in the liver of the obese offspring (hyperacetylation of fetal hepatic tissue), which was associated with the high fat diet (Aagaard-Tillery et al. 2008). Numerous studies suggest that maternal diet in pregnancy can influence child neurodevelopment. For instance maternal adherence to a Mediterranean diet was associated with favorable neurobehavioral outcomes in early childhood and with differences of CpG methylation of imprinted genes, such as *MEG3* and *IGF2* (House et al. 2018). Also the paternal diet before mating, was found to induce intergenerational metabolic reprogramming in a Drosophila model. A 2 days of sugar rich diet intervention in fathers elicits obesity in offspring. Paternal sugar acts desilencing chromatin-state-defined domains in both mature sperm and in offspring embryos (Öst et al. 2014).

Although most of the scientific evidence is based to date on animal models, also epidemiological studies on humans are starting to emerge. Pioneering work relating the influence of maternal lifestyle on embryonic epigenetic programming was performed on a cohort of individuals exposed prenatally to severe food restrictions due to famine in the winter 1944–45 (the so-called "Dutch famine of 1944–45"), which six decades later showed hypomethylation of the gene encoding the insulin-like growth factor 2 (*IGF2*) compared to non-exposed control subjects, or to siblings born before or after that winter, from the same mothers, but not subject to dietary restrictions during pregnancy (Heijmans et al. 2008). Notably, in utero exposure to famine had life-long effects on health, resulting in impaired glucose tolerance and increased risk of obesity, coronary heart disease, atherogenic lipid profile, hypertension, microalbuminuria, schizophrenia, antisocial personality

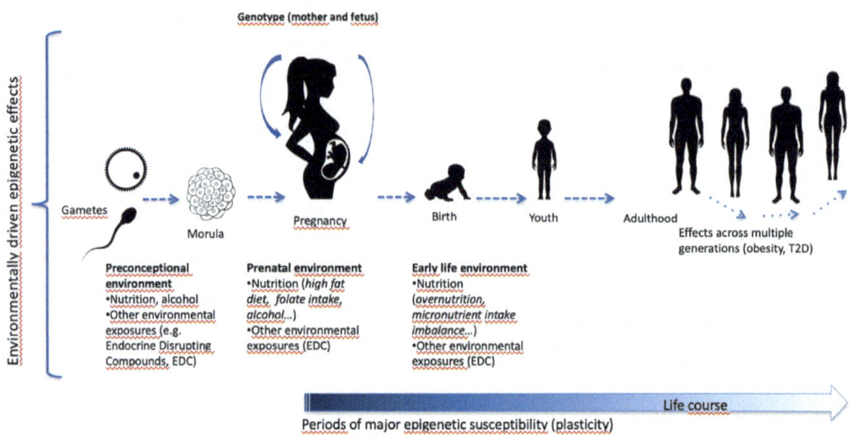

Fig. 3.2 Epigenetic effects from environmental exposures during life course and transgenerational consequences

and affective disorders (Kyle and Pichard 2006). These data are among the few that clearly demonstrate a link between early environmental exposure, specifically nutrition, DNA methylation, and long-term phenotype determination (Fig. 3.2).

3.5 Epigenetic Mechanisms and Nutrients

It is now known that environmental factors, including nutrients, are able to modulate the biochemical reactions underlying the epigenetic modifications. In the previous paragraphs we mentioned the role of folate pathway, but manifold are the influences of environmental factors on the main epigenetic processes such as histone tail modifications and DNA methylation. External factors, including nutrition, exercise and the gut microbiome, regulate histone methylation and acetylation by modulating the intracellular pools of metabolites, including SAM and acetyl-CoA that are employed by histone methyltransferases (HMTs) and histone acetyltransferases (HATs), respectively. The activity of histone demethylases (HDMs) is supported by α-ketoglutarate (αKG), which can be derived from dietary glutamine, and is inhibited by the limited oxygen availability during hypoxia. Ketone bodies and short-chain fatty acids (SCFAs) such as acetate, propionate, and butyrate can provide acyl-CoA precursors for histone acylation, also inhibiting directly the activity of histone deacetylases (HDACs) (Dai et al. 2020).

Furthermore, in a more global vision, recent studies indicate that metabolically regulated epigenetic modifications include a broad spectrum of enzymatic and non-enzymatic modifications on histone, DNA and RNA molecules beyond the "canonical" methylation and acetylation marks (Dai et al. 2020).

Several links exist between epigenetic mechanisms and the intracellular metabolic reactions, and several enzymes involved in the folate metabolic pathway or catalyzing histone tail modifications require FAD or NAD cofactors (Coppedè 2021b). Moreover, several metabolic enzymes that typically localize to the cytoplasm or mitochondria can also be found in the nucleus. These include enzymes required for glycolysis, citric acid cycle, SAM synthesis, and nucleotide synthesis. Inside the nucleus, these enzymes can either supply metabolites to regulate chromatin structure or induce direct modifications of histones and transcription regulators (Li et al. 2018). In addition, the expression levels of several enzymes participating in glucose, lipid or one-carbon metabolism, are regulated by their promoter methylation levels, strengthening evidence of a bidirectional crosstalk between metabolic and epigenetic pathways to tightly regulate gene expression levels according to metabolic demands (Coppedè 2021b). For example, using a cohort of more than 200 elderly individuals, we observed that circulating folate levels correlate with *MTHFR* promoter methylation levels (Tannorella et al. 2015). Moreover, in the same cohort we observed that *MTHFR* methylation levels also correlate with polymorphisms of genes involved in one-carbon metabolism, supporting the need to include genetic data in nutriepigenetic investigations (Coppedè et al. 2019). Investigations involving hundreds or even thousands of individuals have shown that dietary intake of folate, vitamin B12, and other methyl donor compounds correlate with peripheral blood genome-wide methylation levels (Amenyah et al. 2020a). Moreover, also the Mediterranean diet was associated with differential whole genome methylation levels, and in particular with the methylation levels of genes related to inflammation and immunocompetence (Arpon et al. 2016).

3.6 Epigenetic Mechanisms of Antioxidants

Many complex diseases, if not even all those investigated, have been found displaying epigenetic dysregulation. Extending from the paradigm of cancer, epigenetic dysregulation was identified in a plethora of complex disorders including immunological, metabolic (such as obesity and type 2 diabetes) infectious diseases, cardiovascular diseases, and human infertility. Particularly epigenetic regulation is critical for the normal development and functioning of the human brain. Indeed, epigenetic abnormalities have been frequently detected in neurodegenerative diseases, such AD, Parkinson's disease (PD), amyotrophic lateral sclerosis (ALS), and also in psychiatric diseases including schizophrenia, major depressive disorder or post-traumatic stress disorders (Berdasco and Esteller 2019). Evidence for increased oxidative damage in AD and PD neuronal and peripheral tissues was supported by studies in animal models and peripheral blood cells of living patients since the last decade of the past century (Flint Beal 1997; Migliore et al. 2005). Antioxidant therapies have been proposed since long time to slowing the progression and limiting the extent of neuronal cell loss in neurodegenerative diseases. Moreover antioxidant dietary compounds have been demonstrated to have the capacity to combat oxidative stress in many studies involving diverse chronic diseases, besides neurodegenerative diseases, such as cancer, cardiovascular disease, chronic obstructive pulmonary disease, type 2 diabetes (Beetch et al. 2020), and inflammation (Ramos-Lopez et al. 2021). Studies have suggested the antioxidant therapeutics could potentially mimic the physiological actions of the natural antioxidant defense system, resulting in slower aging and prolongation of lifespan.

Many dietary antioxidants, including catechins, flavonoids, anthocyanins, stilbenes and carotenoids, which demonstrate benefits in the prevention and/or support of therapy in chronic diseases are able to reverse altered patterns of DNA methylation in chronic diseases. They remodel the DNA methylation patterns through multiple mechanisms, including regulation of epigenetic enzymes and chromatin remodeling complexes (Remely et al. 2015; Vahid et al. 2015; Beetch et al. 2020).

Among natural antioxidants there are plant-derived polyphenols. For instance resveratrol (3,5,4′-trihydroxystilbene) is a polyphenol present mainly in black grapes and its derivatives, blackberries, peanuts, and peanut products. There is evidence that resveratrol is a potent antioxidant and possesses pleiotropic actions, exerting its activity through various molecular pathways. Experimental data associated with epigenetics changes indicate that resveratrol exhibits the capacity to modulate several neurodegenerative pathways, reducing the expression of genes crucial for age-related diseases. Among these pathways there are inhibition of DNMT activity; activation of SIRT1; regulation of acetylation of histones (H1, H3, H4) and non-histones chromatin protein (FOXO, p53, Ku70, PPAR, PGC- 1α, NF-κβ), regulation of HDAC and HAT activities; modulation of miRNAs, such as the upregulation levels of specific miRNA (for a review see Griñán-Ferré et al. 2021).

3.7 Aging, Epigenetics, Nutrition

Increases in oxidative stress, mitochondrial dysfunction, inflammation, apoptosis as well as epigenetic modifications have been linked to accelerate aging (Griñán-Ferré et al. 2020). An emerging field of epigenetics is represented by the epigenetic clock, which provides a measure of epigenetic age (also known as "biological age"), based on the DNA methylation levels evaluation at 353 CpG sites, called "Horvath epigenetic clock". The DNAmAge clock is computed from some sites that increase and others that decrease methylation with age (Horvath 2013). Investigating how the estimated epigenetic age differs across individuals of the same chronological age can help to determine the impact of endogenous or exogenous stress factors on biological aging and in the onset of diseases (Levine et al. 2018). Perhaps, the most exciting feature of epigenetic biomarkers is that epigenetic changes are reversible, raising the prospect that DNA methylation age estimates might thus be useful for identifying or validating anti-aging interventions. In a pilot randomized clinical trial on a total of 43 healthy adult males of 50–72 years, a potential reversal of epigenetic age using a diet and lifestyle intervention was recently achieved (Fitzgerald et al. 2021). The 8-week treatment program included diet, sleep, exercise and relaxation guidance, and supplemental probiotics and phytonutrients. Both changes in blood biomarkers were obtained (increased mean serum 5-methyltetrahydrofolate and decreased mean triglycerides) together with a DNAmAge of those in the treatment group decreased by an average 1.96 years by the end of the program (Fitzgerald et al. 2021). Other similar randomized clinical trials or cross sectional studies have shown that vitamin D3, folic acid and vitamin B12, carotenoids, fish, and tocopherol, are among dietary factors associated with a reduced DNAmAge (for a review see Amenyah et al. 2020b).

3.8 The Importance of the Gender in Precision Nutrition Medicine

In a study mentioned above significant associations of maternal periconceptional adherence to the Mediterranean diet both with positive neurodevelopmental phenotypes in offspring as well as with differential methylation of CpGs in the control regions of imprinted genes were found (House et al. 2018). However some associations varied by sex: either sex-specific differences in the methylation of imprinted control regions as well as behavioral differences (decreased odds of depression, anxiety, atypical, and Autism Spectrum Disorder found in females) (House et al. 2018).

In general gender differences in food choices, in energy and nutrient intakes are certainly predictable. However these aspects have not yet been fully investigated in the human populations. The need to including a gender dimension in clinical studies and practice is increasingly felt, especially nowadays with a growing interest in precision medicine, which has been defined as "an emerging approach for disease treatment and prevention that takes into account individual variability in genes, environment, and lifestyle for each person" (https://medlineplus.gov/genetics/understanding/precisionmedicine/definition/) and, as such, this approach must necessarily include the gender. As well as until now the majority of drug therapies are not yet optimized for both genders, also in the nutritional field, the modern dietary recommendations are based on studies conducted predominantly among males. The different efficacy of drugs, as likely of nutrients, in women and men is certainly due to biological differences that may be caused by sex-specific gene expression, likely triggered by sex-specific epigenetic modifications. Certainly, there are many factors involved in general in gender differences. A significant proportion of human imprinted genes are realistically involved, as well as altered expression of critical regions, due for instance to a skewed inactivation of the X-chromosome in females, or to the presence of genes that escape to X-inactivation or to a deregulation of miRNAs located on the X-chromosome. Moreover, the expression of specific genes can be modified epigenetically by many intracellular (hormones) or extracellular environmental factors, including dietary factors (Migliore et al. 2021).

3.9 Concluding Remarks

As an important constituent of precision medicine, precision nutrition focuses on the individual rather than groups of people, and assumes that many inherited and chronic metabolic diseases might be prevented or managed through personalized nutritional intervention because each person may have a different response to specific foods and nutrients (Chen and Wang 2016). In the present chapter we discuss several examples showing how inherited genetic variants can result in inter-individual metabolic differences ultimately leading to a different response to foods and nutrients, and how these differences can increase the risk of various diseases in carriers of certain metabolic genotypes. Moreover, we have largely discussed how foods and nutrients are able to modulate the expression levels of hundreds different genes through epigenetic mechanism. These gene-nutrient interactions occur pre-conceptionally, resulting in epigenetic changes in the gametes, as well as during pre- and post-natal life and, in combination with other person's variables, including gender, genetic variants, ethnicity, gut microbiome, environmental exposures, and lifestyles, can either slow the aging process in certain individuals or result in accelerated aging and increased burden of age-related diseases in others. Therefore, nutrigenetics/nutrigenomics and nutriepigenetics/nutriepigenomics offer complementary approaches to better understand the integrated metabolic response to foods and nutrients with the final goal to personalize the diet according to the individual's demands.

References

Aagaard-Tillery KM, Grove K, Bishop J, Ke X, Fu Q, McKnight R, Lane RH (2008) Developmental origins of disease and determinants of chromatin structure: maternal diet modifies the primate fetal epigenome. J Mol Endocrinol 41(2):91–102. https://doi.org/10.1677/JME-08-0025

Amenyah SD, McMahon A, Ward M, Deane J, McNulty H, Hughes CF, Strain JJ, Horigan G, Purvis J, Walsh CP, Lees-Murdock DJ (2020a) Riboflavin supplementation alters global and gene-specific DNA methylation in adults with the MTHFR 677 TT genotype. Biochimie 173:17–26. https://doi.org/10.1016/j.biochi.2020.04.007

Amenyah SD, Ward M, Strain JJ, McNulty H, Hughes CF, Dollin C, Walsh CP, Lees-Murdock DJ (2020b) Nutritional epigenomics and age-related disease. Curr Dev Nutr 4(7). https://doi.org/10.1093/cdn/nzaa097

Arpón A, Riezu-Boj JI, Milagro FI, Marti A, Razquin C, Martínez-González MA, Corella D, Estruch R, Casas R, Fitó M, Ros E, Salas-Salvadó J, Martínez JA (2016) Adherence to Mediterranean diet is associated with methylation changes in inflammation-related genes in peripheral blood cells. J Physiol Biochem 73(3):445–455.https://doi.org/10.1007/s13105-017-0552-6

Barker DJP (2007) The origins of the developmental origins theory. J Intern Med 261:412–417

Barker DJ, Eriksson JG, Forsén T, Osmond C (2002) Fetal origins of adult disease: strength of effects and biological basis. Int J Epidemiol 31(6):1235–1239

Beal FM (1997) Oxidative damage in neurodegenerative diseases. Neuroscientist 3(1):21–27

Beetch M, Harandi-Zadeh S, Shen K, Lubecka K, Kitts DD, O'Hagan HM, Stefanska B (2020) Dietary antioxidants remodel DNA methylation patterns in chronic disease. Br J Pharmacol 177(6):1382–1408. https://doi.org/10.1111/bph.14888

Berdasco M, Esteller M (2019) Clinical epigenetics: seizing opportunities for translation. Nat Rev Genet 20(2):109–127. https://doi.org/10.1038/s41576-018-0074-2
Brennan RO, Mulligan WC (1975) Nutrigenetics: new concepts for relieving hypoglycemia. M. Evans & Co, New York
Brennan L, de Roos B (2021) Nutrigenomics: lessons learned and future perspectives. Am J Clin Nutr 113(3):503–516
Camp KM, Trujillo E (2014) Position of the academy of nutrition and dietetics: nutritional genomics. J Acad Nutr Diet 114(2):299–312. https://doi.org/10.1016/j.jand.2013.12.001
Chen PZ, Wang H (2016) Precision nutrition in the era of precision medicine. Zhonghua Yu Fang Yi Xue Za Zhi 50(12):1036–1042. https://doi.org/10.3760/cma.j.issn.0253-9624.2016.12.004
Chita DS, Tudor A, Christodorescu R, Buleu FN, Sosdean R, Deme SM, Mercea S, Pop Moldovan A, Pah AM, Docu Axelerad A, Docu Axelerad D, Dragan SR (2020) MTHFR gene polymorphisms prevalence and cardiovascular risk factors involved in cardioembolic stroke type and severity. Brain Sci 10(8):476.https://doi.org/10.3390/brainsci10080476
Coppedè F, Grossi E, Migheli F, Migliore L (2010) Polymorphisms in folate-metabolizing genes, chromosome damage, and risk of Down syndrome in Italian women: identification of key factors using artificial neural networks. BMC Med Genomics 3:42. https://doi.org/10.1186/1755-8794-3-42
Coppedè F, Tannorella P, Pezzini I, Migheli F, Ricci G, Caldarazzo lenco E, Piaceri I, Polini A, Nacmias B, Monzani F, Sorbi S, Siciliano G, Migliore L (2012) Folate, homocysteine, vitamin B12, and polymorphisms of genes participating in one-carbon metabolism in late-onset Alzheimer's disease patients and healthy controls. Antioxid Redox Signal 17(2):195–204. https://doi.org/10.1089/ars.2011.4368
Coppedè F, Stoccoro A, Tannorella P, Gallo R, Nicolì V, Migliore L (2019) Association of polymorphisms in genes involved in one-carbon metabolism with MTHFR methylation levels. Int J Mol Sci 20(15):3754. https://doi.org/10.3390/ijms20153754
Coppedè F (2021a) Pregnancy loss and polymorphisms in folic acid genes. In: Vinciguerra M, Sanchez P (eds) Molecular nutrition: mother and infant, 1st edn. Elsevier Inc, pp 317–335
Coppedè F (2021b) One-carbon epigenetics and redox biology of neurodegeneration. Free Radic Biol Med 170:19–33. https://doi.org/10.1016/j.freeradbiomed.2020.12.002
Dai Z, Ramesh V, Locasale JW (2020) The evolving metabolic landscape of chromatin biology and epigenetics. Nat Rev Genet 21(12):737–753. https://doi.org/10.1038/s41576-020-0270-8
da Silva TER, Andrade NL, Cunha DO, Leão-Cordeiro JAB, Vilanova-Costa CAST, Silva AMTC (2018) The FTO rs9939609 polymorphism and obesity risk in teens: evidence-based meta-analysis. Obes Res Clin Pract 12(5):432–437. https://doi.org/10.1016/j.orcp.2018.08.001
Feil R, Fraga MF (2012) Epigenetics and the environment: emerging patterns and implications. Nat Rev Genet 13(2):97–109
Feinberg AP (2018) The key role of epigenetics in human disease prevention and mitigation. N Engl J Med 378(14):1323–1334
Ferguson LR, De Caterina R, Görman U, Allayee H, Kohlmeier M, Prasad C, Choi MS, Curi R, de Luis DA, Gil Á, Kang JX, Martin RL, Milagro FI, Nicoletti CF, Nonino CB, Ordovas JM, Parslow VR, Portillo MP, Santos JL, Serhan CN, Simopoulos AP, Velázquez-Arellano A, Zulet MA, Martinez JA (2016) Guide and position of the international society of nutrigenetics/nutrigenomics on personalised nutrition: part 1—fields of precision nutrition. J Nutrigenet Nutrigenomics 9(1):12–27. https://doi.org/10.1159/000445350
Fitzgerald KN, Hodges R, Hanes D, Stack E, Cheishvili D, Szyf M, Henkel J, Twedt MW, Giannopoulou D, Herdell J, Logan S, Bradley R (2021) Potential reversal of epigenetic age using a diet and lifestyle intervention: a pilot randomized clinical trial. Aging (albany NY) 13(7):9419–9432. https://doi.org/10.18632/aging.202913
Griñán-Ferre' C, Izquierdo V, Pallàs M (2020) Dietary antioxidants, epigenetics, and brain aging: a focus on resveratrol. In: Oxidative stress and dietary antioxidants in neurological diseases. Elsevier, pp 343–357. https://doi.org/10.1016/B978-0-12-817780-8.00022-0
Griñán-Ferré C, Bellver-Sanchis A, Izquierdo V, Corpas R, Roig-Soriano J, Chillón M, Andres-Lacueva C, Somogyvári M, Sőti C, Sanfeliu C, Pallàs M (2021) The pleiotropic neuroprotective

effects of resveratrol in cognitive decline and Alzheimer's disease pathology: from antioxidant to epigenetic therapy. Ageing Res Rev 67:101271. https://doi.org/10.1016/j.arr.2021.101271
Heijmans BT et al (2008) Persistent epigenetic differences associated with prenatal exposure to famine in humans. Proc Natl Acad Sci USA 105:17046–17049
Horvath S (2013) DNA methylation age of human tissues and cell types. Genome Biol 14(10):R115
House JS, Mendez M, Maguire RL, Gonzalez-Nahm S, Huang Z, Daniels J, Murphy SK, Fuemmeler BF, Wright FA, Hoyo C (2018) Periconceptional maternal mediterranean diet is associated with favorable offspring behaviors and altered CpG methylation of imprinted genes. Front Cell Dev Biol 6:107. https://doi.org/10.3389/fcell.2018.00107
James SJ, Pogribna M, Pogribny IP, Melnyk S, Hine RJ, Gibson JB, Yi P, Tafoya DL, Swenson DH, Wilson VL, Gaylor DW (1999) Abnormal folate metabolism and mutation in the methylenetetrahydrofolate reductase gene may be maternal risk factors for Down syndrome. Am J Clin Nutr 70(4):495–501. https://doi.org/10.1093/ajcn/70.4.495
Khalil YA, Rabès JP, Boileau C, Varret M (2021) APOE gene variants in primary dyslipidemia. Atherosclerosis 328:11–22. https://doi.org/10.1016/j.atherosclerosis.2021.05.007
Kyle UG, Pichard C (2006) The Dutch Famine of 1944–1945: a pathophysiological model of long-term consequences of wasting disease. Curr Opin Clin Nutr Metab Care 9(4):388–394. https://doi.org/10.1097/01.mco.0000232898.74415.42
La Merrill M, Torres-Sánchez L, Ruiz-Ramos R, López-Carrillo L, Cebrián ME, Chen J (2012) The association between first trimester micronutrient intake, MTHFR genotypes, and global DNA methylation in pregnant women. J Matern Fetal Neonatal Med 25(2):133–137. https://doi.org/10.3109/14767058.2011.564242
Levine ME, Lu AT, Quach A, Chen BH, Assimes TL, Bandinelli S, Hou L, Baccarelli AA, Stewart JD, Li Y, Whitsel EA, Wilson JG, Reiner AP, Aviv A, Lohman K, Liu Y, Ferrucci L, Horvath S (2018) An epigenetic biomarker of aging for lifespan and healthspan. Aging (albany NY) 10(4):573–591
Li X, Egervari G, Wang Y, Berger SL, Lu Z (2018) Regulation of chromatin and gene expression by metabolic enzymes and metabolites. Nat Rev Mol Cell Biol 19(9):563–578. https://doi.org/10.1038/s41580-018-0029-7
Liu CC, Liu CC, Kanekiyo T, Xu H, Bu G (2013) Apolipoprotein E and Alzheimer disease: risk, mechanisms and therapy. Nat Rev Neurol 9(2):106–118. https://doi.org/10.1038/nrneurol.2012.263
Liu PF, Ding B, Zhang JY, Mei XF, Li F, Wu P, Mei CH, Zhou YF, Chen T (2020) Association between MTHFR C677T polymorphism and congenital heart disease. Int Heart J 61(3):553–561. https://doi.org/10.1536/ihj.19-389
Marcum JA (2020) Nutrigenetics/nutrigenomics, personalized nutrition, and precision healthcare. Curr Nutr Rep 9(4):338–345. https://doi.org/10.1007/s13668-020-00327-z
Martinez JA, Parra MD, Santos JL, Moreno-Aliaga MJ, Marti A, Martinez-Gonzalez MA (2008) Genotype-dependent response to energy-restricted diets in obese subjects: towards personalized nutrition. Asia Pac J Clin Nutr 17(Suppl 1):119–122
Martínez-Frías ML (2008) The biochemical structure and function of methylenetetrahydrofolate reductase provide the rationale to interpret the epidemiological results on the risk for infants with Down syndrome. Am J Med Genet A 146A(11):1477–1482
Migliore L, Fontana I, Colognato R, Coppede F, Siciliano G, Murri L (2005) Searching for the role and the most suitable biomarkers of oxidative stress in Alzheimer's disease and in other neurodegenerative diseases. Neurobiol Aging 26(5):587–595. https://doi.org/10.1016/j.neurobiolaging.2004.10.002
Migliore L, Nicolì V, Stoccoro A (2021) Gender specific differences in disease susceptibility: the role of epigenetics. Biomedicines 9:652. https://doi.org/10.3390/biomedicines9060652
Mullins VA, Bresette W, Johnstone L, Hallmark B, Chilton FH (2020) Genomics in personalized nutrition: can you "eat for your genes"? Nutrients 12(10):3118. https://doi.org/10.3390/nu12103118

Nash AJ, Mandaviya PR, Dib MJ, Uitterlinden AG, van Meurs J, Heil SG, Andrew T, Ahmadi KR (2019) Interaction between plasma homocysteine and the MTHFR c.677C > T polymorphism is associated with site-specific changes in DNA methylation in humans. FASEB J 33(1):833–843. https://doi.org/10.1096/fj.201800400R

Ordovas JM (2009) Genetic influences on blood lipids and cardiovascular disease risk: tools for primary prevention. Am J Clin Nutr 89(5):1509S-1517S. https://doi.org/10.3945/ajcn.2009.27113E

Öst A, Lempradl A, Casas E, Weigert M, Tiko T, Deniz M, Pantano L, Boenisch U, Itskov PM, Stoeckius M, Ruf M, Rajewsky N, Reuter G, Iovino N, Ribeiro C, Alenius M, Heyne S, Vavouri T, Pospisilik JA (2014) Paternal diet defines offspring chromatin state and intergenerational obesity. Cell 159(6):1352–1364. https://doi.org/10.1016/j.cell.2014.11.005

Ramos-Lopez O, Milagro FI, Riezu-Boj JI, Martinez JA (2021) Epigenetic signatures underlying inflammation: an interplay of nutrition, physical activity, metabolic diseases, and environmental factors for personalized nutrition. Inflamm Res 70(1):29–49. https://doi.org/10.1007/s00011-020-01425-y

Remely M, Lovrecic L, de la Garza AL, Migliore L, Peterlin B, Milagro FI, Martinez AJ, Haslberger AG (2015) Therapeutic perspectives of epigenetically active nutrients. Br J Pharmacol 172(11):2756–2768. https://doi.org/10.1111/bph.12854

Tabatabaei RS, Fatahi-Meibodi N, Meibodi B, Javaheri A, Abbasi H, Hadadan A, Bahrami R, Mirjalili SR, Karimi-Zarchi M, Neamatzadeh H (2020) Association of fetal MTHFR C677T polymorphism with susceptibility to neural tube defects: a systematic review and update meta-analysis. Fetal Pediatr Pathol 14:1–17. https://doi.org/10.1080/15513815.2020.1775734

Tannorella P, Stoccoro A, Tognoni G, Petrozzi L, Salluzzo MG, Ragalmuto A, Siciliano G, Haslberger A, Bosco P, Bonuccelli U, Migliore L, Coppedè F (2015) Methylation analysis of multiple genes in blood DNA of Alzheimer's disease and healthy individuals. Neurosci Lett 600:143–147. https://doi.org/10.1016/j.neulet.2015.06.009

Ueland PM, Hustad S, Schneede J, Refsum H, Vollset SE (2001) Biological and clinical implications of the MTHFR C677T polymorphism. Trends Pharmacol Sci 22(4):195–201

Ulvik A, Ueland PM, Fredriksen A, Meyer K, Vollset SE, Hoff G, Schneede J (2007) Functional inference of the methylenetetrahydrofolate reductase 677C > T and 1298A > C polymorphisms from a large-scale epidemiological study. Hum Genet 121(1):57–64. https://doi.org/10.1007/s00439-006-0290-2

Vahid F, Zand H, Nosrat-Mirshekarlou E, Najafi R, Hekmatdoost A (2015) The role dietary of bioactive compounds on the regulation of histone acetylases and deacetylases: a review. Gene 562(1):8–15. https://doi.org/10.1016/j.gene.2015.02.045

Vitolo E, Santini E, Seghieri M, Giannini L, Coppedè F, Rossi C, Dardano A, Solini A (2017) Heterozygosity for the rs696217 SNP in the preproghrelin gene predicts weight loss after bariatric surgery in severely obese individuals. Obes Surg 27(4):961–967. https://doi.org/10.1007/s11695-016-2387-6

Wang Z, Li K, Ouyang L, Iko H, Safi AJ, Gao S (2021) Effects of methylenetetrahydrofolate reductase single-nucleotide polymorphisms on breast, cervical, ovarian, and endometrial cancer susceptibilities. Chronic Dis Transl Med 7(3):169–181. https://doi.org/10.1016/j.cdtm.2021.06.003

Waterland RA, Jirtle RL (2003) Transposable elements: targets for early nutritional effects on epigenetic gene regulation. Mol Cell Biol 23(15):5293–5300. https://doi.org/10.1128/MCB.23.15.5293-5300.2003

Wei H, Zhu Y, Wang T, Zhang X, Zhang K, Zhang Z (2021) Genetic risk factors for autism-spectrum disorders: a systematic review based on systematic reviews and meta-analysis. J Neural Transm (vienna) 128(6):717–734. https://doi.org/10.1007/s00702-021-02360-w

Yamazaki Y, Zhao N, Caulfield TR, Liu CC, Bu G (2019) Apolipoprotein E and Alzheimer disease: pathobiology and targeting strategies. Nat Rev Neurol 15(9):501–518. https://doi.org/10.1038/s41582-019-0228-7

Yi J, Xiao L, Zhou SQ, Zhang WJ, Liu BY (2019) The C677T polymorphism of the methylenetetrahydrofolate reductase gene and susceptibility to late-onset Alzheimer's disease. Open Med (wars) 4(14):32–40. https://doi.org/10.1515/med-2019-0006

Precision Nutrition from the View of the Gut Microbiome

4

Plamena Dikarlo, Isabel Dorst, Olga Moskalenko, and Murad Yateem

Contents

Abstract

There is a growing body of evidence on the importance of the gut microbiome in human health and disease. Diet is a main factor that can alter the microbiome in ways that are beneficial to the host. However, the individual gut microbiome is like a fingerprint: it is unique to each person, and there is no universal recommendation for a healthy diet. This increases the need for microbiome-oriented precision nutrition research. Only in this way can the complex interplay between host, diet, and gut microbiome be understood and, in a next step, put into practice. Here, we summarize the emerging topics in the field of the human gut microbiome and show how to use and generate valuable research results in microbiome-oriented precision nutrition research.

P. Dikarlo (✉) · I. Dorst · O. Moskalenko · M. Yateem
BIOMES NGS GmbH, Schwartzkopffstraße 1, 15745 Halle 21, Wildau, Germany
e-mail: plamena.dikarlo@biomes.world

A. G. Haslberger (ed.), *Advances in Precision Nutrition, Personalization and Healthy Aging*, https://doi.org/10.1007/978-3-031-10153-3_4

4.1 Introduction

"What works for you–unfortunately doesn't work for me!"—facing this challenge on a daily basis, modern medical research was forced to develop new strategies for improving healthcare outcomes. A new field was born—the so-called precision medicine (Ginsburg and Phillips 2018) or P4 medicine—predictive, preventive, personalized, and participatory. It's main goals are to define biomarkers for early diagnosis of chronic degenerative diseases (prediction), identify at-risk individuals, so they can take proper action at early stages (prevention), stratify the patients using their personal metadata (personalization) and incorporate patient-centric data, incl. patient experiences and values into the decision-making process (participation) (Younesi and Hofmann-Apitius 2013).

At the same time, the growing interest in nutrition as a main driver of personal and public health (Nutrition and Public Health 1950; Lenoir-Wijnkoop et al. 2013) turned it into a target of health promotion and therapeutic strategies (Ross et al. 2012).

However, the high complexity of the subject and the lack of one-size-fits-all formula increased the need for integrative solutions in this field. Analogous to precision medicine, the term "precision nutrition" (PN) established itself as a multimodal approach for the generation of personalized nutrition advice, adopting the common methods of precision medicine (Zeisel 2020).

The development of both precision medicine and nutrition is based on collection, analysis, and interpretation of highly diverse biological data, including genomic and other types of omics data from different body sites, vital parameters, information about lifestyle, and personal preferences as well as many others (Schüssler-Fiorenza Rose et al. 2019). In fact, recent advances in DNA sequencing and computational tools for data analysis have enabled the in-depth study of the human genome, as well as the study of the so-called "second genome"—the microbiome (Knight et al. 2018).

Indeed, microbiome research has received increasing attention over the past decade. Although its importance for human health has been well known for a long time (Milestones in Human Microbiota Research 2019), there was no sufficient method for its in-depth analysis. Only with the introduction of the first next-generation high throughput sequencing, was there finally a method to explore this, what was then called "black box". One development led to another, and the microbiome-related research underwent rapid expansion (NIH Human Microbiome Portfolio Analysis Team 2019). However, the first challenge was finding an appropriate definition of the term microbiome to set clear research goals in this field.

According to the Human Microbiome Project Consortium, the human microbiome is the "collection of all the microorganisms living in association with the human body" (Human Microbiome Project 2019). These communities, especially the gut microbiome, mainly consist of bacteria (Qin et al. 2010), and in smaller amounts of archaea, eukaryotes, and viruses. Along with this definition, which describes the microbiome from a host/microorganism point of view, there are also definitions from an ecological (the entire community of microorganisms, inhabiting specific niche, for example, the gut) and a method-driven perspective (the entire genome/proteome/transcriptome/metabolome of these microorganisms). In contrast, the term "microbiota" covers only the microorganisms in a defined environment, without their metabolites or structural elements (Berg et al. 2020; Marchesi and Ravel 2015). In this sense, the term microbiome is the more comprehensive one, reflecting thc functionality of the microorganisms and the interaction with their environment. Therefore, when talking about the microbiome, both the composition and the activity should be considered, and both the genotype and the phenotype of the microbiome should be incorporated for further interpretations.

4.2 The Human Gut Microbiome—(Un)limited Possibilities for Improving Human Health

In a study published in 1977, the total number of procaryotic and eukaryotic microbial cells was estimated to be 10^{14}, which is equal to the number of host cells by factor 10 (estimated to be 10^{13}) (Savage 1977). Later, in 2016, Sender et al. 2016 proposed a new calculation method and showed that the bacterial count is of the same order as the number of human cells. However, both estimations are based on a "standard adult male", so they can vary across sex, age, or health condition (Sender et al. 2016; Sender et al. 2016). More importantly, most of these microbial cells (about 99%) reside in the colon and about half of daily wet stool mass is composed of bacteria (Stephen and Cummings 1980).

These and other findings from the last century (Dworkin et al. 2012; Farré-Maduell et al. 2019) increased the interest in the bacterial communities of the intestine, not only because of their tremendous abundance, but also because of the gut itself. The gut is the largest immunological (GALT—Gut Associated Lymphatic Tissue) and endocrine (EECs—enteroendocrine cells) organ in the human body. It fulfils a chemical, physical and immunological barrier function against pathogens, regulates the nutrient intake and has a dense innervation (ENS—enteric nervous system) (Michael Denbow and Chapter, 2015; Gunawardene et al. 2011; Sasselli et al. 2012). Also, the epithelial cells in a healthy gastrointestinal tract (GIT) undergo high turnover rates (Sender and Milo 2021).

The hypothesis that the gut microbiome may have a role in maintaining the gut homeostasis, encouraged the gut microbiome research, resulting in many new insights into the mechanisms of interaction between intestinal bacteria and the mucus layer (Sicard et al. 2017; Ouwerkerk et al. 2013). Furthermore, gut microbes are involved in the immune regulation (Wells 2011; Miquel et al. 2015; Liu et al.

2017; Lathrop et al. 2011; Peterson et al. 2015), nutrient utilization (Cockburn et al. 2015; Flint et al. 2012), vitamin synthesis (LeBlanc et al. 2013; Cooke et al. 2006), production of neurotransmitters (Strandwitz et al. 2019; Strandwitz 2018), postprandial blood sugar response (Zeevi et al. 2015), calorie extraction (Turnbaugh et al. 2006), and other metabolic processes (see Fig. 4.1). These facts support the assumption of a bi-directional microbiome—host interaction, which is fundamental for the so-called holobiont concept (see Box 1) (Bordenstein and Theis 2015).

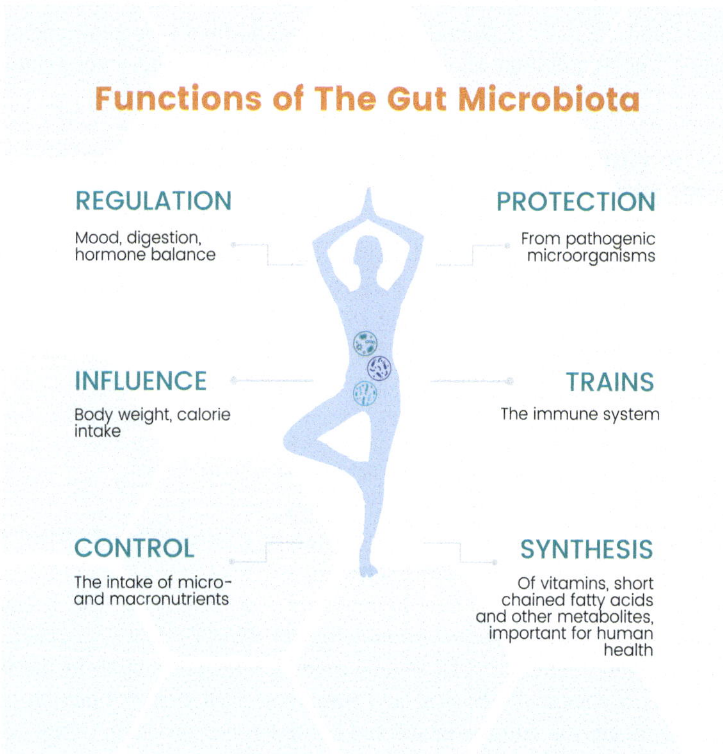

Fig. 4.1 Functions of the gut microbiota

Box 1 Definitions of holobiont, mutualism and commensialism, Alpha- and Beta-diversity and Pan-Microbiome

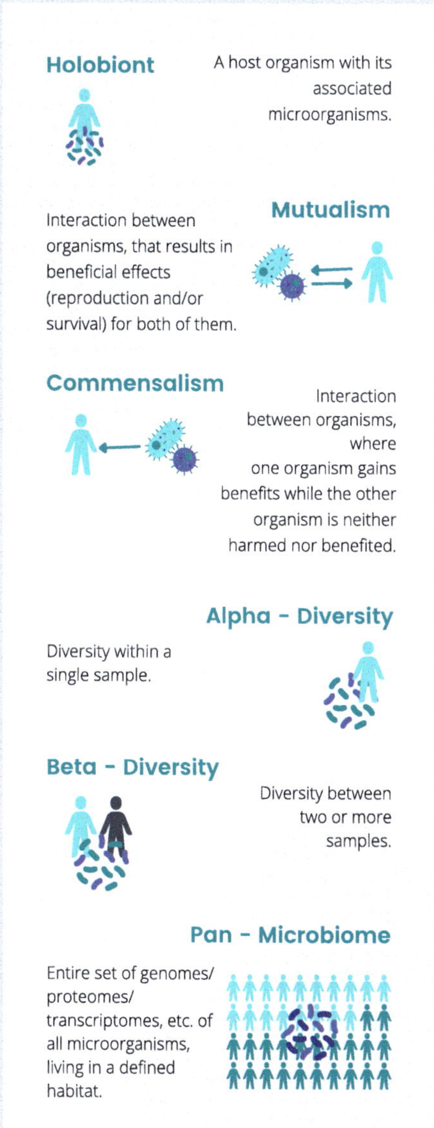

Moreover, the gut microbiome undergoes dynamic shifts in response to changing environmental conditions. Lifestyle modifications in physical activity

(Scheiman et al. 2019; Mohr et al. 2020; Grosicki et al. 2019) or circadian rhythm (Carasso et al. 2021), as well as drug intake (Klünemann et al. 2021; Liou et al. 2019; Jakobsson et al. 2010), alcohol consumption (Ames et al. 2020; Roy et al. 2020), tobacco smoking (Prakash et al. 2021), pet adoption (Levin et al. 2016; Kates et al. 2020), social relationships like partnerships and siblings (Dill-McFarland et al. 2019), travel (Devkota 2020; Rasko 2017), water consumption (Hansen et al. 2018; Zhou et al. 2021), or seasonality (Davenport et al. 2014) are known to have an impact on the gut microbiome composition (see Fig. 4.2). However, the habitual diet is the most prominent factor that shapes and alters the gut microbiome—both short-and long-term (David et al. 2014; Arumugam et al. 2011; Wu et al. 2011; Lilja et al. 2021). Therefore, studying changes in the gut microbiota in response to diet is a promising tool for improved outcomes in precision nutrition research.

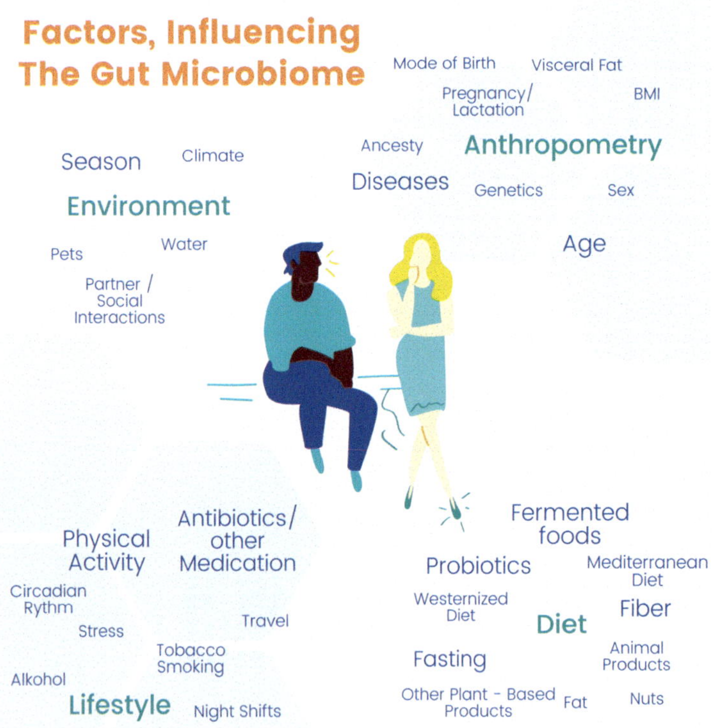

Fig. 4.2 Factors, influencing the gut microbiome

4.3 Exploring the Human Gut Microbiome: Start Low Go Slow—Advances in Microbiome Research

The dynamics of the gut microbiome described above contribute to host genetic adaptation (Bäckhed et al. 2005; Suzuki and Ley 2020) and can extend the host evolutionary potential by shifting host phenotype (Henry et al. 2021). While most of the gut bacteria are commensals, more precisely: mutualists (see Box 1) (Bäckhed et al. 2005), there are also some potentially harmful bacteria, which may have negative impact on the host when their population growth rate goes high (Shin et al. 2015; Baron 1997). These and other findings raise the primary question in gut microbiome research: what is the definition of a healthy gut microbiome?

4.3.1 Eubiosis versus Dysbiosis

While dysbiosis is a gut microbiome pattern linked to a host disease state characterized by unwanted activation of the inflammasome (Zheng et al. 2020; Ley et al. 2006; Manichanh et al. 2006; Cani et al. 2012; Quévrain et al. 2016; Vieira-Silva et al. 2019; Filippis et al. 2021; Bolte et al. 2021) or dominance of other pathological mechanisms (Hertel et al. 2019; Koeth et al. 2019; Ghoshal et al. 2016), the state of "eubiosis", or "healthy state" of the gut is still not clearly defined (Shanahan et al. 2021).

With this purpose in mind, two working groups, provided two different approaches as an answer to this question in 2012. First, Bäckhed et al. 2012 proposed a conceptual interpretation of the healthy microbiome, outlining its two main characteristics: Stability, or the ability of the microbial community to resist change under stress conditions and resilience, or the ability of a community to return to a balance state after experiencing stress-related perturbation (Bäckhed et al. 2012). Second, the Human Microbiome Project Consortium published results of large-scale genomic data of 242 healthy male and female adults from five major areas of the human body, incl. the lower gastrointestinal tract (The Human Microbiome Project Consortium 2012a, 2012b). The main goal of the study was to create a roadmap of the microbial metacommunity in humans and to understand the interplay between host and microbiome in health and disease (Turnbaugh et al. 2007). Another important issue that both groups are focusing on is the *Core Microbiome* as an essential compositional part of the individual and collective microbiome in a given habitat. In theory this "set of genes" is responsible for vital functions of the host and it usually remains stable during life. Shortly before, the results of the MetaHIT project had been published (European Metagenomics of the Human Intestinal Tract project). By analyzing a cohort of 124 Europeans, the scientists were able to identify a total of 1000–1500 species (Qin et al. 2010). Eighteen of these species were present among all samples, including *F. prausnitzii*, *Roseburia intestinalis*, species of *Ruminococcus,* and *Bacteroides* genus—all bacteria, known to be involved in complex carbohydrate degradation (Flint et al. 2012; Townsend et al. 2020). Additionally, a recently published genome-wide association study merged cross-sectional data of 21 different cohorts and showed that from a total of 410 bacterial genera, nine only where present in ≥ 95% of the samples (Kurilshikov et al. 2021). Again, among these bacteria are various complex polysaccharides degrading genera such as *Faecalibacterium*, *Roseburia*, *Ruminococcus*, *Blautia*, and *Bacteroides*, outlining the evolutionary importance of the gut microbiome in nutrients utilization.

Consequently, the next logical question is: If we all have a similar microbiome—*it is necessary to integrate microbiome data into PN research?*

First, the simple prevalence of certain bacteria in the gut does not provide information about its abundance (relative or absolute). As an example, individuals can be clustered into three different enterotypes depending on the relative predominance of only three genera: *Prevotella*, *Ruminococcus*, and *Bacteroides* (Arumugam et al. 2011). The relative predominance of these genera is associated with co-abundance of other bacterial taxa (for instance due to cross-feeding (Rios-Covian et al. 2015)), which confirms the complex qualitative and quantitative interplay within a community and underlines a necessity for studying the whole community. As mentioned before, enterotypes are also linked to long-term dietary patterns, outlining the role of daily diet for shaping the gut microbiome (Fig. 4.2) (Wu et al. 2011). In contrast, further investigations based on quantitative profiling contradicted the taxonomic trade-off between *Bacteroides* and *Prevotella*—enterotype. In addition, another, low-count *Bacteroides* Enterotype (B2) was proposed, which is prevalent in patients with Crohn's disease (Vandeputte et al. 2017). This suggests implementation of different workflows in different clinical settings, dependent on the scientific issue.

Second, nine genera clearly do not represent the whole diversity of the gut microbiome. Also, the perception of certain enterotypes is an ambitious effort to stratify individuals depending on their predominant taxa (Costea et al. 2018). However, these findings do not give information about the whole picture. In contrast, diversity (for example as diversity index) implies overall species richness and evenness of the sample and can be measured with different approaches (DeJong 1975; Hagerty et al. 2020). Diversity is also linked to different health conditions and laboratories parameters (Turnbaugh et al. 2009; Zhernakova et al. 2016), with some controversial outcomes (Jiang et al. 2015; Paulsen et al. 2017). Additionally, other methods exist that also take functional bacterial characteristics into account (gene ontology-based clustering) (Arumugam et al. 2011; Manor et al. 2020).

Third, the same genera, or even the same species are associated with different phenotypes. One prominent example is the *Prevotella* genus. As it seems, they are predominant not only in people with high-carbohydrate diets or diets rich in agricultural products (Wu et al. 2011; Yatsunenko et al. 2012), but also in HIV-positive or RA–patients and are involved in inflammatory processes (Dillon et al. 2014; Scher et al. 2013). Moreover, distinct species within *Prevotella* ssp. compete for similar plant derived polysaccharides (Gálvez et al. 2020). The *Prevotella* copri complex, with currently 106 identified genomes, seems to have a different phenotype in omnivores and vegetarians (Filippis et al. 2019; Tett et al. 2021). A comprehensive review (Tett et al. 2021) discusses in detail the hallmarks of *Prevotella* genus. In fact, these arguments are pointing toward the importance of functional profiling of the bacteria. Along with this, personal metadata and host phenotype should be included into interpretation and subsequent decision-making process.

Fourth, the gut microbiome is changing. There are measurable changes both on a daily basis (Vandeputte et al. 2021) and in a long-term perspective (Faith et al. 2013). On the one hand, the gut microbiome is known to be involved in healthy aging (Biagi et al. 2010a, 2010b; Claesson et al. 2012; Wilmanski et al. 2021). On the other hand, some specific daily fluctuations are linked to unfavorable metabolic alterations of the host (risk of type 2 diabetes) (Reitmeier et al. 2020). Only by consequent measurements, for example as a part of longitudinal studies, we can understand which impact these fluctuations in alpha-diversity may have on human health (see Box 1).

Fifth, the evidence in the microbiome field is expanding. Currently, a lot of associations and reproducible correlations between alteration of gut microbiome and a disease state can be demonstrated, some of them with causal explanation (Quévrain et al. 2016; Koeth et al. 2019), but many of them—without (Turnbaugh et al. 2009; Zhernakova et al. 2016; Dao et al. 2016). Additionally, the phylogenetic diversity of a human gut is still undefined, as results of newly published study are showing (Almeida et al. 2019). Also, together with bacteria and archaea, a small part of the gut microbiome consists of fungi, viruses, phage, etc., which are still not sufficient explored (Nash et al. 2017; Auchtung et al. 2018). Future research goals should clarify the human gut pan-microbiome and the causal relationships between genotype and phenotype (see Box 1).

Sixth, the knowledge about the microbiome connection to other organ systems supports the holistic approach in medicine. "*All disease begins in the gut*" (Hippocrates) but, as we know, it doesn't stay there. New research confirms the existence of gut-brain (Valles-Colomer et al. 2019), gut-liver (Jian et al. 2021 Jan; Zhang et al. 2021), gut-heart (Koeth et al. 2019; Wang et al. 2011), and other types of gut—organ axes (Arrieta et al. 2015; Salem et al. 2018). This again points toward the important role that PN plays in the support and maintenance of the general health of each individual.

Seventh, understanding inter- and intraindividual metabolic heterogeneity is one of the main goals of precision medicine and precision nutrition (Zeisel 2020). Further investigation of the compositional and functional spectrum of the microbiome may be the missing part, which completes the big picture (hologenome).

All these and other facts are rising the next key questions in PN research: *how to conduct a gut microbiome targeted study in order to gain valuable results?*

4.4 The Microbiome Study in PN Research—Important Aspects (see Figs. 4.3 and 4.4, Box 2)

Despite rapid advances in high throughput sequencing (HTS) technologies and software development for analysis of metagenomic data, study outcomes in microbiome research have relative low reproducibility. A variety of different methods and high discrepancy in workflow between the laboratories may lead to different results, even for the same samples. Furthermore, various sources of bias can negatively affect the final outcome. In fact, each step in this multimodal process, from sample collection to completion of a microbiome profile, is simultaneously a potential source of bias (Hiergeist and Gessner 2018). Therefore, one of the current efforts in the microbiome field is to understand why these failures happen, as well as to propose strategies for error management and error prevention.

Box 2 Definitions of OTU (Operational Taxonomic Unit), ASV, 16s rDNA sequencing, WGS and RNA sequencing

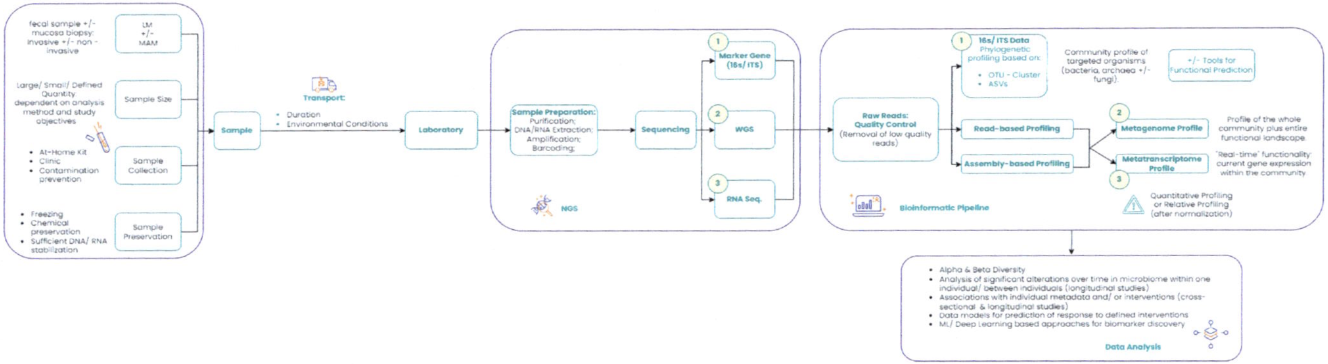

Fig. 4.3 Process flow: from sample to report

Fig. 4.4 Conducting a gut microbiome-oriented study in PN research: decision-making process

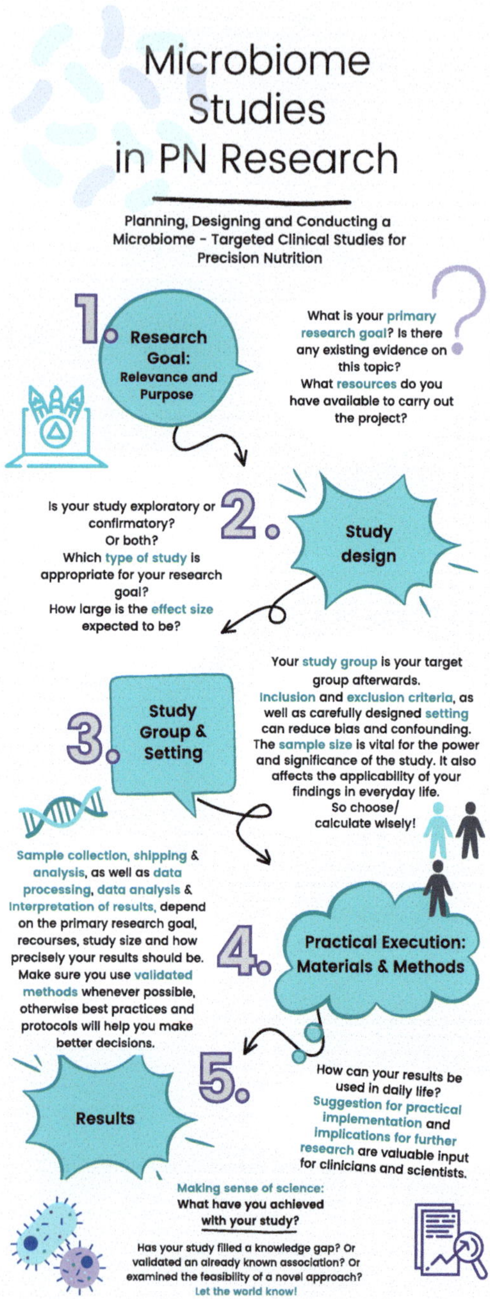

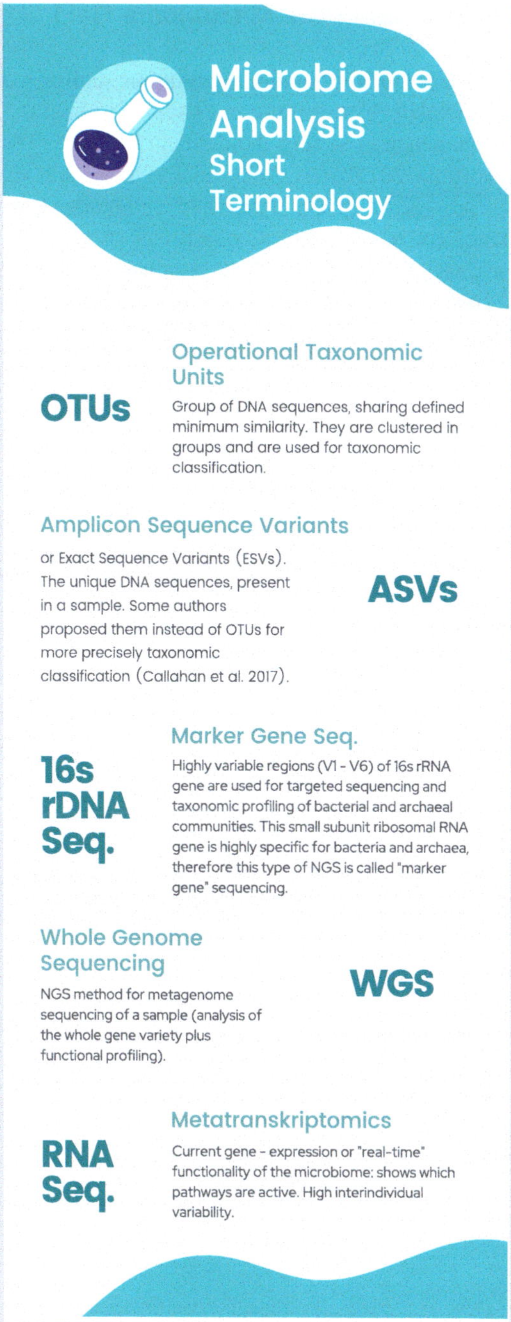
Microbiome Analysis
Short Terminology
OTUs
Operational Taxonomic Units
Group of DNA sequences, sharing defined minimum similarity. They are clustered in groups and are used for taxonomic classification.
Amplicon Sequence Variants
or Exact Sequence Variants (ESVs). The unique DNA sequences, present in a sample. Some authors proposed them instead of OTUs for more precisely taxonomic classification (Callahan et al. 2017).
ASVs
16s rDNA Seq.
Marker Gene Seq.
Highly variable regions (V1 - V6) of 16s rRNA gene are used for targeted sequencing and taxonomic profiling of bacterial and archaeal communities. This small subunit ribosomal RNA gene is highly specific for bacteria and archaea, therefore this type of NGS is called "marker gene" sequencing.
Whole Genome Sequencing
NGS method for metagenome sequencing of a sample (analysis of the whole gene variety plus functional profiling).
WGS
RNA Seq.
Metatranskriptomics
Current gene - expression or "real-time" functionality of the microbiome: shows which pathways are active. High interindividual variability.

4.4.1 Setting Standards in the Microbiome Field

With the increasing interest in the human microbiome and microbiome research, projects like HMP/MetaHit Consortium or The Microbiome Quality Control project, took on the challenge of setting standards in this area (The Human Microbiome Project Consortium 2012b)[1] [2]. Additionally, new methods, protocols, and bioinformatic tools are regularly evaluated and summarized in comprehensive reviews with guidance for best practices, which can support a researcher in the decision-making process (Knight et al. 2018; Allaband et al. 2019; Shakya et al. 2019; Whon et al. 2021). However, these reviews are not clearly linked to microbiome studies in precision nutrition research, so here we want to emphasize again the most important points in a PN-context:

1. Sample

There is lots of evidence about the differences between so-called mucosa associated microbiome (MAM) and luminal microbiome (LM). Because of the variable environmental conditions, microbes with distinct metabolism are located in the lumen or in the mucosal surface of the intestine (Ringel et al. 2015; Eckburg et al. 2005). Both luminal and mucosa associated microbiome are undergoing alterations in a disease state, for instance obesity, inflammatory bowel disease or liver cirrhosis (Dong et al. 2017; Derrien et al. 2004; Everard et al. 2013; Shen et al. 2021). Nevertheless, because of the feasibility and non-invasive sampling, most of the current evidence is based on analysis of fecal samples. In contrast, to study MAM, a biopsy from a defined intestinal area should be taken. Usually this is performed during endoscopy, after intake of laxatives, which can lead to systematic bias of the results (Tropini et al. 2018). Rectal swabs have been suggested as an alternative, but in this case only a rectal sample is analysed, without other bowel regions (Shen et al. 2021). Other suggested sampling techniques are laser capture microdissection (LCM) or luminal brushing (Tang et al. 2020). In summary, to understand the microbiome–mucosal barrier interactions and the influence of nutrition on intestinal communities, it is vital to study the compositional and functional characteristics of the MAM, along with LM. Research areas of great interest are diseases with evidence of altered intestinal epithelial homeostasis such as colon cancer, inflammatory bowel disease (IBD), celiac disease, and many others (Ramezani et al. 2021; Gitter et al. 2001; Schumann et al. 2012). In any case, the study of the MAM in PN research is not an either/or question but will be an additional *MAM—test useful for my findings*—question.

[1] https://www.sanger.ac.uk/resources/downloads/bacteria/metahit/.
[2] https://www.mbqc.org/.

Other details of the sample collection that may affect the results are sample size (depends mainly on analysis method) and preservation/storage/shipping conditions (conservation/freezing of the sample is needed to prevent further growth of facultative aerobic bacteria (McDonald et al. 2018)).

2. Sequencing Method

For phylogenetic and functional profiling of the gut microbiome three types of omics technologies are commonly used: metataxonomics (or marker gene sequencing, such as 16 s rRNA-gene sequencing), metagenomics (whole genome sequencing or WGS), and metatranscriptomics (see Info box 1). The three of them have positives and negatives. Which of them should be used in certain case depends on the study goal and the resources available. For an instance, **16s rDNA Sequencing** is suitable for confirmatory studies, focusing on phylogenetic changes in the bacterial community, as well as overall diversity (alpha and beta) (Knight et al. 2018). Consider the following case: Scientists want to prove the effect of a novel diet with whole-foods on overweight and obese individuals. A control group with a similar structure in terms of anthroposophical characteristics and everyday habits is also planned. Based on the literature, the scientists decide to measure the change in the ratio of *Firmicutes* to *Bacteroidetes* (or F to B ratio as a primary research goal) (Turnbaugh et al. 2006; Ley et al. 2006). They want to find out whether these phyla change over time. Additionally, they want to measure alpha-diversity (Turnbaugh et al. 2009) and see if other phyla, e.g., proteobacteria, are also affected. In this case, 16s rDNA sequencing is perfectly adequate, first because of the low cost pro run. In this way, more resources can be used elsewhere, e.g., for recruiting more participants. Secondly, because the results have hight correlation rate with genomic content when done correctly (Knight et al. 2018). This is made possible by a variety of quality control tools, based on 16s rDNA data, as well as by regularly updated databases for 16s rDNA sequences (only if the technical implementation has been carried out correctly) (Quast et al. 2013; Parks et al. 2018).

Turning back to the thought experiment: after analyzing the data, the results are not sufficient. After testing for normal distribution, the Wilcoxon rank sum test (non-parametric) is used to test for significance (Pan 2021). Unfortunately, there are no significant changes in the F to B ratio between intervention and control group. However, the scientists can see a positive tendency in favor of the novel diet. In addition, the group of responders has a higher *Prevotella* to *Bacteroides* ratio, whereas in the non-responder group *Bacteroides* genus dominates over *Prevotella* genus within *Bacteroidetes* phylum (Roager et al. 2014; Hjorth et al. 2019).

Therefore, the researchers conduct a multivariate analysis (ANOVA) showing that the "responder" group is more likely to be associated with a high variable plant-based diet (more than 30 different plant-based products per week), high in fiber, than a medium variable (20–30 different plant-based products) or a low variable (less than 20 different plant-based products per week)—lower in fiber. To predict the functional potential of the samples, an additional bioinformatic tool is implemented (PICRUSt2) (Douglas et al. 2020). The researchers want to know which substrates exactly promote the higher *Prevotella* abundance, so they decide to use MetaCyc as a reference database (Caspi et al. 2014). After running the analysis, an increased carbohydrate metabolism can be shown in both groups (responder and non-responder in comparison to control). To find out how genera abundances correlate with different diet carbohydrates, a Spearman correlation analysis is performed. While the abundance of *Prevotella* genus is related to simple and complex carbohydrates, the abundance of *Bacteroides* genus is very slightly related to complex carbohydrates and plant derivates (Wu et al. 2011).

These results are very intriguing, so the researcher decide to go further and to use **metagenomic (whole genome sequencing)** approach to achieve high taxonomic resolution on species level (which is not sufficient with 16s rDNA seq.), as well as to build a potential functional profile of the samples. EggNOG mapper (Cantalapiedra et al. 2021) is used for functional annotation and CAZy database (Drula et al. 2021) is used for detailed classification of the carbohydrate-active enzymes, present in the samples. In addition, the participants' metadata and daily food frequency questionnaires are carefully examined with nutrition software allowing the different types of dietary fiber to be categorized and attributed to changes in the gut microbiome. In-depth correlation analysis of CAZy enzyme families and abundance of bacterial genera shows two cluster of glycosyl hydrolases—*Prevotella*-annotated and *Bacteroides*-annotated glycosyl hydrolases. Furthermore, the distinct CAZy profile of responder with high abundance of *Prevotella* copri complex suggests metabolic adaptation of different *Prevotella* species to different substrates from the diet (Aakko et al. 2020). In order to understand this interaction in real time, the scientists perform a **metatranscriptomic (RNA seq)** analysis after a tightly controlled dietary intervention with daily changes in dietary fiber composition and amount in the participants' diets. In this way, the combined metatranscriptomic and metagenomic analysis shows the viability in gene expression on strain level within the *Prevotella* complex in response to different fibers and in context of the individual microbiome composition. Additionally, the scientists note an enrichment of gene expression of the beneficial bacteria *Bacteroides thetaiotaomicron* after pectin-rich meals (Townsend et al. (2020); Porter et al. 2018). Furthermore, the scientists use this data to train a machine learning model for prediction of metabolic response to different fiber types so that they can create personalized advice.

In the end, the researchers conclude that their novel whole-foods diet is beneficial for individuals with initially higher *Prevotella* to *Bacteroides* rates, although for optimal results the exact composition of dietary fiber should be customized based on the individual's metagenomic profile using a predictive model (Oh and Zhang 2020).

This hypothetical case shows how a possible decision-making process can proceed, and in which cases the different sequencing techniques can be applied. It is important to note the differences in the results obtained with the various methods and the subsequent use of the data without compromising scientific correctness. Furthermore, this case underlines the diversity of bioinformatics tools, which enables a high interoperability of the data obtained. In this way, even more information can be extracted from the data.

3. Bioinformatic Pipeline

In fact, various criteria can be used to evaluate the different bioinformatic pipelines in microbiome research. Apart from the sequencing method, criteria such as:

- reference database (is it updated or not, size, manual),
- graphic interface and/or command line,
- system requirements (e.g., RAM, size of pipeline/tool),
- time (how long does it take to run an analysis),
- single or paired reads,
- costs (open source or paid),
- options for customization and
- ease of implementation
- influence the final decision.

For example, one important part of the pipeline is the reference database for taxonomic profiling of the sample. Features such as the number of reference sequences and taxonomies describe the size of the database. Even more important, however, is the regular update rate of the database. In fact, Greengenes is the largest database for 16s rDNA, but it has not been updated since 2013 (McDonald et al. 2012). As already mentioned, there are still many undiscovered bacteria species (Almeida et al. 2019). Therefore, other databases such as GTDB 89 or Silva 138 are used more frequently today because of their regular updates (Quast et al. 2013; Parks et al. 2018). Furthermore, implementation of additional tools for functional prediction such as PICRUSt, or databases like CAZy are working with different datasets (input/output). Therefore, when planning a study, consideration should be given to what type of data is needed for the intended analysis (16s rDNA/metagenomic/metaproteomic data, etc.).

4. Study Population

Even before we ask ourselves what (sample) and how (method) to test, we need to know who (study population) we are testing. As discussed above, the gut microbiome composition and functionality are influenced by various lifestyle factors, as well as by different anthropometrical factors such as body mass index (BMI), age, sex, and genetics (Biagi et al. 2010a, 2010b; Claesson et al. 2012; Wilmanski et al. 2021; Cuesta-Zuluaga et al. 2019; Scepanovic et al. 2019; Nie et al. 2020; Goodrich et al. 2014). The selection of a highly homogeneous group of people, as well as strongly controlled study setting can reduce bias and confounding, however, strict dietary plans and comprehensive food frequency questionnaires in interventional studies are often a burden for the participants and can reduce the feasibility and external evidence of the study. This can lead to a lot of missing data, which prevents meaningful data analysis, or to results that are not relevant at all, so that they cannot be put into practice. Therefore, efforts should be made to develop and validate a customer-friendly data collection tools (e.g., wearables and apps) and to devise strategies for comprehensive data analysis. Alternatively, different nutritional scores (HEI (Krebs-Smith et al. 2019), MEDAS (Hebestreit et al. 2017), etc.) are being developed to summarize the results and form different clusters, but their practicality is limited as they can show a trend but not enable a highly personalized recommendation. Nevertheless, they are an important component of retrospective studies in which dietary habits are based only on the information provided by the participants. This leads to the next challenge in microbiome-oriented PN research, namely

5. Big Data Analysis

High throughput sequencing technologies generate large, complex, and multi-dimensional datasets. As already discussed, they are generated by 16S rRNA Amplicon, metagenomics, metatranscriptomics, or other omics technologies, for instance, metaproteomics. As a result, this data is put into tables with read counts or relative abundances. The predominant data structure in microbiome projects is a feature table, that displays the distribution of different features (e.g., OTUs/ASVs/taxonomic levels, etc.) to samples in the form of counts or relative abundance (Callahan et al. 2017). Typical characteristics of microbiome data are:

- Compositionality: are made up of relative proportions of a defined unity (library size) (Gloor et al. 2017).
- High Dimensionality: Mostly due to a high number of OTUs or taxa (Xia et al. 2018).
- Sparsity: absence of numerous taxa across samples results in an abundance of zeros within feature tables (Pan 2021).

These characteristics pose particular challenges for the processing, visualization, and statistical analysis of microbiome data. As discussed before, the aim of microbiome-oriented PN research is to investigate metabolic heterogeneity, based on the relationship between environmental conditions, interventions, clinical, and biological parameters, as well as microbiome composition and phenotype. For this purpose, various methods can be used. Classical methods, for instance different parametric and non-parametric tests such as t-test, ANOVA (parametric) or Wilcoxon rank sum test, Kruskal-Wallis test (non-parametric) can be used to compare continuous variables between groups (alpha-, beta-diversity-indices, relative abundance of taxa). Here, the significance is mostly determined by q-values (*p*-values that are adjusted for multiple hypotheses testing). Additionally, multivariate methods can be used to examine the association between the microbiome and covariates on a compositional community level. The benefit is, that multiple variables can be examined simultaneously, focusing on the whole composition, not on a single taxa. In addition, naturally occurring interaction in a microbiome dataset can be identified. These methods are based on a distance metric (e.g., Unifrac, Bray-Curtis dissimilarity), e.g.:

- PERMANOVA (like ANOVA but for multivariate data).
- ANOSIM.
- Mantel's test.

Further strategy for analyzing microbiome data is the compositional analysis. Due to the compositional nature of microbiome data, common statistical methods can lead to spurious correlations. Therefore, Aitchison proposed a log-ratio transformation, to account for the issue of a fixed sample size (Aitchison 1986). After log-ratio transformation, other multivariate methods can be applied to compositional data. In fact, Aitchison distance (Euclidean distance of centred log-ratio transformed abundances) can serve as a more robust alternative to other beta-diversity metrics, such as Bray-Curtis dissimilarity, weighted, or unweighted Unifrac.

4.4.2 Visualizing Methods in Microbiome Research

Another important aspect of big data analysis is data visualization. In fact, visualization methods can strongly influence decision-making processes and accelerate discovery rates (Park et al. 2021).

In case of microbiome research, differences in microbial diversity between a multitude of samples cannot be visual by a standard scatterplot, as a microbial feature table consists of many dimensions. Therefore, to reduce dimensions and to graphically display microbiome data, different ordination methods can be applied. Ordination methods try to find axes in a multidimensional space that account for the maximum amount of variance in a dataset. Most used visualization methods are:

- Principal coordinate analysis (PCoA): Uses distance matrix as input. Distance matrix can be calculated with any given beta-diversity metric (Jaccard-Indey, Bray-Curtis, Unifrac).
- Principal component analysis (PCA): Uses feature table directly as input. The input can also be log-ratio corrected to account for compositionality.

Of course, this is only a brief overview of the most used data analysis methods in microbiome research. More sophisticated statistical models such as machine learning or deep learning techniques have been developed over the years and enable the analysis of huge data sets. A recent comprehensive review discusses the widely used methods in precision nutrition research (Kirk et al. 2021).

4.5 From the Clinical Trial to the Personal Recommendation: Putting the Individual Pieces of the Puzzle Together

All the steps mentioned so far have one goal: the generation of a personalized recommendation. A common approach is to take the data of healthy people as a reference and see if and where the individual gut flora deviates greatly. In this way, the targets of the planned intervention are defined. Personal metadata, diseases, allergies, etc., must be included in the decision-making process. However, personalization is just one aspect of the precision nutrition. Personalized solutions from a purely medical perspective are often not feasible in practice, due to unwillingness of the consumers to carry out the intervention. Various factors may play a role here, such as lack of discipline and motivation, but also inadequate guidance from health professionals, a conflict with personal preferences such as diet (vegan/vegetarian), religious requirements (kosha, no alcohol, etc.), and many others. Therefore, participation, or integration of the consumer's needs in the decision-making process could be a powerful tool to enhance adherence to the intervention. This can be pursued through questionnaires or personal interviews, but the important point is to understand the consumer's personal goals and wishes in order to find an appropriate solution to their problem and prioritize the individual steps together with them. This type of consumer-centered approach aims to achieve better results with long-lasting positive behavioral changes. Additionally, as the gut microbiome tests are a snapshot, close monitoring is required to ensure regular evaluation of the measures taken and to be able to intervene if necessary.

4.6 Conclusion

Insights into the complex interplay between host, environment and microorganism are constantly improving with advances in technology and software development. In addition, tools for data collection, processing, and analysis are becoming more precise, enabling better health outcomes at lower costs. Consumer-centered approaches view each person as a unique being with own values and preferences and seek personalized solutions with high compliance for optimal health outcomes. However, there is much more to be done. Looking at precision nutrition as a process rather than a target of perfect stratification, one of the main goals in this field is to optimize and validate the different processes and force improvements without putting consumers at risk. With flexibility, constant evaluation, and appropriate error management, this is possible.

References

Aakko J, Pietilä S, Toivonen R et al (2020) A carbohydrate-active enzyme (CAZy) profile links successful metabolic specialization of Prevotella to its abundance in gut microbiota. Sci Rep 10:12411. https://doi.org/10.1038/s41598-020-69241-2

Aitchison J (1986) The statistical analysis of compositional data. Chapman and Hall, London

Allaband C, McDonald D, Vázquez-Baeza Y, Minich JJ, Tripathi A, Brenner DA, Loomba R, Smarr L, Sandborn WJ, Schnabl B, Dorrestein P, Zarrinpar A, Knight R (2018) Microbiome 101: studying, analyzing, and interpreting gut microbiome data for clinicians. Clin Gastroenterol Hepatol 17(2):218–230. https://doi.org/10.1016/j.cgh.2018.09.017. Epub 18 Sep 2018. PMID: 30240894; PMCID: PMC6391518

Almeida A, Mitchell AL, Boland M et al (2019) A new genomic blueprint of the human gut microbiota. Nature 568(7753):499–504. https://doi.org/10.1038/s41586-019-0965-1

Ames NJ, Barb JJ, Schuebel K, Mudra S, Meeks BK, Tuason RTS, Brooks AT, Kazmi N, Yang S, Ratteree K, Diazgranados N, Krumlauf M, Wallen GR, Goldman D (2020) Longitudinal gut microbiome changes in alcohol use disorder are influenced by abstinence and drinking quantity. Gut Microbes 11(6):1608–1631. https://doi.org/10.1080/19490976.2020.1758010. PMID: 32615913; PMCID: PMC7527072

Arrieta MC, Stiemsma LT, Dimitriu PA, Thorson L, Russell S, Yurist-Doutsch S, Kuzeljevic B, Gold MJ, Britton HM, Lefebvre DL, Subbarao P, Mandhane P, Becker A, McNagny KM, Sears MR, Kollmann T (2015) CHILD study investigators, Mohn WW, Turvey SE, Finlay BB. Early infancy microbial and metabolic alterations affect risk of childhood asthma. Sci Transl Med 7(307):307ra152. https://doi.org/10.1126/scitranslmed.aab2271. PMID: 26424567

Arumugam M, Raes J, Pelletier E et al (2011) Enterotypes of the human gut microbiome. Nature 473:174–180. https://doi.org/10.1038/nature09944

Auchtung TA, Fofanova TY, Stewart CJ, Nash AK, Wong MC, Gesell JR, Auchtung JM, Ajami NJ, Petrosino JF (2018) Investigating colonization of the healthy adult gastrointestinal tract by Fungi. mSphere 3(2):e00092-18. https://doi.org/10.1128/mSphere.00092-18. PMID: 29600282; PMCID: PMC5874442

Bäckhed F, Ley RE, Sonnenburg JL, Peterson DA, Gordon JI (2005) Host-bacterial mutualism in the human intestine. Science 307(5717):1915–1920. https://doi.org/10.1126/science.1104816. PMID: 15790844

Bäckhed F, Fraser CM, Ringel Y, Sanders ME, Sartor RB, Sherman PM, Versalovic J, Young V, Finlay BB (2012) Defining a healthy human gut microbiome: current concepts, future directions, and clinical applications. Cell Host Microbe 12(5):611–622. https://doi.org/10.1016/j.chom.2012.10.012

Baron EJ (1997) Bilophila wadsworthia: a unique Gram-negative anaerobic rod. Anaerobe. 3(2–3):83–6. https://doi.org/10.1006/anae.1997.0075. PMID: 16887567

Berg G, Rybakova D, Fischer D et al (2020) Microbiome definition re-visited: old concepts and new challenges. Microbiome 8:103. https://doi.org/10.1186/s40168-020-00875-0

Biagi E, Nylund L, Candela M, Ostan R, Bucci L, Pini E et al (2010a) Through ageing, and beyond: gut microbiota and inflammatory status in seniors and Centenarians. PLoS ONE 5(5):e10667. https://doi.org/10.1371/journal.pone.0010667

Biagi E, Nylund L, Candela M, Ostan R, Bucci L et al (2010b) Correction: through ageing, and beyond: gut microbiota and inflammatory status in seniors and Centenarians. PLoS ONE 5(6). https://doi.org/10.1371/annotation/df45912f-d15c-44ab-8312-e7ec0607604d

Bolte LA, Vich Vila A, Imhann F et al (2021) Long-term dietary patterns are associated with pro-inflammatory and anti-inflammatory features of the gut microbiome. Gut 70:1287–1298

Bordenstein SR, Theis KR (2015) Host biology in light of the microbiome: ten principles of holobionts and hologenomes. PLoS Biol 13(8):e1002226. https://doi.org/10.1371/journal.pbio.1002226. PMID: 26284777; PMCID: PMC4540581

Callahan B, McMurdie P, Holmes S (2017) Exact sequence variants should replace operational taxonomic units in marker-gene data analysis. ISME J 11:2639–2643. https://doi.org/10.1038/ismej.2017.119

Cani PD, Osto M, Geurts L, Everard A (2012) Involvement of gut microbiota in the development of low-grade inflammation and type 2 diabetes associated with obesity. Gut Microbes 3(4):279–288. https://doi.org/10.4161/gmic.19625

Cantalapiedra CP, Hernández-Plaza A, Letunic I, Bork P, Huerta-Cepas J (2021) eggNOG-mapper v2: functional annotation, orthology assignments, and domain prediction at the metagenomic scale. Mol Biol Evol 38(12):5825–5829. https://doi.org/10.1093/molbev/msab293

Carasso S, Fishman B, Lask LS, Shochat T, Geva-Zatorsky N, Tauber E (2021) Metagenomic analysis reveals the signature of gut microbiota associated with human chronotypes. FASEB J 35:e22011. https://doi.org/10.1096/fj.202100857RR

Caspi R, Altman T, Billington R, Dreher K, Foerster H, Fulcher CA, Holland TA, Keseler IM, Kothari A, Kubo A, Krummenacker M, Latendresse M, Mueller LA, Ong Q, Paley S, Subhraveti P, Weaver DS, Weerasinghe D, Zhang P, Karp PD (2014) The MetaCyc database of metabolic pathways and enzymes and the BioCyc collection of pathway/genome databases. Nucleic Acids Res 42(D1):D459–D471. https://doi.org/10.1093/nar/gkt1103

Claesson M, Jeffery I, Conde S et al (2012) Gut microbiota composition correlates with diet and health in the elderly. Nature 488:178–184. https://doi.org/10.1038/nature11319

Cockburn DW, Orlovsky NI, Foley MH et al (2015) Molecular details of a starch utilization pathway in the human gut symbiont Eubacterium rectale. Mol Microbiol 95(2):209–230. https://doi.org/10.1111/mmi.12859

Cooke G, Behan J, Costello M (2006) Newly identified vitamin K-producing bacteria isolated from the neonatal faecal flora. Microb Ecol Health Dis 18(3–4):133–138. https://doi.org/10.1080/08910600601048894

Costea PI, Hildebrand F, Arumugam M et al (2018) Enterotypes in the landscape of gut microbial community composition. Nat Microbiol 3:8–16. https://doi.org/10.1038/s41564-017-0072-8

Dao MC, Everard A, Aron-Wisnewsky J et al (2016) Akkermansia muciniphila and improved metabolic health during a dietary intervention in obesity: relationship with gut microbiome richness and ecology. Gut 65:426–436

Davenport ER, Mizrahi-Man O, Michelini K, Barreiro LB, Ober C, Gilad Y (2014) Seasonal variation in human gut microbiome composition. PLoS ONE 9(3):e90731. https://doi.org/10.1371/journal.pone.0090731. PMID: 24618913; PMCID: PMC3949691

David L, Maurice C, Carmody R et al (2014) Diet rapidly and reproducibly alters the human gut microbiome. Nature 505:559–563. https://doi.org/10.1038/nature12820

De Filippis F, Pasolli E, Tett A, Tarallo S, Naccarati A, De Angelis M, Neviani E, Cocolin L, Gobbetti M, Segata N, Ercolini D (2019) Distinct genetic and functional traits of human intestinal Prevotella copri strains are associated with different habitual diets. Cell Host Microbe 25(3):444-453.e3. https://doi.org/10.1016/j.chom.2019.01.004. Epub 21 Feb 2019. PMID: 30799264

De Filippis F, Paparo L, Nocerino R, Della Gatta G, Carucci L, Russo R, Pasolli E, Ercolini D, Berni Canani R (2021) Specific gut microbiome signatures and the associated pro-inflamatory functions are linked to pediatric allergy and acquisition of immune tolerance. Nat Commun 12(1):5958. https://doi.org/10.1038/s41467-021-26266-z. PMID: 34645820; PMCID: PMC8514477

de la Cuesta-Zuluaga J, Kelley ST, Chen Y, Escobar JS, Mueller NT, Ley RE, McDonald D, Huang S, Swafford AD, Knight R, Thackray VG (2019) Age- and sex-dependent patterns of gut microbial diversity in human adults. mSystems 4(4):e00261-19. https://doi.org/10.1128/mSystems.00261-19. PMID: 31098397; PMCID: PMC6517691

DeJong TM (1975) A comparison of three diversity indices based on their components of richness and evenness. Oikos 26:222–227. https://doi.org/10.2307/3543712

Derrien M, Vaughan EE, Plugge CM, de Vos WM (2004) Akkermansia muciniphila gen. nov., sp. nov., a human intestinal mucin-degrading bacterium. Int J Syst Evol Microbiol 54(Pt 5):1469–1476. https://doi.org/10.1099/ijs.0.02873-0. PMID: 15388697

Devkota S (2020) The gut microbiome during acute lifestyle transition. Nat Med 26:1013–1015. https://doi.org/10.1038/s41591-020-0980-7

Dill-McFarland KA, Tang ZZ, Kemis JH et al (2019) Close social relationships correlate with human gut microbiota composition. Sci Rep 9:703. https://doi.org/10.1038/s41598-018-37298-9

Dillon S, Lee E, Kotter C et al (2014) An altered intestinal mucosal microbiome in HIV-1 infection is associated with mucosal and systemic immune activation and endotoxemia. Mucosal Immunol 7:983–994. https://doi.org/10.1038/mi.2013.116

Dong LN, Wang JP, Liu P, Yang YF, Feng J, Han Y (2017) Faecal and mucosal microbiota in patients with functional gastrointestinal disorders: correlation with toll-like receptor 2/toll-like receptor 4 expression. World J Gastroenterol 23(36):6665–6673. https://doi.org/10.3748/wjg.v23.i36.6665

Douglas GM, Maffei VJ, Zaneveld JR et al (2020) PICRUSt2 for prediction of metagenome functions. Nat Biotechnol 38:685–688. https://doi.org/10.1038/s41587-020-0548-6

Drula E, Garron M-L, Dogan S, Lombard V, Henrissat B, Terrapon N (2021) The carbohydrate-active enzyme database: functions and literature. Nucleic Acids Res gkab1045. https://doi.org/10.1093/nar/gkab1045

Dworkin M, Gutnick D (2012) Sergei Winogradsky: a founder of modern microbiology and the first microbial ecologist. FEMS Microbiol Rev 36(2):364–379. https://doi.org/10.1111/j.1574-6976.2011.00299.x

Eckburg PB, Bik EM, Bernstein CN, Purdom E, Dethlefsen L, Sargent M, Gill SR, Nelson KE, Relman DA (2005) Diversity of the human intestinal microbial flora. Science 308(5728):1635–8. https://doi.org/10.1126/science.1110591. Epub 14 Apr 2005. PMID: 15831718; PMCID: PMC1395357

Everard A, Belzer C, Geurts L et al (2013) Cross-talk between Akkermansia muciniphila and intestinal epithelium controls diet-induced obesity. Proc Natl Acad Sci USA 110(22):9066–9071. https://doi.org/10.1073/pnas.1219451110

Faith JJ, Guruge JL, Charbonneau M et al (2013) The long-term stability of the human gut microbiota. Science 341(6141):1237439. https://doi.org/10.1126/science.1237439

Farré-Maduell E, Casals-Pascual C (2019) The origins of gut microbiome research in Europe: from Escherich to Nissle. Hum Microbiome J 14:100065. ISSN: 2452-2317. https://doi.org/10.1016/j.humic.2019.100065

Flint HJ, Scott KP, Duncan SH, Louis P, Forano E (2012) Microbial degradation of complex carbohydrates in the gut. Gut Microbes 3(4):289–306. https://doi.org/10.4161/gmic.19897

Gálvez EJC, Iljazovic A, Amend L, Lesker TR, Renault T, Thiemann S, Hao L, Roy U, Gronow A, Charpentier E, Strowig T (2020) Distinct polysaccharide utilization determines interspecies competition between intestinal Prevotella spp. Cell Host Microbe 28(6):838-852.e6. https://doi.org/10.1016/j.chom.2020.09.012. Epub 27 Oct 2020. PMID: 33113351

Ghoshal U, Shukla R, Srivastava D, Ghoshal UC (2016) Irritable bowel syndrome, particularly the constipation-predominant form, involves an increase in Methanobrevibacter smithii, which is associated with higher methane production. Gut Liver 10(6):932–938. https://doi.org/10.5009/gnl15588

Ginsburg GS, Phillips KA (2018) Precision medicine: from science to value. Health Aff (millwood) 37(5):694–701. https://doi.org/10.1377/hlthaff.2017.1624

Gitter AH, Wullstein F, Fromm M, Schulzke JD (2001) Epithelial barrier defects in ulcerative colitis: characterization and quantification by electrophysiological imaging. Gastroenterology 121(6):1320–1328. https://doi.org/10.1053/gast.2001.29694. PMID: 11729111

Gloor GB, Macklaim JM, Pawlowsky-Glahn V, Egozcue JJ (2017) Microbiome datasets are compositional: and this is not optional. Front Microbiol 8:2224. Published 15 Nov 2017. https://doi.org/10.3389/fmicb.2017.02224

Goodrich JK, Waters JL, Poole AC et al (2014) Human genetics shape the gut microbiome. Cell 159(4):789–799. https://doi.org/10.1016/j.cell.2014.09.053

Grosicki GJ, Durk RP, Bagley JR (2019) Rapid gut microbiome changes in a world-class ultramarathon runner. Physiol Rep 7(24):e14313. https://doi.org/10.14814/phy2.14313. PMID: 31872558; PMCID: PMC6928244

Gunawardene AR, Corfe BM, Staton CA (2011) Classification and functions of enteroendocrine cells of the lower gastrointestinal tract. Int J Exp Pathol 92(4):219–231. https://doi.org/10.1111/j.1365-2613.2011.00767.x

Hagerty SL, Hutchison KE, Lowry CA, Bryan AD (2020) An empirically derived method for measuring human gut microbiome alpha diversity: demonstrated utility in predicting health-related outcomes among a human clinical sample. PLoS ONE 15(3):e0229204. https://doi.org/10.1371/journal.pone.0229204

Hansen TH, Thomassen MT, Madsen ML et al (2018) The effect of drinking water pH on the human gut microbiota and glucose regulation: results of a randomized controlled cross-over intervention. Sci Rep 8:16626. https://doi.org/10.1038/s41598-018-34761-5

Hebestreit K, Yahiaoui-Doktor M, Engel C et al (2017) Validation of the german version of the mediterranean diet adherence screener (MEDAS) questionnaire. BMC Cancer 17:341. https://doi.org/10.1186/s12885-017-3337-y

Henry LP, Bruijning M, Forsberg SKG et al (2021) The microbiome extends host evolutionary potential. Nat Commun 12:5141. https://doi.org/10.1038/s41467-021-25315-x

Hertel J, Harms AC, Heinken A, Baldini F, Thinnes CC, Glaab E, Vasco DA, Pietzner M, Stewart ID, Wareham NJ, Langenberg C, Trenkwalder C, Krüger R, Hankemeier T, Fleming RMT, Mollenhauer B, Thiele I (2019) Integrated analyses of microbiome and longitudinal metabolome data reveal microbial-host interactions on sulfur metabolism in Parkinson's disease. Cell Rep 29(7):1767–1777.e8. https://doi.org/10.1016/j.celrep.2019.10.035. PMID: 31722195; PMCID: PMC6856723

Hiergeist A, Gessner A (2018) Clinical implementation of high-throughput sequencing. In: Haller D (eds) The gut microbiome in health and disease. Springer, Cham. https://doi.org/10.1007/978-3-319-90545-7_19

Hjorth MF, Blædel T, Bendtsen LQ et al (2019) Prevotella-to-Bacteroides ratio predicts body weight and fat loss success on 24-week diets varying in macronutrient composition and dietary fiber: results from a post-hoc analysis. Int J Obes 43:149–157. https://doi.org/10.1038/s41366-018-0093-2

Human Microbiome Project (2019) https://hmpdacc.org. Accessed 15 Oct 2019

Jakobsson HE, Jernberg C, Andersson AF, Sjölund-Karlsson M, Jansson JK, Engstrand L (2010) Short-term antibiotic treatment has differing long-term impacts on the human throat and gut microbiome. PLoS ONE 5(3):e9836. https://doi.org/10.1371/journal.pone.0009836. PMID: 20352091; PMCID: PMC2844414

Jian C, Luukkonen P, Sädevirta S, Yki-Järvinen H, Salonen A (2021) Impact of short-term overfeeding of saturated or unsaturated fat or sugars on the gut microbiota in relation to liver fat in obese and overweight adults. Clin Nutr 40(1):207–216. https://doi.org/10.1016/j.clnu.2020.05.008. Epub 16 May 2020. PMID: 32536582

Jiang H, Ling Z, Zhang Y, Mao H, Ma Z, Yin Y, Wang W, Tang W, Tan Z, Shi J, Li L, Ruan B (2015) Altered fecal microbiota composition in patients with major depressive disorder. Brain Behav Immun 48:186–194. https://doi.org/10.1016/j.bbi.2015.03.016. Epub 13 Apr 2015. PMID: 25882912

Kates AE, Jarrett O, Skarlupka JH, Sethi A, Duster M, Watson L, Suen G, Poulsen K, Safdar N (2020) Household pet ownership and the microbial diversity of the human gut microbiota. Front Cell Infect Microbiol 10:73. https://doi.org/10.3389/fcimb.2020.00073. PMID: 32185142; PMCID: PMC7058978

Kirk D, Catal C, Tekinerdogan B (2021) Precision nutrition: a systematic literature review. Comput Biol Med 133:104365. https://doi.org/10.1016/j.compbiomed.2021.104365

Klünemann M, Andrejev S, Blasche S et al (2021) Bioaccumulation of therapeutic drugs by human gut bacteria. Nature 597:533–538. https://doi.org/10.1038/s41586-021-03891-8

Knight R, Vrbanac A, Taylor BC et al (2018) Best practices for analysing microbiomes. Nat Rev Microbiol 16:410–422. https://doi.org/10.1038/s41579-018-0029-9

Koeth RA, Lam-Galvez BR, Kirsop J, Wang Z, Levison BS, Gu X, Copeland MF, Bartlett D, Cody DB, Dai HJ, Culley MK, Li XS, Fu X, Wu Y, Li L, DiDonato JA, Tang WHW, Garcia-Garcia JC, Hazen SL (2018) l-Carnitine in omnivorous diets induces an atherogenic gut microbial pathway in humans. J Clin Invest 129(1):373–387. https://doi.org/10.1172/JCI94601. Epub 10 Dec 2018. PMID: 30530985; PMCID: PMC6307959

Krebs-Smith SM, Pannucci TE, Subar AF et al (2018) Update of the healthy eating index: HEI-2015. J Acad Nutr Diet 118(9):1591–1602. Published correction appears in 2019 Aug 20. https://doi.org/10.1016/j.jand.2018.05.021

Kurilshikov A, Medina-Gomez C, Bacigalupe R et al (2021) Large-scale association analyses identify host factors influencing human gut microbiome composition. Nat Genet 53:156–165. https://doi.org/10.1038/s41588-020-00763-1

Lathrop SK, Bloom SM, Rao SM, Nutsch K, Lio CW, Santacruz N, Peterson DA, Stappenbeck TS, Hsieh CS (2011) Peripheral education of the immune system by colonic commensal microbiota. Nature 478(7368):250–254. https://doi.org/10.1038/nature10434. PMID: 21937990; PMCID: PMC3192908

Le Roy CI, Wells PM, Si J, Raes J, Bell JT, Spector TD (2020) Red wine consumption associated with increased gut microbiota α-diversity in 3 independent cohorts. Gastroenterology 158(1):270-272.e2. https://doi.org/10.1053/j.gastro.2019.08.024. Epub 28 Aug 2019. PMID: 31472153

LeBlanc JG, Milani C, de Giori GS, Sesma F, van Sinderen D, Ventura M (2013) Bacteria as vitamin suppliers to their host: a gut microbiota perspective. Curr Opin Biotechnol 24(2):160–168. https://doi.org/10.1016/j.copbio.2012.08.005. Epub 30 Aug 2012. PMID: 22940212

Lenoir-Wijnkoop I, Jones P, Uauy R, Segal L, Milner J (2013) Nutrition economics—food as an ally of public health. Br J Nutr 109(5):777–784. https://doi.org/10.1017/S0007114512005107

Levin A, Sitarik A, Havstad S et al (2016) Joint effects of pregnancy, sociocultural, and environmental factors on early life gut microbiome structure and diversity. Sci Rep 6:31775. https://doi.org/10.1038/srep31775

Ley R, Turnbaugh P, Klein S et al (2006) Human gut microbes associated with obesity. Nature 444:1022–1023. https://doi.org/10.1038/4441022a

Lilja S, Stoll C, Krammer U et al (2021) Five days periodic fasting elevates levels of longevity related Christensenella and Sirtuin expression in humans. Int J Mol Sci 22(5):2331. Published 26 Feb 2021. https://doi.org/10.3390/ijms22052331

Liou JM, Chen CC, Chang CM, Fang YJ, Bair MJ, Chen PY, Chang CY, Hsu YC, Chen MJ, Chen CC, Lee JY, Yang TH, Luo JC, Chen CY, Hsu WF, Chen YN, Wu JY, Lin JT, Lu TP, Chuang EY, El-Omar EM, Wu MS (2019) Taiwan gastrointestinal disease and helicobacter consortium. Long-term changes of gut microbiota 2019 Oct antibiotic resistance, and metabolic parameters after Helicobacter pylori eradication: a multicentre, open-label, randomised trial. Lancet Infect Dis 19(10):1109–1120. https://doi.org/10.1016/S1473-3099(19)30272-5. PMID: 31559966

Liu MY, Yang ZY, Dai WK, Huang JQ, Li YH, Zhang J, Qiu CZ, Wei C, Zhou Q, Sun X, Feng X, Li DF, Wang HP, Zheng YJ (2017) Protective effect of Bifidobacterium infantis CGMCC313-2 on ovalbumin-induced airway asthma and β-lactoglobulin-induced intestinal food allergy mouse models. World J Gastroenterol 23(12):2149-2158. https://doi.org/10.3748/wjg.v23.i12.2149. PMID: 28405142; PMCID: PMC5374126.g

Manichanh C, Rigottier-Gois L, Bonnaud E, Gloux K, Pelletier E, Frangeul L, Nalin R, Jarrin C, Chardon P, Marteau P, Roca J, Dore J (2006) Reduced diversity of faecal microbiota in Crohn's disease revealed by a metagenomic approach. Gut 55(2):205–11. https://doi.org/10.1136/gut.2005.073817. Epub 27 Sep 2005. PMID: 16188921; PMCID: PMC1856500

Manor O, Dai CL, Kornilov SA et al (2020) Health and disease markers correlate with gut microbiome composition across thousands of people. Nat Commun 11:5206. https://doi.org/10.1038/s41467-020-18871-1

Marchesi JR, Ravel J (2015) The vocabulary of microbiome research: a proposal. Microbiome 3:31. https://doi.org/10.1186/s40168-015-0094-5

McDonald D, Price M, Goodrich J et al (2012) An improved Greengenes taxonomy with explicit ranks for ecological and evolutionary analyses of bacteria and archaea. ISME J 6:610–618. https://doi.org/10.1038/ismej.2011.139

McDonald D, Hyde E, Debelius JW et al (2018) American gut: an open platform for citizen science microbiome research. mSystems. 3(3):e00031-18. Published May 15 2018. https://doi.org/10.1128/mSystems.00031-18

Michael Denbow D (2015) Chapter 14—Gastrointestinal anatomy and physiology. In: Scanes CG (eds) Sturkie's avian physiology (Sixth edition). Academic Press, pp 337–366. ISBN: 9780124071605. https://doi.org/10.1016/B978-0-12-407160-5.00014-2

Milestones in human microbiota research (2019) https://www.nature.com/immersive/d42859-019-00041-z/index.html

Miquel S, Leclerc M, Martin R, Chain F, Lenoir M, Raguideau S, Hudault S, Bridonneau C, Northen T, Bowen B, Bermúdez-Humarán LG, Sokol H, Thomas M, Langella P (2015) Identification of metabolic signatures linked to anti-inflammatory effects of Faecalibacterium prausnitzii. mBio 6(2):e00300-15. https://doi.org/10.1128/mBio.00300-15. PMID: 25900655; PMCID: PMC4453580

Mohr AE, Jäger R, Carpenter KC et al (2020) The athletic gut microbiota. J Int Soc Sports Nutr 17:24. https://doi.org/10.1186/s12970-020-00353-w

Nash AK, Auchtung TA, Wong MC et al (2017) The gut mycobiome of the human microbiome project healthy cohort. Microbiome 5(1):153. Published 25 Nov 2017. https://doi.org/10.1186/s40168-017-0373-4

Nie X, Chen J, Ma X, Ni Y, Shen Y, Yu H, Panagiotou G, Bao Y (2020) A metagenome-wide association study of gut microbiome and visceral fat accumulation. Comput Struct Biotechnol J 18:2596–2609. https://doi.org/10.1016/j.csbj.2020.09.026. PMID: 33033580; PMCID: PMC7528071

NIH Human Microbiome Portfolio Analysis Team (2019) A review of 10 years of human microbiome research activities at the US National Institutes of Health, Fiscal years 2007–2016. Microbiome 7:31. https://doi.org/10.1186/s40168-019-0620-y

Nutrition and public health (1950) Nutr Rev 8(8):236–237. https://doi.org/10.1111/j.1753-4887.1950.tb02458.x

Oh M, Zhang L (2020) DeepMicro: deep representation learning for disease prediction based on microbiome data. Sci Rep 10:6026. https://doi.org/10.1038/s41598-020-63159-5

Ouwerkerk JP, de Vos WM, Belzer C (2013) Glycobiome: bacteria and mucus at the epithelial interface. Best Pract Res Clin Gastroenterol 27(1):25–38. https://doi.org/10.1016/j.bpg.2013.03.001. PMID: 23768550

Pan AY (2021) Statistical analysis of microbiome data: the challenge of sparsity. Curr Opin Endocr Metab Res 19:35–40. ISSN: 2451-9650. https://doi.org/10.1016/j.coemr.2021.05.005

Park S, Bekemeier B, Flaxman A, Schultz M (2021) Impact of data visualization on decision-making and its implications for public health practice: a systematic literature review. Inform Health Soc Care. https://doi.org/10.1080/17538157.2021.1982949

Parks D, Chuvochina M, Waite D et al (2018) A standardized bacterial taxonomy based on genome phylogeny substantially revises the tree of life. Nat Biotechnol 36:996–1004. https://doi.org/10.1038/nbt.4229

Paulsen JA, Ptacek TS, Carter SJ et al (2017) Gut microbiota composition associated with alterations in cardiorespiratory fitness and psychosocial outcomes among breast cancer survivors. Support Care Cancer 25:1563–1570. https://doi.org/10.1007/s00520-016-3568-5

Peterson DA, Planer JD, Guruge JL, Xue L, Downey-Virgin W, Goodman AL, Seedorf H, Gordon JI (2015) Characterizing the interactions between a naturally primed immunoglobulin A and its conserved Bacteroides thetaiotaomicron species-specific epitope in gnotobiotic mice. J Biol Chem 290(20):12630–49. https://doi.org/10.1074/jbc.M114.633800. Epub 20 Mar 2015. PMID: 25795776; PMCID: PMC4432283

Porter NT, Luis AS, Martens EC (2018) Bacteroides thetaiotaomicron. Trends Microbiol 26(11):966–967. https://doi.org/10.1016/j.tim.2018.08.005

Prakash A, Peters BA, Cobbs E, Beggs D, Choi H, Li H, Hayes RB, Ahn J (2021) Tobacco smoking and the fecal microbiome in a large, multi-ethnic cohort. Cancer Epidemiol Biomarkers Prev 30(7):1328–1335. https://doi.org/10.1158/1055-9965.EPI-20-1417. Epub 21 May 2021. PMID: 34020999; PMCID: PMC8254769

Qin J, Li R, Raes J et al (2010) A human gut microbial gene catalogue established by metagenomic sequencing. Nature 464:59–65. https://doi.org/10.1038/nature08821

Quast C, Pruesse E, Yilmaz P et al (2013) The SILVA ribosomal RNA gene database project: improved data processing and web-based tools. Nucleic Acids Res 41(Database issue):D590–D596. https://doi.org/10.1093/nar/gks1219

Quévrain E, Maubert MA, Michon C, Chain F, Marquant R, Tailhades J, Miquel S, Carlier L, Bermúdez-Humarán LG, Pigneur B, Lequin O, Kharrat P, Thomas G, Rainteau D, Aubry C, Breyner N, Afonso C, Lavielle S, Grill JP, Chassaing G, Chatel JM, Trugnan G, Xavier R, Langella P, Sokol H, Seksik P (2016) Identification of an anti-inflammatory protein from Faecalibacterium prausnitzii, a commensal bacterium deficient in Crohn's disease. Gut 65(3):415–425. https://doi.org/10.1136/gutjnl-2014-307649. Epub 4 Jun 2015. PMID: 26045134; PMCID: PMC5136800

Ramezani S, Parkhideh A, Bhattacharya PK, Farach-Carson MC, Harrington DA (2021) Beyond colonoscopy: exploring new cell surface biomarkers for detection of early, heterogenous colorectal lesions. Front Oncol 11:657701. https://doi.org/10.3389/fonc.2021.657701. PMID: 34290978; PMCID: PMC8287259

Rasko DA (2017) Changes in microbiome during and after travellers' diarrhea: what we know and what we do not. J Travel Med 24(suppl_1):S52–S56. https://doi.org/10.1093/jtm/tax017

Reitmeier S, Kiessling S, Clavel T, List M, Almeida EL, Ghosh TS, Neuhaus K, Grallert H, Linseisen J, Skurk T, Brandl B, Breuninger TA, Troll M, Rathmann W, Linkohr B, Hauner H, Laudes M, Franke A, Le Roy CI, Bell JT, Spector T, Baumbach J, O'Toole PW, Peters A, Haller D (2020) Arrhythmic gut microbiome signatures predict risk of type 2 diabetes. Cell Host Microbe 28(2):258-272.e6. https://doi.org/10.1016/j.chom.2020.06.004. Epub 2 Jul 2020. PMID: 32619440

Ringel Y, Maharshak N, Ringel-Kulka T, Wolber EA, Balfour Sartor R, Carroll IM (2015) High throughput sequencing reveals distinct microbial populations within the mucosal and luminal niches in healthy individuals. Gut Microbes 6(3):173–181. https://doi.org/10.1080/19490976.2015.1044711

Rios-Covian D, Gueimonde M, Duncan SH, Flint HJ, de los Reyes-Gavilan CG (2015) Enhanced butyrate formation by cross-feeding between Faecalibacterium prausnitzii and Bifidobacterium adolescentis. FEMS Microbiol Lett 362(21):fnv176. https://doi.org/10.1093/femsle/fnv176. Epub 28 Sep 2015. PMID: 26420851

Roager HM, Licht TR, Poulsen SK, Larsen TM, Bahl MI (2014) Microbial enterotypes, inferred by the Prevotella-to-Bacteroides ratio, remained stable during a 6-month randomized controlled diet intervention with the new nordic diet. Appl Environ Microbiol 80(3):1142–1149. https://doi.org/10.1128/AEM.03549-13

Ross AC, Caballero BH, Cousins RJ, Tucker KL, Ziegler TR (2012) Modern nutrition in health and disease: eleventh edition. Wolters Kluwer Health Adis (ESP)

Salem I, Ramser A, Isham N, Ghannoum MA (2018) The gut microbiome as a major regulator of the gut-skin axis. Front Microbiol 9:1459. https://doi.org/10.3389/fmicb.2018.01459. PMID: 30042740; PMCID: PMC6048199

Sasselli V, Pachnis V, Burns AJ (2012) The enteric nervous system. Dev Biol 366(1):64–73. https://doi.org/10.1016/j.ydbio.2012.01.012. Epub 24 Jan 2012. PMID: 22290331

Savage DC (1977) Microbial ecology of the gastrointestinal tract. Annu Rev Microbiol 31:107–133. https://doi.org/10.1146/annurev.mi.31.100177.000543. PMID: 334036

Scepanovic P, Hodel F, Mondot S et al (2019) A comprehensive assessment of demographic, environmental, and host genetic associations with gut microbiome diversity in healthy individuals. Microbiome 7:130. https://doi.org/10.1186/s40168-019-0747-x

Scheiman J, Luber JM, Chavkin TA et al (2019) Meta-omics analysis of elite athletes identifies a performance-enhancing microbe that functions via lactate metabolism. Nat Med 25:1104–1109. https://doi.org/10.1038/s41591-019-0485-4

Scher JU, Sczesnak A, Longman RS, Segata N, Ubeda C, Bielski C, Rostron T, Cerundolo V, Pamer EG, Abramson SB, Huttenhower C, Littman DR (2013) Expansion of intestinal Prevotella copri correlates with enhanced susceptibility to arthritis. Elife 2:e01202. https://doi.org/10.7554/eLife.01202. PMID: 24192039; PMCID: PMC3816614

Schumann M, Kamel S, Pahlitzsch M-L, Lebenheim L, May C, Krauss M, Hummel M, Daum S, Fromm M, Schulzke J-D (2012) Defective tight junctions in refractory celiac disease. Ann N Y Acad Sci 1258:43–51. https://doi.org/10.1111/j.1749-6632.2012.06565.x

Schüssler-Fiorenza Rose SM, Contrepois K, Moneghetti KJ et al (2019) A longitudinal big data approach for precision health. Nat Med 25:792–804. https://doi.org/10.1038/s41591-019-0414-6

Sender R, Milo R (2021) The distribution of cellular turnover in the human body. Nat Med 27:45–48. https://doi.org/10.1038/s41591-020-01182-9

Sender R, Fuchs S, Milo R (2016) Are we really vastly outnumbered? Revisiting the ratio of bacterial to host cells in humans. Cell 164(3):337–340. https://doi.org/10.1016/j.cell.2016.01.013. PMID: 26824647

Sender R, Fuchs S, Milo R (2016) Revised estimates for the number of human and bacteria cells in the body. PLoS Biol 14(8):e1002533. https://doi.org/10.1371/journal.pbio.1002533. PMID: 27541692; PMCID: PMC4991899

Shakya M, Lo CC, Chain PSG (2019) Advances and challenges in metatranscriptomic analysis. Front Genet 10:904. https://doi.org/10.3389/fgene.2019.00904. PMID: 31608125; PMCID: PMC6774269

Shanahan F, Ghosh TS, O'Toole PW (2021) The healthy microbiome-what is the definition of a healthy gut microbiome? Gastroenterology 160(2):483–494. https://doi.org/10.1053/j.gastro.2020.09.057. Epub 27 Nov 2020. PMID: 33253682

Shen T-C, Daniel SG, Patel S, Kaplan E, Phung L, Lemelle-Thomas K, Chau L, Herman L, Trisolini C, Stonelake A, Toal E, Khungar V, Kyle Bittinger K, Reddy R, Wu GD (2021) The mucosally-adherent rectal microbiota contains features unique to alcohol-related cirrhosis. Gut Microbes 13:1. https://doi.org/10.1080/19490976.2021.1987781

Shin NR, Whon TW, Bae JW (2015) Proteobacteria: microbial signature of dysbiosis in gut microbiota. Trends Biotechnol 33(9):496–503. https://doi.org/10.1016/j.tibtech.2015.06.011. Epub 22 Jul 2015. PMID: 26210164

Sicard JF, Le Bihan G, Vogeleer P, Jacques M, Harel J (2017) Interactions of intestinal bacteria with components of the intestinal mucus. Front Cell Infect Microbiol 7:387. Published 5 Sep 2017. https://doi.org/10.3389/fcimb.2017.00387

Stephen A, Cummings J (1980) The microbial contribution to human fecal mass. J Med Microbiol 13:45–56

Strandwitz P (2018) Neurotransmitter modulation by the gut microbiota. Brain Res 1693(Pt B):128–133. https://doi.org/10.1016/j.brainres.2018.03.015

Strandwitz P, Kim KH, Terekhova D et al (2019) GABA-modulating bacteria of the human gut microbiota. Nat Microbiol 4:396–403. https://doi.org/10.1038/s41564-018-0307-3

Suzuki TA, Ley RE (2020) The role of the microbiota in human genetic adaptation. Science. 370(6521):eaaz6827. https://doi.org/10.1126/science.aaz6827. PMID: 33273073

Tang Q, Jin G, Wang G, Liu T, Liu X, Wang B, Cao H (2020) Current sampling methods for gut microbiota: a call for more precise devices. Front Cell Infect Microbiol 10:151. https://doi.org/10.3389/fcimb.2020.00151. PMID: 32328469; PMCID: PMC7161087

Tett A, Pasolli E, Masetti G et al (2021) Prevotella diversity, niches and interactions with the human host. Nat Rev Microbiol 19:585–599. https://doi.org/10.1038/s41579-021-00559-y

The Human Microbiome Project Consortium (2012a) Structure, function, and diversity of the healthy human microbiome. Nature 486:207–214. https://doi.org/10.1038/nature11234

The Human Microbiome Project Consortium (2012b) A framework for human microbiome research. Nature 486:215–221. https://doi.org/10.1038/nature11209

Townsend GE, Han W, Schwalm ND, Hong X, Bencivenga-Barry NA, Goodman AL, Groisman EA (2020) A master regulator of Bacteroides thetaiotaomicron gut colonization controls carbohydrate utilization and an alternative protein synthesis factor. mBio 11(1):e03221-19. https://doi.org/10.1128/mBio.03221-19

Tropini C, Moss EL, Merrill BD, Ng KM, Higginbottom SK, Casavant EP, Gonzalez CG, Fremin B, Bouley DM, Elias JE, Bhatt AS, Huang KC, Sonnenburg JL (2018) Transient osmotic perturbation causes long-term alteration to the gut microbiota. Cell 173(7):1742-1754.e17. https://doi.org/10.1016/j.cell.2018.05.008. PMID: 29906449; PMCID: PMC6061967

Turnbaugh PJ, Ley RE, Mahowald MA, Magrini V, Mardis ER, Gordon JI (2006) An obesity-associated gut microbiome with increased capacity for energy harvest. Nature 444(7122):1027–1031. https://doi.org/10.1038/nature05414. PMID: 17183312

Turnbaugh P, Ley R, Hamady M et al (2007) The human microbiome project. Nature 449:804–810. https://doi.org/10.1038/nature06244

Turnbaugh P, Hamady M, Yatsunenko T et al (2009) A core gut microbiome in obese and lean twins. Nature 457:480–484. https://doi.org/10.1038/nature07540

Valles-Colomer M, Falony G, Darzi Y et al (2019) The neuroactive potential of the human gut microbiota in quality of life and depression. Nat Microbiol 4:623–632. https://doi.org/10.1038/s41564-018-0337-x

Vandeputte D, Kathagen G, D'hoe K et al (2017) Quantitative microbiome profiling links gut community variation to microbial load. Nature 551:507–511. https://doi.org/10.1038/nature24460

Vandeputte D, De Commer L, Tito RY et al (2021) Temporal variability in quantitative human gut microbiome profiles and implications for clinical research. Nat Commun 12(1):6740. Published 18 Nov 2021. https://doi.org/10.1038/s41467-021-27098-7

Vieira-Silva S, Sabino J, Valles-Colomer M et al (2019) Quantitative microbiome profiling disentangles inflammation- and bile duct obstruction-associated microbiota alterations across PSC/IBD diagnoses. Nat Microbiol 4:1826–1831. https://doi.org/10.1038/s41564-019-0483-9

Wang Z, Klipfell E, Bennett B et al (2011) Gut flora metabolism of phosphatidylcholine promotes cardiovascular disease. Nature 472:57–63. https://doi.org/10.1038/nature09922

Wells JM (2011) Immunomodulatory mechanisms of lactobacilli. Microb Cell Fact 10(Suppl 1):S17. https://doi.org/10.1186/1475-2859-10-S1-S17. Epub 30 Aug 2011. PMID: 21995674; PMCID: PMC3231924

Whon TW, Shin NR, Kim JY et al (2021) Omics in gut microbiome analysis. J Microbiol 59:292–297. https://doi.org/10.1007/s12275-021-1004-0

Wilmanski T, Diener C, Rappaport N et al (2021) Gut microbiome pattern reflects healthy ageing and predicts survival in humans. Nat Metab 3:274–286. https://doi.org/10.1038/s42255-021-00348-0

Wu GD, Chen J, Hoffmann C, Bittinger K, Chen YY, Keilbaugh SA, Bewtra M, Knights D, Walters WA, Knight R, Sinha R, Gilroy E, Gupta K, Baldassano R, Nessel L, Li H, Bushman FD, Lewis JD (2011) Linking long-term dietary patterns with gut microbial enterotypes. Science. 334(6052):105–108. https://doi.org/10.1126/science.1208344. Epub 1 Sep 2011. PMID: 21885731; PMCID: PMC3368382

Xia Y, Sun J, Chen D-G (2018) Statistical analysis of microbiome data with R. Springer, Singapore. ISBN: 978-981-13-4645-3. https://doi.org/10.1007/978-981-13-1534-3

Yatsunenko T, Rey F, Manary M et al (2012) Human gut microbiome viewed across age and geography. Nature 486:222–227. https://doi.org/10.1038/nature11053

Younesi E, Hofmann-Apitius M (2013) From integrative disease modeling to predictive, preventive, personalized and participatory (P4) medicine. EPMA J 4:23. https://doi.org/10.1186/1878-5085-4-23

Zeevi D, Korem T, Zmora N, Israeli D, Rothschild D, Weinberger A, Ben-Yacov O, Lador D, Avnit-Sagi T, Lotan-Pompan M, Suez J, Mahdi JA, Matot E, Malka G, Kosower N, Rein M, Zilberman-Schapira G, Dohnalová L, Pevsner-Fischer M, Bikovsky R, Halpern Z, Elinav E, Segal E (2015) Personalized nutrition by prediction of glycemic responses. Cell 163(5):1079–1094. https://doi.org/10.1016/j.cell.2015.11.001. PMID: 26590418

Zeisel SH (2020) Precision (Personalized) nutrition: understanding metabolic heterogeneity. Annu Rev Food Sci Technol 25(11):71–92. https://doi.org/10.1146/annurev-food-032519-051736. Epub 13 Jan 2020. PMID: 31928426

Zhang X, Coker OO, Chu ES, Fu K, Lau HCH, Wang YX, Chan AWH, Wei H, Yang X, Sung JJY, Yu J (2021) Dietary cholesterol drives fatty liver-associated liver cancer by modulating gut microbiota and metabolites. Gut 70(4):761–774. https://doi.org/10.1136/gutjnl-2019-319664. Epub 21 Jul 2020. PMID: 32694178; PMCID: PMC7948195

Zheng D, Liwinski T, Elinav E (2020) Inflammasome activation and regulation: toward a better understanding of complex mechanisms. Cell Discov 6:36. https://doi.org/10.1038/s41421-020-0167-x

Zhernakova A, Kurilshikov A, Bonder MJ, Tigchelaar EF, Schirmer M, Vatanen T, Mujagic Z, Vila AV, Falony G, Vieira-Silva S, Wang J, Imhann F, Brandsma E, Jankipersadsing SA, Joossens M, Cenit MC, Deelen P, Swertz MA; LifeLines cohort study, Weersma RK, Feskens EJ, Netea MG, Gevers D, Jonkers D, Franke L, Aulchenko YS, Huttenhower C, Raes J, Hofker MH, Xavier RJ, Wijmenga C, Fu J (2016) Population-based metagenomics analysis reveals markers for gut microbiome composition and diversity. Science 352(6285):565–569. https://doi.org/10.1126/science.aad3369. Epub 28 Apr 2016. PMID: 27126040; PMCID: PMC5240844

Zhou K, Liu W, Chen Z, Yang D, Qiu Z, Feng H, Li C, Jin M, Li J, Xu Q, Shen Z (2021) The effect of different drinking water in culture medium on feces microbiota diversity. J Water Health 19(2):267–277. https://doi.org/10.2166/wh.2020.075. PMID: 33901023

Personalized Nutrition for Healthy Aging, A Review

5

Angelika Pointner and Alexander G. Haslberger

Contents

A. Pointner · A. G. Haslberger (✉)
Department for Nutritional Sciences, University of Vienna, Vienna, Austria
e-mail: Alexander.haslberger@univie.ac.at

A. G. Haslberger (ed.), *Advances in Precision Nutrition, Personalization and Healthy Aging*, https://doi.org/10.1007/978-3-031-10153-3_5

Abstract

Aging is a multifactorial biological process manifested by different characteristic changes at the molecular and cellular level affecting multiple physiological functions and immune competence. While varying enormously between individuals, aging and its underlying mechanisms increase the susceptibility to many diseases. Research focusing on individual aspects of disease pathologies and aging has shown the huge importance of epigenetic regulation and microbial metabolites beyond hereditary genetic dispositions. Epigenetic mechanisms, but also microbiota and their metabolites reflect impacts of lifestyle and nutrition. Epigenetics and microbiota therefore provide some of the most accurate biomarkers of healthy of premature aging such as the epigenetic clock. These developments suggest that the 'one-size-fits-all' concept in medicine as well as in nutrition is no longer sufficient to narrow the gap between health span and life span. Integrating and combining data from different platforms (genome-DNA sequence, transcriptome, proteome, metabolome, and epigenome) leads to a better understanding of the basis of complex diseases and paves the way toward personalized medicine and personalised nutrition. Consumer organizations will have to find a delicate balance between safety, use of modern concepts of precise nutrition as well as quite divergent expectation of different consumer groups.

5.1 Healthy Aging

Aging is a complex multifactorial biological process manifested by a gradual decline of normal physiological functions. Although different characteristic age-dependent changes at the molecular and cellular level are well known to date, process and rates vary enormously between individuals. This has a central importance, when it comes to human health. Aging and its underlying mechanisms increase the susceptibility to many diseases, including cancer, metabolic disorders, such as diabetes, cardiovascular disorders, and neurodegenerative diseases. Aging-associated diseases may be a reason why recently aging is addressed as a curable, preventable disease in the extended life span debate by some scientists (Faragher 2015; Gems 2011).

Modern molecular biology has summarized major molecular mechanisms, hallmarks of aging, which determine biological aging. These hallmarks are especially: genomic instability, telomere attrition, epigenetic alterations, loss of proteostasis, deregulated nutrient sensing, mitochondrial dysfunction, cellular senescence, stem cell exhaustion, and altered intercellular communication (Fig. 5.1; López-Otín et al. 2013). Presumably, especially, aging microbiota should be considered as an additional hallmark. Cellular senescence, reflecting organismal senescence, is a specialized process resulting in an irreversible growth arrest and possibly evolved

Fig. 5.1 Hallmarks of aging (López-Otín et al. 2013)

as an endogenous anticancer mechanism. Although it is essential in some physiological events, it proves to be detrimental in a variety of age-related diseases. With advancing age, senescent cells accumulate in tissues and organs, and subsequently promote the aging process, largely through their complex senescence-associated secretory phenotype (SASP) displaying highly inflammatory activities (Campisi 2013).

5.1.1 Genetics and Healthy Aging

Research into longevity and in particular in genetic factors driving healthy aging has progressed rapidly in recent years. Although genome-wide association studies (GWAS) have emerged recently indicating several gene loci correlated to key human aging traits (Melzer et al. 2020), the genetic basis has been mainly studied in model organisms and resulted in a considerably number of genes and mutations with relevance for lifetime. Especially, the nematode *Caenorhabditis elegans* with a single-gene mutation in Daf-2 became famous for an enhanced lifetime (Braeckman and Vanfleteren 2007; Kenyon et al. 1993). Daf-2 encodes a tyrosine kinase receptor in *C. elegans* and is homolog in sequence and structure to the human receptor protein that responds to insulin. **Daf**-2 normally controls many other genes, which in turn regulate a variety of physiological processes at different stages in life (Pal and Tyler 2016a).

There is evidence that **insulin signaling** is involved in longevity as the insulin pathway plays a role in diseases like diabetes and cancer. Insulin resistance at the cellular level is a key feature of type II diabetes. Mutations in the pathways of insulin/IGF-1 receptors have been associated with the dysregulation of growth in cancers. The risk of both, diabetes and cancer, increases with age. Possibly Daf-2 controls aging because it controls many other genes involved in aging (Adams 2008).

Although previous studies may have overestimated the extent of genetic heritability of longevity, many common age-related disorders such as cardiovascular disease, Alzheimer's disease, or type-2 diabetes have a substantial heritable component. Furthermore, GWAS have also analyzed genetic variants across groups of older individuals and younger control individuals and identified various loci associated with longevity such as in the APOE, GPR78, or the FOXO3A gene (Melzer et al. 2020).

5.1.2 Epigenetics and Healthy Aging

Among the hallmarks of aging, epigenetic alterations represent crucial mechanisms which seem to interfere with most other hallmarks. A large number of studies have shown that epigenetic processes not only accompany but strongly affect aging and age-related diseases. By definition, 'epigenetics represents the **reversible heritable mechanisms that occur without any alteration of the underlying DNA sequence**' (Fig. 5.2; Fymat 2017). Epigenetic mechanisms play a crucial role in normal development and function of the organism as they regulate the accessibility and therefore activity of genes.

Epigenetics connects the genotype with the phenotype and provides an explanation why the pattern of aging is different between two genetically identical individuals, such as identical twins (Brunet and Berger 2014a; Martin 2005; Sargent 2010). Although longevity studies on the human population have shown that genetic factors may explain a fraction (16–30%) of the differences in life spans, the majority of the remainder of variation is thought to have arisen through epigenetic drift during lifetime (Herskind et al. 1996; Muñoz-Najar and Sedivy 2011; Poulsen et al. 2007; Morris et al. 2019).

Various environmental stimuli, including diet or lifestyle, cause differential alterations of epigenetic information. However, epigenetic patterns are reversible

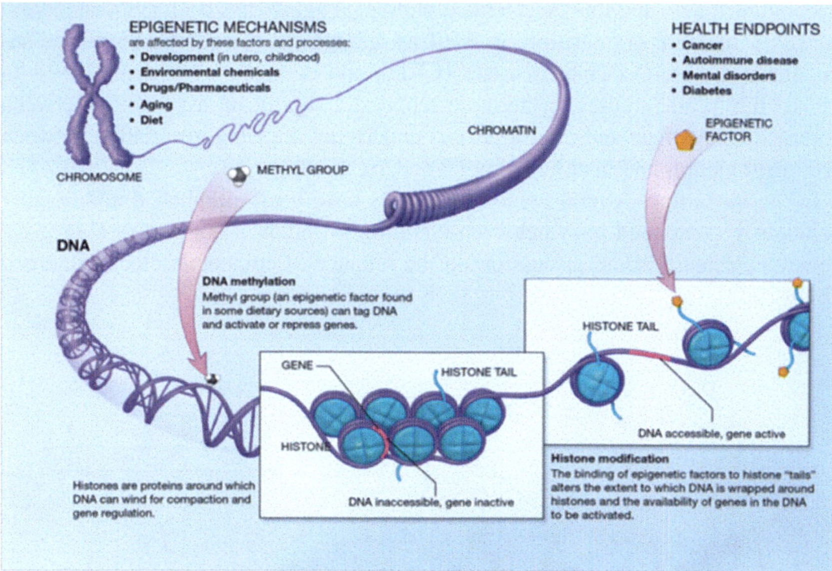

Fig. 5.2 Epigenetic mechanisms (Fymat 2017)

and therefore constitute a potential target for therapeutic and/or nutritional interventions, in contrast to genetic changes, which are mostly still irreversible. Accordingly, understanding the role of epigenetic changes that happen during aging is a promising way to delay aging and age-related diseases (Pal and Tyler 2016b).

Types of epigenetic information: There are different types of epigenetic regulation occurring at various levels including DNA methylation, chromatin remodeling, and transcription of noncoding RNAs (ncRNAs). (Brunet and Berger 2014b; Feser and Tyler 2011; Gelato and Fischle 2008; Lazarus et al. 2013; O'Sullivan and Karlseder 2012). All together, they comprise our epigenome which provides the molecular basis for the regulation of gene expression and DNA stability.

5.1.2.1 The Epigenetic Clocks

Among all epigenetic mechanisms, DNA methylation is the most researched epigenetic mark. Changes in DNA methylation patterns have been shown to occur with advancing age and thus may be a fundamental mechanism that drives human aging. This specific age-dependent modifications in the genome have been recently used for age evaluation, developing so called epigenetic clocks (Fig. 5.3). Referred to specifically as 'DNA methylation age' (DNAmAge), they provide an accurate estimate of age across a range of tissues, and at different stages of life, and are some of the most promising **biomarkers of aging**. DNAmAge has also permitted the identification of individuals who show substantial deviations from their actual chronological age, and this 'accelerated biological aging' has been associated with unhealthy lifestyle or nutrition, as well as age-related disorders as frailty, cancer, diabetes, cardiovascular diseases (CVD), and dementia (Woodcox 2018). In the last few years, studies furthermore reported a significant association between increased DNAmAge and mortality risk, underlying the close correlation between DNA methylation and healthy aging (Woodcox 2018).

In the meantime, several epigenetic clocks have been established additionally including proteins and enzymes even claiming to indicate aging risks (Fig. 5.4). However, recently, some discussion on the reliance of epigenetic clocks emerged (Drew 2022).

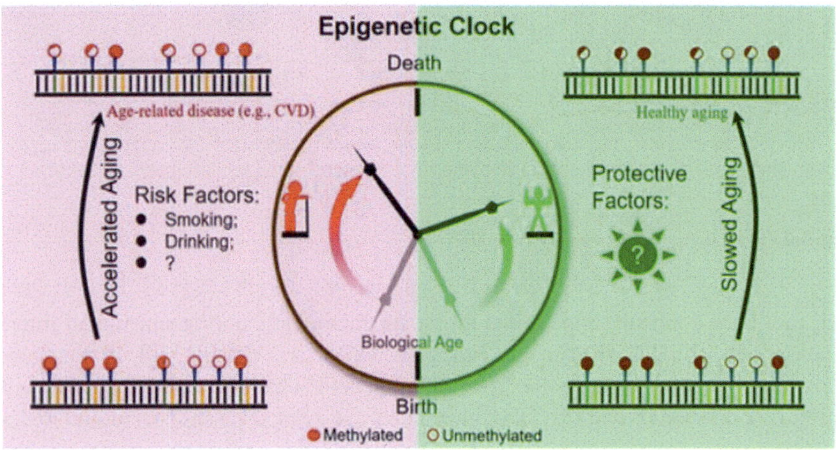

Fig. 5.3 Epigenetic clocks adopted from (Pal and Tyler 2016a; Xiao et al. 2019)

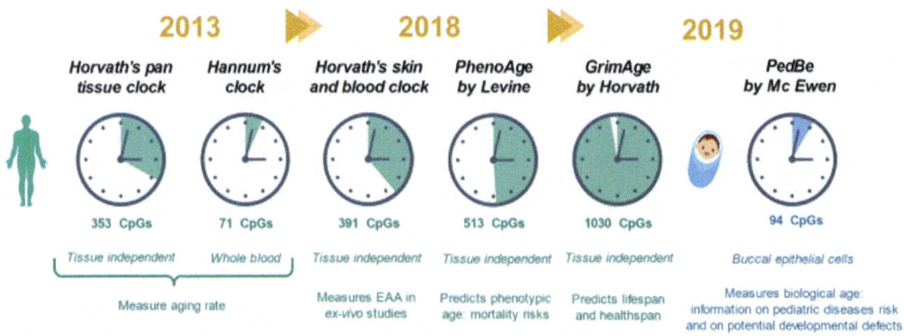

Fig. 5.4 Development of epigenetic clocks from the assessment of the biological age to the prediction of life span (Horvath 2013; Hannum et al. 2013; Klutstein et al. 2022; Topart et al. 2020)

5.1.3 Histones and Healthy Aging

Changes in chromatin structure are a common feature during aging. The main protein components of chromatin are histones, which bind to DNA and function as 'anchors' around which the strands are wound (Bowman and Poirier 2015). Consequently, modifications in histones also provide a major regulator for chromatin functions affecting crucial processes including cell-cycle progression, DNA replication, and transcription as well as tissue specification.

Histone tails extend from the nucleosomal core into the nucleoplasm and constitute sites of posttranslational modifications (Fenley et al. 2018) In general, histones can undergo multiple types of modifications, among which acetylation, methylation, phosphorylation, and ubiquitination are the best known. The different modifications can even 'communicate' and influence each other's presence (Molina-Serrano et al. 2019). Moreover, changes of chromatin state seem to interfere with other epigenetic processes such as DNA methylation (Pal and Tyler 2016a; Fig. 5.5).

Acetyltransferases (HATs) and histone deacetylases (HDACs) alter the positive charge of the histone resulting in a modified affinity between histones and DNA, and thus undertake a critical task in regulating gene expression (Legube and Trouche 2003).

One of the earliest models of epigenetic aging was the '**heterochromatin loss model of aging**' (Bernadotte et al. 2016; Haithcock et al. 2005; Wilson 2005). This model suggested that 'the loss of heterochromatin that accompanies aging leads to changes in global nuclear architecture and the expression of genes residing in those regions, directly or indirectly causing aging and cellular senescence'. Increasing evidence supports this theory, and the gradual loss of heterochromatin

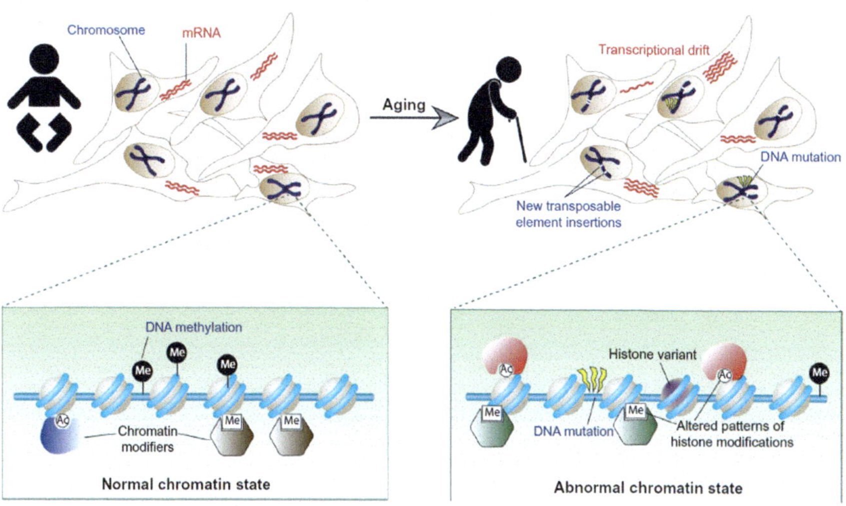

Fig. 5.5 Epigenetic changes during aging (Pal and Tyler 2016a)

during lifetime has been observed across many model organisms to date (Pal and Tyler 2016a; Anton and Leeuwenburgh 2013).

Consequently, there are projects focusing on life span expansion strategies through histone modulation. In this regard, research focusing on histone acetylation could show that treatment with histone HDAC inhibitors shortens life span, whereas chemical activation or overexpression of SIR2 or **sirtuins has the ability to extend life span** in different models (Anastasiou and Krek 2006; Dillin and Kelly 2007; Finkel et al. 2009; Guarente 2011; Guarente and Guarente 2007; Haigis and Sinclair 2010; Hall et al. 2013; Longo and Kennedy 2006; Morris 2013). Plant derived sirtuins have drawn specific attention in mimicking fasting and its beneficial effects on health, and potentially longevity. (https://doi.org/10.31989/ffhd.v10i10.752).

5.1.4 Noncoding RNAs (NcRNAs) and Aging

NcRNAs are the most recent players in the epigenetics field, influencing obviously all biological processes. It is now widely accepted that approximately 60–90% of the human genome is transcribed. Noncoding RNAs (ncRNAs) are RNA molecules that are not translated; thus, they do not have any apparent protein-coding roles. NcRNAs differ both in length and mechanisms of biogenesis and functions. The role of ncRNAs covers a wide range of functions within various cellular processes: They affect biogenesis and stability of other RNAs, regulate gene expression and chromatin packaging, and are involved in multiple other physiological and also pathological processes, including senescence (Fig. 5.6; Sidler et al. 2017; Wagner 2019).

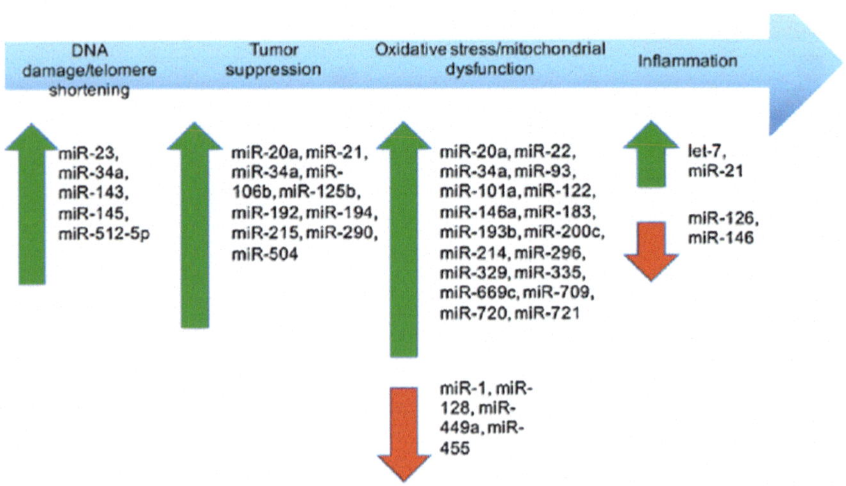

Fig. 5.6 miRNAS, aging, senescence (Williams et al. 2017)

Disruption of ncRNA function has been implicated in numerous disease such as cancer, neurodegenerative disorders, cardiovascular disorders, and aging (Huarte 2015; Li et al. 2019; Ren et al. 2018; Toomey et al. 2016; Soriano-Tárraga et al. 2019). One of the most prominent examples in the area of aging is microRNA (miRNA) lin-4 and its pro-aging target miRNA lin-14. Loss of function of lin-4 shortens life span, whereas overexpression of lin-4 extends life span. In contrast, knocking down lin-14 extends life span. However, the majority of miRNAs are down-regulated with age (Goodall et al. 2019; Huan et al. 2018; Kinser and Pincus 2020; McGregor and Seo 2016; Soriano-Tárraga et al. 2019; Abdelmohsen et al. 2013; Szafranski et al. 2015).

Until recently, most of the studies focused on short ncRNAs such as miRNAs spanning that between 18 and 24 nucleotides, but the functional importance of long ncRNAs (lncRNAs) with transcript lengths over 200 nucleotides is attracting growing interest, as they are described to be more complex and act more specific than other ncRNAs. Meanwhile, thousands of lncRNAs have been identified; however, most of their functions and their role in the pathophysiology of diseases are still a subject of investigation. However, increasing evidence suggests their regulating role in senescence-associated processes (Jin et al. 2019).

5.1.5 Aging of the Immune System (I.S.) and Epigenetics

Aging is associated with reduced immune functions which lead to a higher to infections. Changes in both, the innate and adaptive immune system, occur in aging. This reduction of the immune competence is associated with low-grade chronic inflammation (**inflammaging**) and the senescence-associated inflammatory phenotype, characterized by high levels of circulating cytokines and chemokines. This inflammatory state accelerates the age-related immune dysfunction. Stem cell diversity also reduces during lifetime with an exponential increase in the occurrence of clonal hematopoiesis (Keenan and Allan 2019).

The **thymus**, where T-cells develop, begins to involute at puberty due to age-related changes that affect both T-cell progenitors and the thymic microenvironment. Decreased hematopoietic activity in the bone marrow means that B-cell lymphopoiesis also decreases with age. **DNA damage** promotes cellular senescence. **DNA repair** mechanisms and clearance mechanisms of damaged cells by the innate immune system are impaired with aging. The reduced clearance of senescent cells results in an accumulation of genomically damaged and inflammatory cells in all tissues of the body, including the immune system. Age-associated changes in DNA methylation have been reported in a number of human immune cell types including monocytes, $CD4^+$ and $CD8^+$ T-cells, and stem cells. Many of these changes appear to be cell type-specific (Fig. 5.7; Allis and Jenuwein 2016; Briceño et al. 2016; Busslinger and Tarakhovsky 2014; Cambier 2005; Chambers et al. 2007; Keenan and Allan 2019).

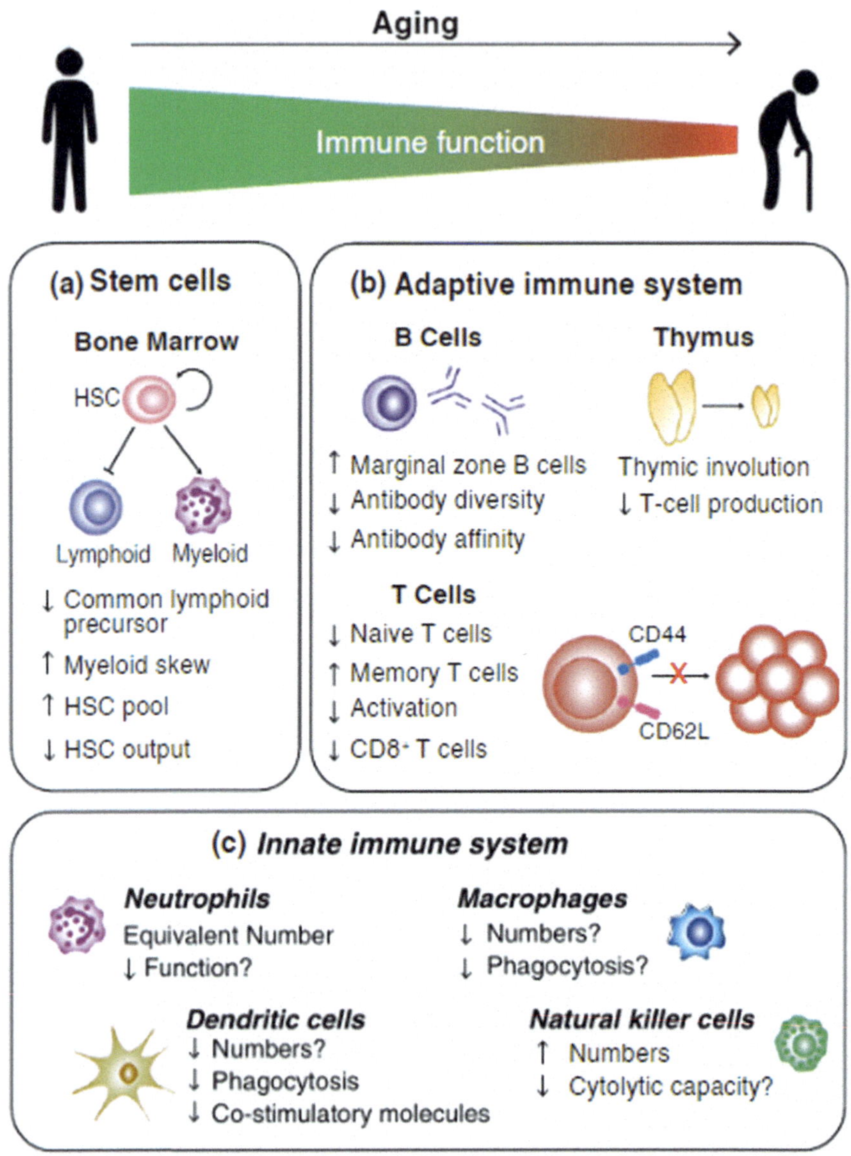

Fig. 5.7 Dysfunctions of the aged I.S. (Keenan and Allan 2019)

5.1.6 Neurodegenerative Diseases, Aging, and Epigenetics

One frequent problem of human aging is neuroinflammation, neurodegeneration, and reduced brain function (Gruendler et al. 2020). Premature death of neurons is considered to be a major aspect of neurodegenerative diseases. Accumulating evidence indicates a correlation of epigenetic mechanisms and neurodegenerative disorders. Several transcription factors have been identified to be involved in epigenetic remodeling in the brain. The transcriptional repressor REST (Repressor Element-1 Silencing Transcription factor), e.g., is a key modulator of the neuronal epigenome (Basavarajappa and Subbanna 2021; Zhang et al. 2019).

Studies of neurodegenerative processes often address the regulation by ncRNAs. Whereas, some of the miRNAs correlate with neuroprotection; others clearly contribute toward neurodegenerative diseases and/or aging. miR-107 and miR-650 are even under discussion as markers for Alzheimer disease (AD).

There are higher levels of miR-34 in samples collected from **AD** patients. The pro-survival factor BCL2 and the anti-aging deacetylase SIRT1 both are seen as targets of miR-34, and the expression of SIRT1 correlates inversely with miR-34 expression, revealing a potential mechanism for miR-34 function in the aged brain. Similarly, another miRNA, miR-144, seems to be enriched in aged brains and may also contribute to age-associated neurodegeneration through down-regulation of key protective factors.

Histone modifications and DNA methylation have also been linked to functions of glial cells and the neuronal cell death in neurodegenerative diseases and impaired cognition, associated with neurodegeneration (Hwang et al. 2017).

5.1.7 Microbiota and Healthy Aging

The human gut microbiome is composed of several different phyla, including **Bacteroidetes, Firmicutes, and Actinobacteria** (Eckburg et al. 2005). It plays a central role in many physiological and immunological processes, e.g., in the defense against pathogens and in the development of immune and intestinal barrier functions (West et al. 2015). Furthermore, it is involved in many aspects of **the metabolism**, including the production of bile acids, lipids, vitamins, choline, and polyamines (Nicholson et al. 2012).

Microbiota are especially involved in the breakdown of indigestible polysaccharides (fermentation of resistant starch, oligosaccharides, and inulin), whereby energy for the host is obtained from foods that are absorbed but not digested by the host (Flint et al. 2012; Holscher 2017).

According to the **16S ribosomal DNA sequencing** data of faecal samples, individual gut microbiota show distinct profiles, and this inter-individual variation is greater in older adults. Longitudinally, gut microbiota of healthy adults are relatively stable after establishment early in life, within three years after birth. Recently, it became clear that the community structure of the intestinal microbiota and metabolism of the mother shifts during pregnancy (Koren et al. 2012), and that microbes colonize the amniotic fluid, the umbilical blood cord, and the placenta, indicating maternal **microbial colonization of the fetus** in utero (D'argenio 2018) with important consequences for healthy further development. The '1000 days of life' theory developed by this findings of the prenatal establishment of GI microbiota and their interaction with the prenatal immune system and the epigenetic system. This time spanning roughly between conception and one's second birthday has been established to be a unique period of opportunity when the foundations of optimum health, growth, and neurodevelopment across the life span are established. Therefore, maternal nutrition and maternal microbiota have a central influence on these developments (Fuhler 2020).

Microbiota respond to diets, the social and physical environment, lifestyle, and aging mainly via metabolites, especially short chain fatty acids (SCFAs), which strongly correspond **with the epigenome**. SCFAs are also a main element of the 'gut–brain axis' which connects the gut microbiome with the central nervous system. Recently, it was shown that bacterial cell wall derived muropeptides from the gut could reach the brain and are recognized by the pattern recognition receptor NOD2, thus regulating neurons this way. This suggests that the brain may sense changes in gut bacteria as a measure of food intake (Gabanyi et al. 2022).

In humans, diets have been shown to affect microbiota diversity rapidly. Diets containing higher amounts of carbohydrates derived from cooked grains contribute to higher numbers of bifidobacterium. Diets providing higher amounts of fiber from fruits lead to enrichment of Lachnospira (Wilmanski et al. 2021). Consumption of a plant protein diet, based on glycated pea proteins, significantly increases the levels of commensal lactobacilli and bifidobacteria and elevates short-chain fatty acid production in humans (Szczyrek et al. 2021).

The long-term consumption of complex carbohydrates, especially dietary fibers, has been shown to promote the *Prevotella* genus (Simpson and Campbell 2015; Vinke et al. 2017).

In contrast, a high intake of dietary fat (mainly saturated fatty acids) is associated with reduced microbiota richness and diversity in both adults and infants (Wolters et al. 2019). The consumption of omega-3 polyunsaturated fatty acids (PUFAs) leads to an increased abundance of several butyrate-producing bacteria, in line with the known anticancer and anti-inflammatory effects of omega-3 PUFAs (Freitas and Campos 2019) In humans, a long-term animal protein-rich diet is associated with the Bacteroides enterotype (de Moraes et al. 2017; Fan et al. 2020).

Changes in the GI microbiota distribution have crucial consequences on the metabolites of the microbiota. **SCFAs** are products of the breakdown of dietary fibers by the mostly anaerobic gut microbiota. They enter the circulation from the gut and have multiple beneficial roles in energy metabolism. Acetate is discussed to reduce serum cholesterol and triglyceride levels, propionate can lower glucose levels (Byrne et al. 2015; Hernández et al. 2019; Wolever et al. 1996) Butyrate can increase insulin sensitivity and modulate the expression of crucial mediators of the gut–brain axis, which ae also involved in the regulation of appetite (Byrne et al. 2015; Fluitman et al. 2018; Wolever et al. 1996; Silva et al. 2020).

SCFAs play their regulatory role by binding to different G-coupled protein receptors (GCPRs). The free fatty acid receptor 3 (FFAR3), for example, has been linked to regulating insulin secretion and appetite, which suggests the importance of SCFAs in satiety and energy balance control. Butyrate and propionate are non-competitive HDAC inhibitors. This exhibits their role in epigenetic gene regulation by influencing histone acetylation. SCFAs are also involved in regulating immunity, by suppressing the production of proinflammatory mediators and enhancing the release of anti-inflammatory cytokines (Li et al. 2017; Fig. 5.8). A negative correlation between promotor methylation of the FFAR3 gene and BMI was shown. Both type-2 diabetic and obese individuals showed lower methylation in comparison to lean controls.

To what amount SCFAs are transported from the gut to the brain is still under discussion as the fast binding of SCFAs to their G-protein-coupled receptors (GPR41, GPR34) makes the assumption of SCFA production and SCFA blood concentrations, e.g., from analysis in feces difficult. Also, a large number of other biologically active metabolites are produced by GI microbiota, such as gamma aminobutyric acid (GABA) which plays a key role in anxiety and depression disorders microbiota (Dhakal et al. 2012; Duranti et al. 2020), and **GABA** was shown to diminish with increasing **age** (Mazzoli and Pessione 2016).

Also, bile acids, which relate the GI-microbiome, absorption, distribution, metabolism, and excretion of nutrients and sense the intestinal contents, been shown to be reduced with aging (Frommherz et al. 2016; Shulpekova et al. 2022).

Studies of fecal samples from individuals in different age groups suggest drastic age-related changes in the gut microbiota composition and diversity (Hippe et al.

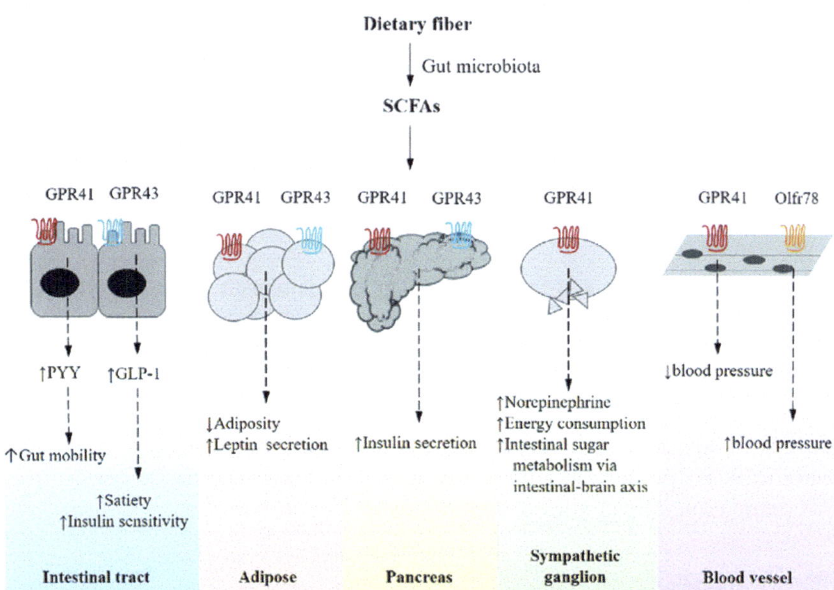

Fig. 5.8 Short-chain fatty acid (SCFA)-receptor-mediated pathways and their effects on host energy metabolism in peripheral tissues. Gut microbes can ferment dietary fiber into SCFAs, which induce an array of G-protein-coupled receptor-mediated signaling pathways that are essentially implicated in host energy homeostasis in multiple tissues (Li et al. 2017)

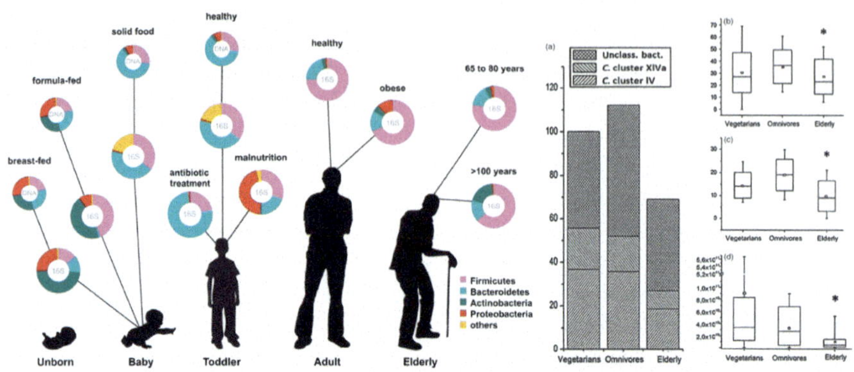

Fig. 5.9 Microbiota structure and producers of SCFAs in aging (Hippe et al. 2011)

2011; Fig. 5.9). The gut microbiota of the elderly becomes more variable with advancing age. For instance, the three bacterial families in the core microbiota are less abundant in older age groups, while certain health-associated species become more abundant in older age. In older age (over ~80 years), healthy individuals show continued microbial drift toward a unique compositional state, whereas this drift is absent in less healthy individuals. Retaining a high bacteroides dominance into older age, or having a low gut microbiome uniqueness measure, predicts a decreased survival fitness (Wilmanski et al. 2021).

Reflecting changes of microbiota structure strong age-related changes in the metagenomics of SCFA production have been observed. Frequencies of genes encoding SCFA production and those involved in carbohydrate breakdown decrease, while genes involved in protein breakdown increase. The reduced frequency of genes for short-chain fatty acid production is also associated with frailty. Thus, the short-chain fatty acids strongly modulate healthy aging (Den Besten et al. 2013).

Recently, results of metagenomic data analysis demonstrate the importance of the high abundance of *Akkermansia* and the butyrate biosynthesis pathway in aging. Oral administration of *Akkermansia* ameliorated the senescence-related phenotype in the intestinal systems in aged mice and extended the health span (Shin et al. 2021).

In conclusion, diet regulates microbiota composition and is the biggest contributor to SCFA production. Therefore, personalized nutrition has to benefit from incorporating microbiota information into their recommendations.

5.1.8 Individual-Specific Aging

The conventional medical concept of 'one-size-fits-all' to our aging population is widely considered to be no longer sufficient to narrow the gap between health span and life span. (Garmany et al. 2021; Olshansky 2018) A striking argument

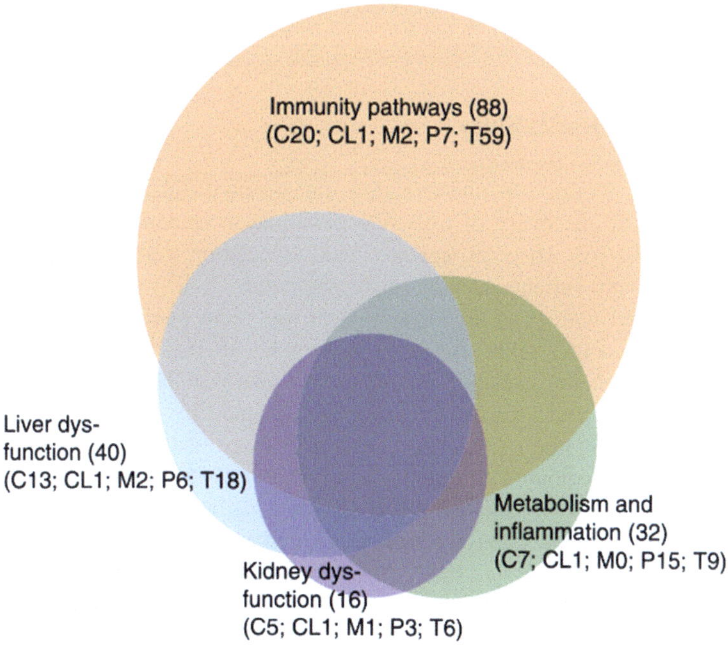

Fig. 5.10 Ageotypes, adapted from Ahadi et al. (2020)

for the use of personalized approaches in the aging biology comes from the omics-wide analysis of markers within the hallmarks of aging and the identifications of personal-specific mechanisms which drive accelerated aging. Recently, individual metabolism-specific aging patterns have been identified by deep longitudinal profiling. These so called ageotypes may allow to assess personal aging on a molecular level, also reflecting the individual lifestyle and medical history (Fig. 5.10; Ahadi et al. 2020). Thus, adapted and personalized interventions on this basis could be more beneficial in preventing and treating age-related disorders (Ordovas et al. 2018).

5.2 Ways to Personalization

An organism's complex trait composition is the product of the interaction of genetic, epigenetic, and environmental impacts. Figure 5.10 Genetic mechanisms controlling complex traits are 'Mendelian' in their transmission pattern. However, epigenetic contributors, by contrast, do not follow these Mendelian inheritance principles. Epigenetically driven phenotypic variation (EPV) explains why the same genotype can yield distinct phenotypes. EPV results predominantly from developmental responses to the environment, termed 'phenotypic plasticity'. Studies suggest that phenotypic plasticity is triggered preferentially during critical windows of high sensitivity including embryonic development and growth phases (Panzeri and Pospisilik 2018).

5.2.1 Missing Heritability

Multiple studies focused on understanding the genotype–phenotype interaction, and moreover, the discovery of causative mutations is underlying different diseases. During the past 16 years, genome-wide association studies have identified hundreds of genetic variants such as single nucleotide (SNPs) or allelic polymorphisms associated with complex physiological traits and diseases. Most variants identified so far confer relatively small increments in risk and explain only a small proportion of the expected heritable fraction (Manolio et al. 2009). This discrepancy between the amount of variance explained by specific genetic factors identified in genome-wide association and twin-estimated heritability is also known as 'missing heritability', a perpetuating problem of human genetic research. Therefore, many different explanations for the 'missing heritability' have been proposed: (1) that complex traits are highly polygenic and affected by many rare variants; (2) that twin studies have overestimated heritability (Youngid 2019). However, recently, it has become increasingly clear, that epigenetic programs, which are systematically missed by conventional DNA sequencing, may account for a significant fraction of the 'missing heritability'.

Environmental or nutritional factors interact with complex traits, such as those associated with height, stature, or metabolism of food. Furthermore, some of these factors have been shown to be epigenetically inherited (Lacal and Ventura 2018). Environmental factors can include aspects such as stress, but also microbiome composition. In this case, the influence is reciprocal: Our nutrition and genetics shape the composition of microbiota, and thus our own environment. Either way, interactions between genes and lifestyle exist in human, and importantly, a given genetic susceptibility is modifiable by lifestyle (Riggs and Porter 1996). Epigenetics so bridges the gap between genotype and phenotype and paves a way for a new understanding of the interplay of genes and environment including nutrition, lifestyle, or social environment. Thus, environment can influence the phenotypic variation directly on a molecular level. The ability of environmental epigenetics to alter phenotypic and genotypic variation, directly, can also significantly impact natural selection. Neo-Lamarckian concepts can thus facilitate neo-Darwinian evolution (Haslberger 2009; Skinner 2015).

5.2.2 Markers Enable a Personalized Pre- and Intervention

Biomarkers are a central tool to assess biological changes due to age, disease, or treatment in a patient and thus play a crucial role for personalized health prevention. As the main focus is to establish correlations between changes of biomarkers and diseases, research into biomarkers started to be developed primarily in medical areas addressing cancer or neurological disorders. However, experiences in these fields were rapidly translated to other areas, especially metabolic diseases, and are applied to prevention and treatment options such as personalized nutrition. The main benefits of most biomarkers are that they are non-invasive (blood samples) and have the potential to offer early personalized screening for diseases. Although the focus of personalized medicine is on genomic research, the exposome, all non-genetic internal and external influences that are determining a person's health play a crucial role in disease development (DeBord et al. 2016).

In the context of cancer research, there are various biomarkers addressing the different levels of carcinogenesis: starting with genetic markers to altered gene expression, protein status and the metabolic level. Genetic mutation markers are mostly relevant in hereditary diseases, whereas epigenetic-based markers have a function in both, inherited and sporadic diseases. Epigenetic biomarkers can be advantageous as they provide information on the patient's genetic and environmental background (García-Giménez et al. 2017). Whereas e.g. metabolic markers mostly reflect short time responses to stimuli, epigenetic markers condense effects from longer periods.

In general, **epigenetic biomarkers** address the mechanisms of epigenetics, CpG methylation, histone modification, and noncoding RNAs. Especially, the various forms of **noncoding RNAs** are increasingly used in the area of complex diseases, or for metabolic or aging-related mechanisms. Analysis of sets of miRNAs is usually done using next-generation sequencing (NGS) platforms.

Recently, **cell-free DNA** (cfDNA) offers an exciting new class of biomarkers. cfNDA, which is often released into the blood after tissue destruction received a special importance as biomarkers in the area of early detection, staging, and specification of cancer (Huang et al. 2021; Fig. 5.11).

Liquid biopsy markers include circulating tumor cells (CTCs), tumor-derived cell-free DNA (ctDNA) carrying tumor-specific mutations or CpG methylation, as well as tumor-specific miRNAs. Liquid biopsy provides specific information about tumor sub-populations, enabling personalized therapies (*i.e.*, precision medicine) (Snow et al. 2019).

While liquid biopsy is becoming a routine analysis in precision medicine for cancer, cfDNA is started to be used in areas of autoimmunity, metabolic pathogenesis (Bronkhorst et al. 2019), and to detect age-related epigenetic changes (Epigenetics of Aging Poster activemotif.com).

Epigenetic markers are more and more used as reliable, stable markers condensing nutrition and lifestyle effects over longer time periods. They can so be explored to predict the responsiveness to dietary prescriptions and guide effects of nutrition and lifestyle modifications. Analysis of epigenetic marks detected, e.g., the effect of nutritional treatments on weight loss and changes in metabolic profiles (Samblas et al. 2019). Furthermore, methylation levels of circadian genes correlated with the magnitude of weight loss and circulating blood lipids after a nutritional program based on a Mediterranean dietary pattern (Samblas et al. 2016). Methylation patterns of appetite-regulatory genes were also found to be associated with the success in weight loss or the risk of weight regain (Crujeiras et al. 2013).

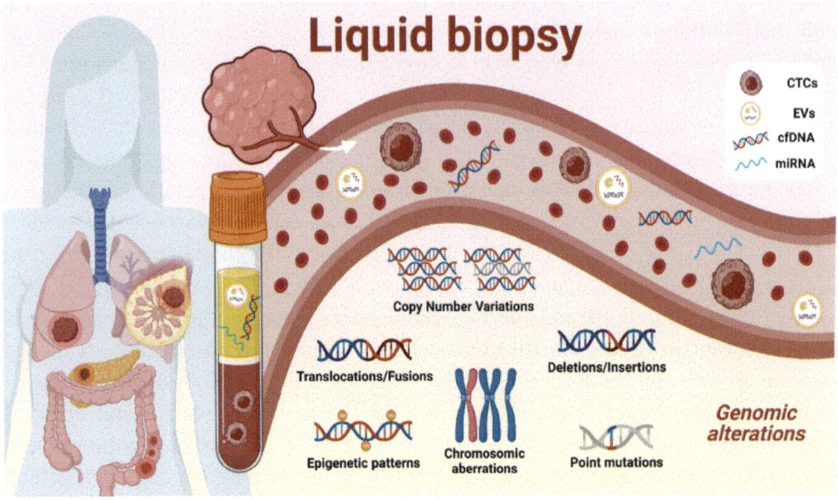

Fig. 5.11 Liquid biopsy obtained from peripheral blood is composed of different tumoral components such as circulating tumor cells (CTCs), circulating cell-free DNA (cfDNA) and miRNAs. These elements can be isolated for the identification of various tumor-specific genomic aberrations including point mutations, copy number variations, structural rearrangements, or epigenetic patterns (Palacín-aliana et al. 2021)

Reductions of body fat and serum lipids were related to changes in the methylation status of genes involved in inflammatory response and fatty acid metabolism (Panchal and Brown 2020).

A specific importance in the analysis of dietary patterns and cancer metabolism is given to **metabolomic markers**. Studies applied multivariate methods to identify panels of metabolites that discriminate between high and low scores within or between dietary patterns. This represents an important approach because a panel of metabolites would be expected to best capture the multidimensionality of complex dietary patterns (Guasch-Ferre et al. 2018).

Interestingly, based on metabolite profiles from urine and plasma, including creatinine, branched-chain amino acids, and sarcosine, it was possible to identify metabolite patterns which classify participants according to sex with >90% accuracy (Rist et al. 2017).

5.3 Developments of Precision Medicine

Hippocrates already emphasized the importance of individualizing medical care, proclaiming 'It is more important to know what sort of person has a disease than to know what sort of disease a person has'. From this point of view, personalized medicine seems to have always been within the scope of medical practice and research (Baird 1990; Juengst 2000; Pokorska-Bocci et al. 2014).

The idea of personalized medicine was first introduced by Roger William in the 1950s. It became more attainable only in the early 2000s, when the human genome was mapped, and scientists could study the subtle individual genomic differences.

Recent genetic knowledges allow to anticipate drug response, enable more targeted therapies, and can help to choose more beneficial treatment in particular cases, whereas conventional medicine can only develop blockbusters. https://www.technologynetworks.com/drug-discovery/articles/the-future-of-pharma-beyond-blockbusters-332181.

The hope is that prescriptions could be tailored to an individual's specific molecular characteristics, and that this more accurate prescription will replace the current 'one-size-fits-all' paradigm of drug development and usage.

Precision medicine is certainly driven by improvements in cancer biology. Personalized cancer therapy is a treatment strategy centered on the ability to predict which patients are more likely to respond to specific (cancer) therapies. This approach is founded upon the idea that tumor biomarkers are associated with patient prognosis and tumor response to therapy (Fig. 5.12; Henry and Hayes 2012; *How Are Biomarkers Used to Treat Cancer? | MD Anderson Cancer Center*, n.d.; Shigeyasu et al. 2017). Integration of markers from these omics level is a central development in precision medicine (Fig. 5.13; Nebbioso et al. 2018).

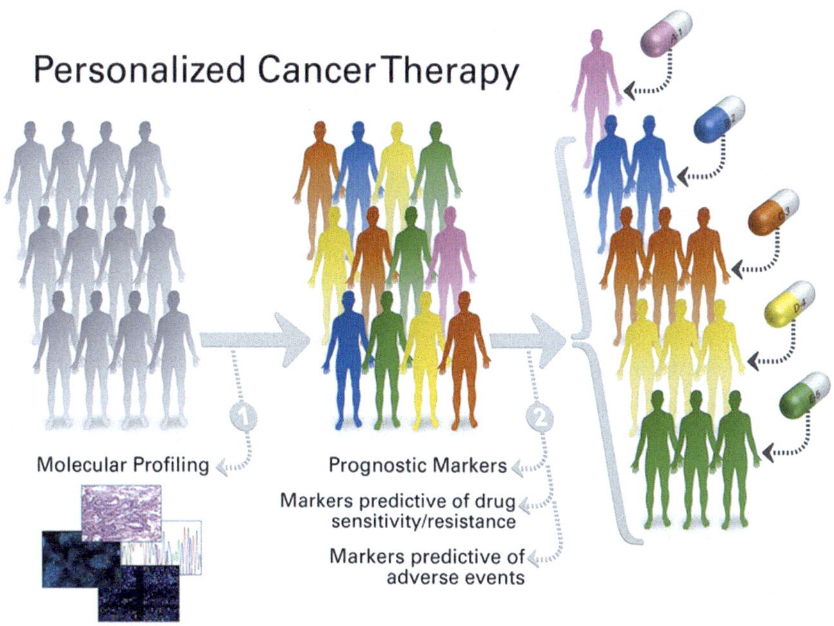

Fig. 5.12 Personalized cancer therapy and markers. http://www.novomics.com/eng/Research/precision.asp

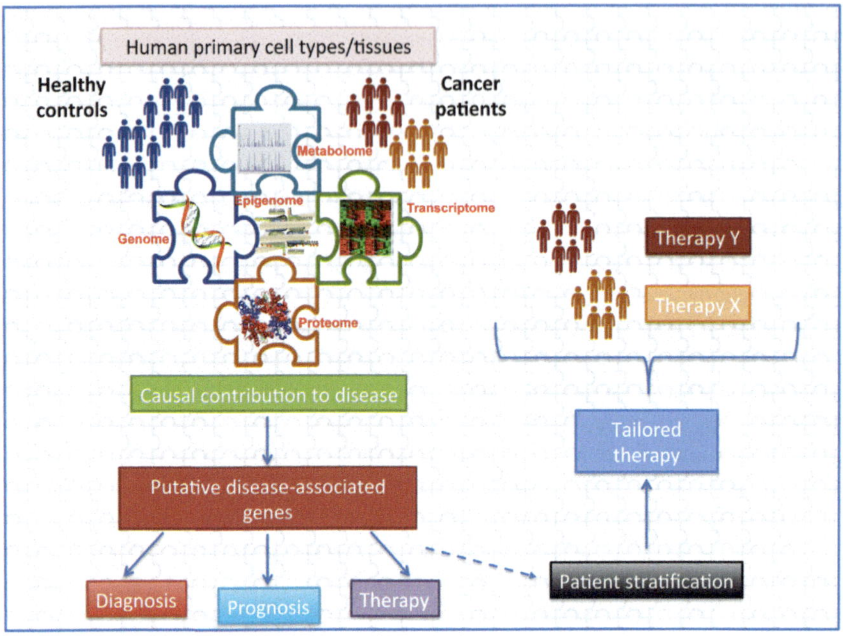

Fig. 5.13 Integrating and combining data from different platforms (genome-DNA sequence, transcriptome, proteome, metabolome, and epigenome) lead to a better understanding of the basis of cancer and paves the way toward personalized medicine (Nebbioso et al. 2018)

Ethical and Social Aspects

The fast increase in the understanding of molecular disease mechanisms is rapidly changing the practice of health care from reactive to preventive. Traditional medicine could mostly only offer a delayed intervention on an already existing disease; continues testing using molecular markers and the increased use of wearable health devices will allow a centralized collection of personal health information, bioinformatic analysis, and the prediction of risks of developing diseases. This can help to prevent the development of diseases by taking appropriate preventive actions (Lu et al. 2020; Ming et al. 2020; Vijayan et al. 2021).

Unfortunately, personalized medicine is also associated with an increased risk of discrimination. The improved possibility of dividing up the population into groups (persons with an identified disease risk, good vs. non-responders to interventions), which often coincides with ethnic different disease susceptibilities, may lead to increase inequalities in access to health care. This would be particularly concerning in insurance-based health care. This could lead to call into question the principles of fairness, solidarity, and justice on which many healthcare systems are based. It could further increase inequalities which are already faced by some socially disadvantaged groups (Menzel 2012).

The strongly increased possibilities to acquire many disease-specific predictive information could lead to create a new class of individuals: the 'pre-patients' or 'potential patients'. This would challenge the standard concept of disease, since health would not be opposed to illness, but to the risk of illness, which would be the new concern of medical practice.

5.4 Development of Personalized Precision Nutrition

In parallel to the development of precision medicine, precision nutrition follows similar concepts to improve serious metabolic problems. With nutrition-related diseases such as diabetes mellitus type II and obesity on the rise, it is for many experts unclear if the current efforts made to combat these diseases are effective or will be effective in the future.

Dietary guidelines are an extremely important instrument, but even a very recent overview '**A Global Review of Food-Based Dietary Guidelines**' (Herforth et al. 2019) addresses the problem of important regional disparities, where certain food groups, e.g., meat, particularly red meat, are treated differently across countries. Another problem of **dietary guidelines** is that they reflect the daily intake required to meet the **nutrient requirements** of 97.5% of the healthy population, meaning they are not geared toward specific individual needs or toward sick individuals. The European Food Safety Authority (EFSA) acknowledges that 'the physiological requirement varies between individuals dependent upon genetic and epigenetic **differences**, age, sex [and], physiological state (Agostoni et al. 2010)'.

The field of **personalized preventative nutrition** tries to incorporate these factors to better advise patients on their nutritional needs (Biesiekierski et al. 2019) So, personalized precision nutrition is based on the idea that individualizing nutritional advice, products, or services will be more effective than more generic approaches (de Toro-Martín et al. 2017; Fig. 5.14).

Personalization can be based on: 'Biological evidence of differential responses to foods/nutrients dependent on genotypic or phenotypic characteristics and the analysis of current behavior, preferences, barriers, and objectives and subsequent delivery of interventions, which motivate and enable each person to make appropriate changes to his or her eating pattern' (Adams et al. 2020; Biesiekierski et al. 2019; de Toro-Martín et al. 2017; Ordovas et al. 2018).

Presumably, personalized nutrition developed from nutrigenetics. SNPs have been used to identify a patient's individual genetic need and disease risk and provide the basis to build individualized diets and therapies. Indeed, single nucleotide variants are by far the most widely studied genetic variation in the field of precision nutrition. The risks (usually expressed as odds ratios) associated with common alleles are <2.0, and for continuous traits such as body mass index (BMI) usually <0.1 standard deviation (SD), e.g., the most strongly associated variants associated with type-2 diabetes (the variant in *TCF7L2*) and coronary heart disease (the variant near *CDKN2A/B*) confer risks, of approximately 1.4 per risk allele. The variants most strongly associated with BMI and height do so with per allele effects of 0.1

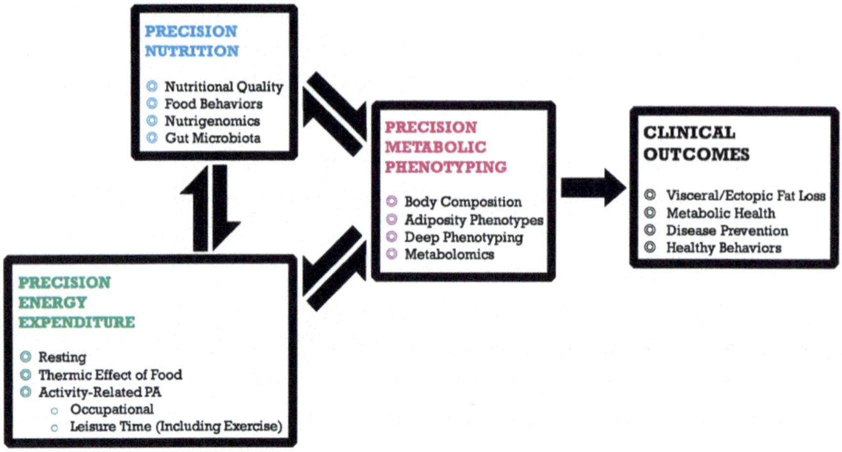

Fig. 5.14 Precision nutrition features and their relationships. PA: Physical activity (de Toro-Martín et al. 2017)

SD (approximately 0.4 kg/m^2) and 1 cm, respectively. Most of the variants confer much smaller effects than these examples (Frayling 2014).

Many studies have combined information from multiple associated SNPs, but even when combined, the variants rarely provided sufficient statistical power to offer any predictive value, e.g., the 40 strongest type-2 diabetes variants have a receiver operator curve (ROC) area under the curve (AUC) value of 0.63, 'where 0.5 is the same as flipping a coin and 0.8 is considered clinically useful'. However, there are some common diseases where directly genotyping sets of common variants could be useful to individual patients (Frayling 2014).

In this sense, several SNPs have been associated with common chronic diseases through interactions with the intakes of macro and micronutrients or with the consumption of particular foods and dietary patterns. Common variants in genes regulating homocysteine metabolism, such as methylenetetrahydrofolate reductase (MTHFR), and methionine synthase (MTR), have been linked to increased risk for breast cancer in individuals with low intakes of folate, vitamin B6, and vitamin B12 (Jiang-hua et al. 2014). Also, vitamin D status can be influenced by several polymorphisms in vitamin D pathway genes. SNPs in the vitamin D receptor (VDR) gene affecting vitamin D availability have been associated with osteoporosis predisposition (Banjabi et al. 2020). SNPs in genes encoding lipid proteins such as apolipoprotein C3 (APOC3) and apolipoprotein A1 (APOA1) conferred a higher risk of metabolic syndrome in subjects with a Western dietary patter (Hosseini-Esfahani et al. 2015). Significant interactions between the genetic risk score and diets on metabolic traits have been shown (Alsulami et al. 2020).

However, the solely use of SNPs only for personalized nutrition has been critically discussed because they are mainly based on correlation between gene and disease, and they do not include interactions between genes and environmental factors, 'and lastly they fail to explain their heritability' (Khalilisamani et al. 2022; López-Cortegano and Caballero 2019; Zuk et al. 2012).

5.4.1 Personalized Nutrition and Nutriepigenetics

Way of nutrition as well as multiple bioactive food components has been shown to influence epigenetic modifications and thus modulate gene expression. This is an important field what nutriepigenetics investigates. It studies the interaction between nutrients and epigenetic modification, questioning what effect they may have on health outcomes. Unfortunately, literature not always differentiates between the term nutrigenetic, nutrigenomic or nutriepigenetics, and nutriepigenomics in a harmonized way. Often, literature even includes both genetic and epigenetics under nutrigenomics.

Complex interactions among nutritional factors and DNA methylation, covalent histone modifications, and noncoding RNAs, including microRNAs (miRNAs), have been observed in obesity, dyslipidemia, T2DM, NAFLD, cancer, and CVD. The anti-inflammatory effects of consuming a Mediterranean diet were related to hypermethylation of proinflammatory genes (Casas et al. 2016; Padin et al. 2019; Stromsnes et al. 2021). The administrations of polyunsaturated fatty acids positively modulated the expression of several miRNAs, which suppressed oncogenic and lipogenic genes (Moradi Sarabi et al. 2018). The anticancer properties of resveratrol, epigallocatechin-3-gallate, curcumin, sulforaphane, and genistein have been associated with some epigenetic modifications including hypomethylation and acetylation of tumor suppressor genes and an increase in miRNAs targeting oncogenes. Apple polyphenols prevented diet-induced obesity by regulating the methylation status of genes involved in lipid metabolism (Boqué et al. 2013). Curcumin was found to exert protective effects against liver injury and heart failure through modulating DNA methylation patterns and histone modifications of key genes (Hassan et al. 2019).

Calorie restriction, high fat, low protein, single nutrient (SN) conditions are sensed by the cell through signaling pathways like TOR, Ras, AMPK, or PI3K/AKT, promoting changes on the epigenome. Histone modifications can be so seen as an intersection between diet and longevity (Molina-Serrano et al. 2019; Fig. 5.15).

Different micronutrient deficiencies such as folate, vitamin A, vitamin B, potassium, iron, and selenium correlated with hypermethylation of tumor suppressor genes (Cuenca-Micó and Aceves 2020). Dietary methionine restriction could improve insulin resistance, glucose homeostasis, oxidative stress, lipid metabolism, and autophagy in diabetes (Yin et al. 2018).

Multiple bioactive nutrients are known for their epigenetic activities (Bautista-García et al. 2017; Fig. 5.16). The use of epigenetic-active food ingredients in a preventative personalized nutrition therefore may provide an easy and inexpensive approach to prevent and potentially cure certain diseases (Tiffon 2018).

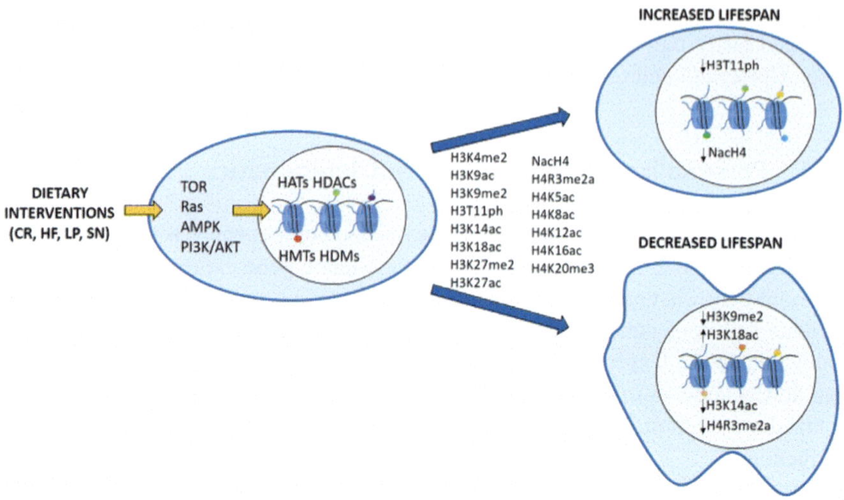

Fig. 5.15 Schematic representation depicting the hypothesis that histone modifications act as an intermediate between diet and longevity. Calorie restriction (CR), high fat (HF), low protein (LP), single nutrient (SN) conditions are sensed by the cell through signaling pathways like TOR, Ras, AMPK, or PI3K/AKT, promoting changes on the epigenome (Molina-Serrano et al. 2019)

The reversible feature of epigenetic marks has encouraged the design of specific nutritional interventions targeting epigenetic alterations that might have a significant impact on preventing and treating human chronic diseases (Tollefsbol 2014). Based on this evidence, the introduction of epigenetic-active dietary compounds into the diet could serve as an effective strategy for reducing metabolic and aging-associated comorbidities (Asif et al. 2020; Mahmoud 2022).

5.4.2 Personalized Nutrition and Gene Expression

Food ingredients but also specific diets can directly affect gene expression, e.g., by interaction with signaling pathways. But, food ingredients may be also be metabolized, and intermediates may affect gene expression or alter cell signaling pathways involved in gene expression (Berná et al. 2014).

As seen for pharmaceuticals, metabolization of food ingredients has been shown to be highly individual, e.g., with respect to genetics, ethnicity, physical exercise, structure of GI-microbiota or age (Chaleckis et al. 2016; Kastenmüller et al. 2015; Martin et al. 2019; Olafuyi et al. 2021; Sato et al., n.d., 2003).

Even effects of ambient conditions such as cool temperature were shown to alter brown fat and human metabolism (*Cool Temperature Alters Human Fat and Metabolism | National Institutes of Health (NIH)*, n.d.).

On the other hand, gene expression profiles have been used to predict the responsiveness to nutritional treatments. It has been reported that, prior to the

Bioactive nutriments	Natural sources	Antineoplastic effects	Epigenetic mechanisms of action
Folate,	spinach, asparagus, beans, peas, lentils, almonds	Anti-cancer, chemoprevention of malignant transformation	Regulation of SAM/SAH ratio, DNMT and MBD expression; regulation of tumor supressor miRNAs and oncogenic miRNAs
Retinoic acid	Mango, papaya, carrots, spinach, sweet potatoes	Anti-cancer, differentiating, pro-apoptotic	Regulation of DNMTs expression and activity, regulation of miRNAs targeting DNMTs; regulation of tumour suppressor miRNAs and oncogenic miRNAs; GNMT regulation; histone acetylation
Vitamin D3	Sun exposure, fish, fish liver oils	Anti-cancer, differentiating, pro-apoptotic	Regulation of DNMTs expression and enzyme activity; regulation of histone acetylation; regulation of oncogenic miRNAs
Resveratrol	Grapes, mulberries, apricots, pineapples, peanuts	Anti-cancer, antioxidant, anti-angiogenesis, pro-apoptotic	Regulation of DNMTs expression and enzyme activity; activation of deacetylase SIRT1 and p300 HAT; down-regulation of UHRF1; regulation of miRNAs
EGCG	Green tea	Anti-cancer, antioxidant, anti-angiogenesis, pro-apoptotic	Regulation of SAM/SAH ratio by COMT-mediated reactions; direct inhibition of DNMTs by binding to catalytic domain of the enzyme; regulation of tumour suppressor miRNAs
Curcumin	Spice turmeric	Anti-cancer, antioxidant, protects against heart failure	Direct inhibition of DNMTs by binding to catalytic domain of the enzyme; inhibition of HDACs and p300 HAT; regulation of tumour suppressor miRNAs and oncogenic miRNAs

Fig. 5.16 Epigenomic roles of bioactive nutrients (Bautista-García et al. 2017)

consumption of a low-fat diet, adipose gene expression profiling was able to differentiate responders from non-responders, as well as serve as a weak predictor of subjects predisposed to lose weight (Mutch et al. 2007). Also, the analysis of gene expression in subcutaneous adipose tissue revealed that genes regulating fatty acid metabolism, citric acid cycle, oxidative phosphorylation, and apoptosis were differentially regulated during a low-calorie diet between weight maintainers and weight regainers after weight loss (Armenise et al. 2017; Mutch et al. 2011). Adipose tissue transcriptome even reflects variations between subjects with continued weight loss and subjects regaining weight 6 mo after caloric restriction independent of energy intake (Márquez-Quiñones et al. 2010).

5.4.3 Personalized Nutrition and Microbiota-Epigenetic Interactions

The fast technological progress in the analysis of GI microbiota, their structure, and functions have severely influenced the understanding of personal mechanisms of metabolisms. Zeevi et al. conducted various studies on the prediction of glycemic response in combination with microbiota composition and personalized nutrition (Fig. 5.17). They showed that the post-prandial glucose response (PPGR) varied significantly in individuals when consuming the same standardized

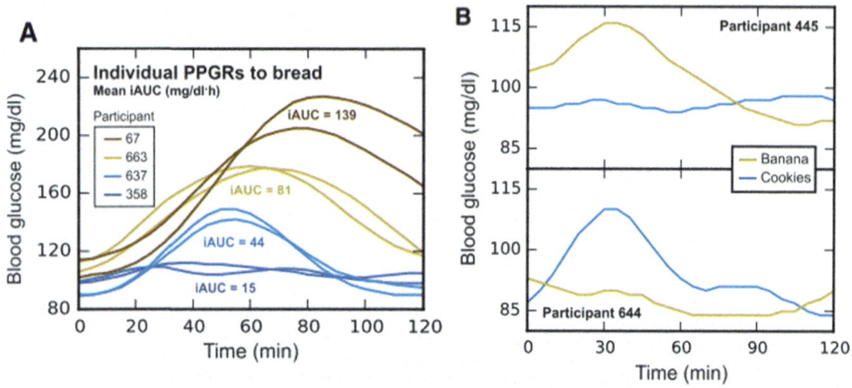

Fig. 5.17 **a** High variability in PPGRs between individuals consuming the same meal; **b** Opposite PPGRs in different individuals consuming the same meal (Zeevi et al. 2015)

meal. Opposite PPGR was also found in different individuals consuming the same meal. The different PPGRs were associated with individual microbiota composition. Taxa such as the *Enterobacteriaceae* were positively linked to PPGR; the same taxa have been associated with poor glycemic control and metabolic syndrome. An algorithm containing the patients' clinical and microbial data could predict PPGRs and proved that personalized nutrition interventions can improve glycemic responses. Until now, many groups conclude that personalized dietary interventions including microbiota data and predictive algorithms allow health professionals to more precisely tailor interventions toward their patients' individual needs (de Toro-Martín et al. 2017; Ordovas et al. 2018).

Personal Variability of Gut Microbiomes

The α-diversity (intra-individual) is a predictor of the extent of microbiota composition change upon the short-term consumption of different protein sources (red meat, white meat, and nonmeat sources) in healthy subjects. Importantly, changes are also highly variable between individuals, without strong population-level trends (Lozupone et al. 2012; Yeoh et al. 2019).

Although response to fibers has a common trend within the general population, heterogeneous and highly personalized shifts in the human microbiota have also been detected in response to carbohydrates, including dietary fiber (Leeming et al. 2019; Modrackova et al. 2021), resistant starches, and carbohydrate-containing prebiotics. Consumption of a high-fiber weight-stabilization or weight-loss diet in obese individuals affects the intestinal microbiota composition with significant interpersonal variation (Salonen 2014; Wang et al. 2016). Targeted prebiotics may be a way to alter the obese gut microbiome in humans (She et al. 2021).

Although fecal butyrate levels generally increase upon indigestible carbohydrate consumption, the response also varies widely among individuals (McOrist et al. 2011). The microbiome response to dietary carbohydrates can be predicted from

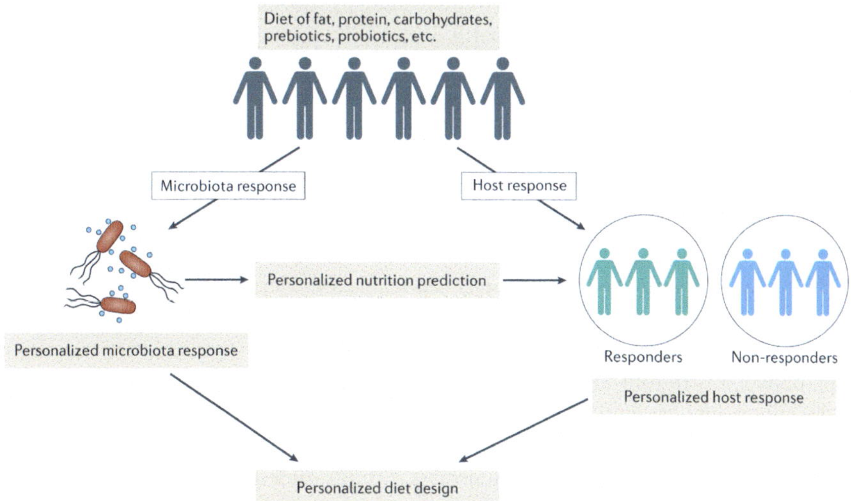

Fig. 5.18 Diet changes the gut microbiome composition and function in a person-specific manner, which is associated with the specific pre-intervention microbiome profile. Diet also results in highly individualized variation in host responses (for example, glycemic response), which can be accurately predicted by the host's unique microbiome signatures. By utilizing both aspects, personalized nutritional strategies can be developed in order to modify an individual's microbiome and further improve the response to a specific diet (Kolodziejczyk et al. 2019)

the baseline microbial diversity and the role of the gut microbiome in predicting responses to diet can be used for the development of precision nutrition models (Hughes et al. 2019; Jardon et al. 2022). Especially, the level of particular bacterial species may be a predictor of the response to particular diets (Kolodziejczyk et al., n.d.; Fig. 5.18).

5.5 Omics Approaches and Data Integration

Experiences from various OMIC layers, such as genetic, epigenetic, gene expression, metabolic or microbiota, have been shown to give information about personal responses to foods and nutrition as well as personal requirements for diets. In many situations, there will also be interactions between molecular pathways involving elements of these layers.

Recent advances in the development of high-throughput sequencing and different 'omics' technologies now allow quantitative analyses of biological molecules at multiple levels, and how they change with aging. The use of next-generation sequencing has paved the way for data integration: Through the same technology, it is today possible to investigate genetics, different aspects of genomics or protein binding (ChipSeq), and the transcriptome (RNAseq).

Different information can so be represented by similar data formats and data sources: This offered opportunities for further development of the omics and a boost to develop new integration methods and approaches (Dato et al. 2021).

High-throughput approaches require proper data normalization and data quality control. e.g., the understanding of the aging process is not possible without taking into account molecular interactions (or '**interactome**') occurring among different components of the biological system (proteins, lipids, nucleic acids, etc.) as well as how they affect physiological parameters associated with aging (i.e., 'physiome') (Randhawa and Kumar 2021).

Multi-layered networks have been proposed as a powerful tool used to establish the necessary connection between different types of information: It does provide a natural way to represent the structure of a biological system, and the relationships between different layers in the network may represent effects which cannot be described just by statistical correlations (Lee et al. 2020). Tensor decomposition has also been proposed as a powerful method to infer relationships between different biological descriptors. A tensor is a multi-dimensional array (Kolda and Bader 2009). Machine learning (ML) approaches have been proven to be extremely powerful in the re-analysis of large datasets collected in the past, allowing an unprecedented capacity for data integration, and providing new insights. The use of feature selection and a combination of support vector machines (SVMs) and random forest (RF) allowed to mine the combined datasets of different aging population studies and enabled the integration of lifestyle, laboratory, and clinical data (Fig. 5.19; Dato et al. 2021; Gomez-Cabrero et al. 2021).

5.5.1 Translation of Personalized Precision Nutrition into Praxis

Personalized nutrition is based on the idea that individualizing nutritional advice, products, or services will be more effective than more generic approaches. Personalized nutrition can be applied in two broad areas: Firstly, for the dietary management of people with specific diseases or those who need special nutritional support, for example, in special life phases like pregnancy or old age, and, secondly, for the development of more effective interventions for improving public health. Individuals may also wish to use personalized nutrition to achieve personal goals/ambitions that are less directly related to health, for example, to deal with preferences for and/or dislikes of specific foods, to attempt to achieve a desired body size or shape, or for competitive sports (Ordovas et al. 2018).

Unlike with medication, dietary changes require individuals to make daily, sometimes hourly, choices. The adoption of these lifestyle changes, including changes in dietary patterns, is highly dependent on effective collaboration with participants who are being helped to take responsibility for their behavior, and, ultimately, health. Increasing technology is available that can motivate healthy eating. However, such applications usually adopt a 'one-size-fits-all' approach that is biased toward specific cultures or population subgroups. More emphasis is needed to develop behavioral approaches that will best motivate particular individual

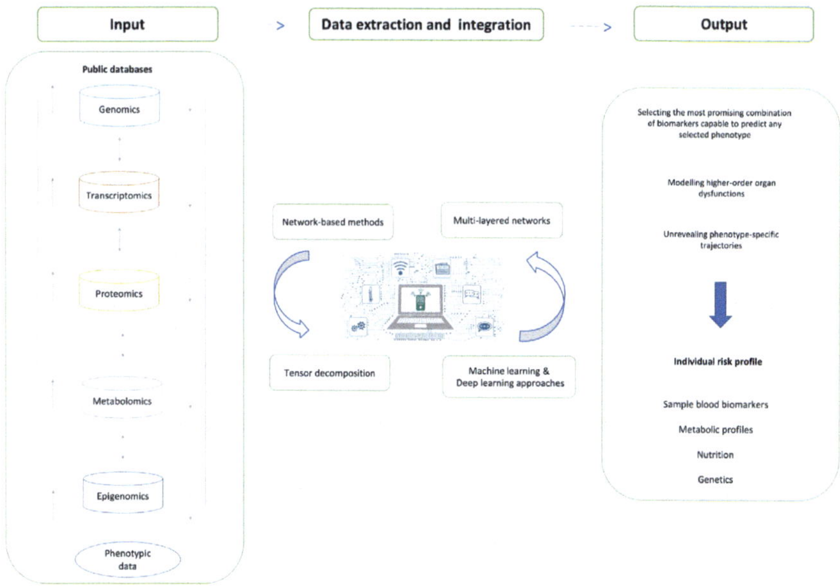

Fig. 5.19 Data integration in aging research. A schematic representation of the process of data integration from public databases and other sources in aging and age-related diseases. The main data sources are represented in the 'input' panel (Dato et al. 2021)

and cultural groups. There may be benefits in moving from a decision framework based on health professionals' perspectives of effectiveness to one of shared decision-making. An intervention based on shared decision-making between the provider and the recipient becomes personalized and may increase acceptance and adherence (Ordovas et al. 2018).

The **Food4Me Study** was a EU funded randomized controlled trial (RCT) involving >1600 participants from seven European countries. The project studied the effects of internet-based personalized nutrition advice on lifestyle changes. Their recent 6-month randomized control trail compared behavioral changes in individuals who received personalized nutrition (PN) advice or conventional dietary advice. Personalized nutrition advice was further categorized in individual diet intake, individual diet intake and phenotypic data and individual diet intake, phenotypic genotypic data and medical practice.

The study asked two key questions: Is personalized nutrition more effective in changing diet than a conventional one-size-fits-all approach? Does the basis used for personalization matter? (With particular interest in the benefit of personalization based on phenotypic and genotypic characteristics).

After 6 months, the answer was clear. Personalization of dietary advice assisted and/or motivated consumers to eat a healthier diet and follow a healthier lifestyle (in comparison with 'impersonal' (conventional) dietary advice) (Celis-Morales et al. 2017).

The 'Healthy Eating Index' (Colby et al. 2020; Kasper et al. 2016) was used as the global measure of 'healthfulness' of eating patterns, and change was measured after 3 and 6 months.

Personal nutrition was not only more effective in improving lifestyle changes compared to conventional advice but also more effective than a conventional one-size-fits-all approach.

Personal precision nutrition aims to prevent and manage chronic diseases by tailoring dietary interventions or recommendations to one or a combination of an individual's genetic and epigenetic background, microbiota structure, metabolic profile, and environmental exposures. With adequate scientific evidence, nutritional tests can provide an early screening opportunity and could increase the demand for health-related consultations and screenings, which will help prevent disease Fig. 5.20. (https://doi.org/10.3390/jcm8071065).

In addition, mobile apps and wearable devices facilitate real-time assessment of dietary intake and provide feedback which can improve glycemic control and diabetes management. By integrating these technologies with big data analytics, precision nutrition has the potential to provide personalized nutrition guidance for more effective prevention and management of complex metabolic diseases such as type-2 diabetes (Wang and Hu 2018; Fig. 5.21).

Further, developments of personalized nutrition include the education of health professionals to correctly interpret genetic and epigenetic data, creating ways to motivate positive behavior change in patients and to correctly implementing personalized nutrition into medical practice. Ethical, and legal guidelines, as well as

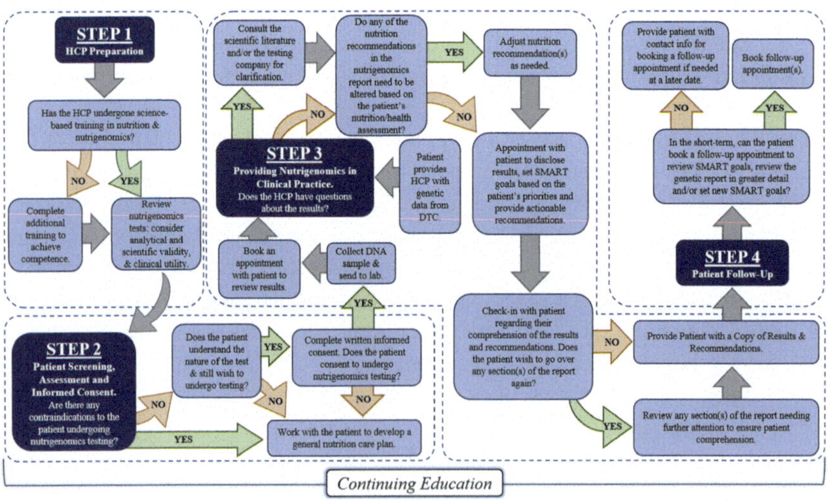

Fig. 5.20 Guiding global best practice in personalized nutrition based on genetics: the development of a nutrigenomics care map. DTC direct-to-consumer; HCP health care provider; SMART specific, measurable, attainable, relevant, and time-based (Horne et al. 2022)

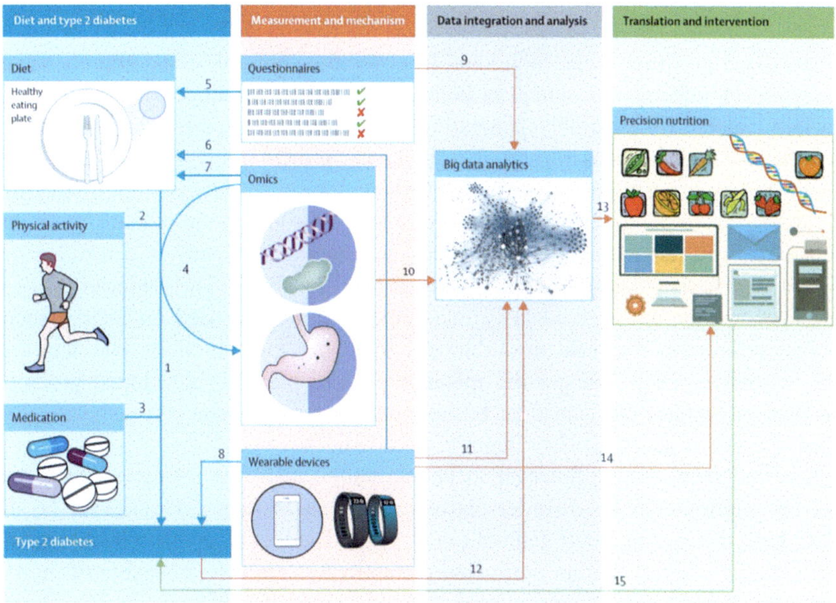

Fig. 5.21 Precision nutrition for prevention and management of type-2 diabetes (Wang and Hu 2018)

standardized regulations for tests will need to be established (Fig. 5.22; Ferguson et al. 2016).

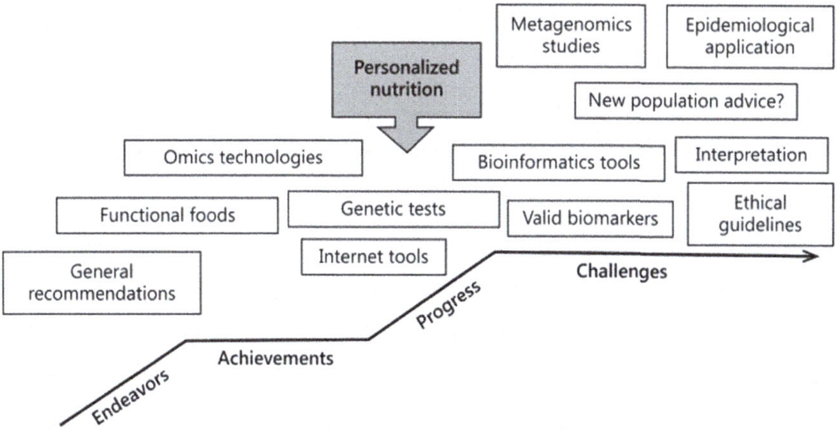

Fig. 5.22 Achievements already made and challenges faced by personalized nutrition (Ferguson et al. 2016)

5.5.2 Personalization or Stratification, Metabotypes

Personalized nutrition at the individual level requires not only the comprehensive collection of information, which is both costly and demanding, but also models that are capable of accurately generating personalized advice for the individual. A somehow easier approach may be to personalize diets at the group level. More recent studies have suggested that individuals may be grouped according to unique metabolic responses to foods and dietary changes. Grouping individuals based on similarities in their metabolic phenotype—that is metabotypes—is a novel concept, and different definitions have been used (Fig. 5.23). The underlying idea behind metabotyping is to identify metabolic phenotypes based on factors such as diet, anthropometric measures, clinical parameters, metabolomics data, and the gut microbiota. Diets for specific metabotypes are presently developed predicting that an optimal diet can then be tailored to fit each metabotype specifically (Garcia-Perez et al. 2020; Hillesheim et al. 2020).

It is under discussion, however, if people with a high risk for certain diseases such as cardiometabolic disorders have special metabotypes (Grabowski et al. 2020; Hillesheim et al. 2020; Palmnäs et al. 2020; Riedl et al. 2020). The results of some studies suggest that an optimized metabotype approach is capable of delivering targeted nutritional counselling to healthy adults and is very comparable to individualized counselling.

The next step would be to determine whether the optimized metabotype approach is effective in changing diet quality (Hillesheim and Brennan 2020). Clearly, metabotyping and group-based nutrition wants to enter conventional population-based guidelines and personalized nutrition.

In conclusion, the way to personalized precision nutrition developed from simple genetic SNP testing to addition of lifestyle and environmental data, epigenetic, microbiota, and metabolic markers already reflecting impacts of lifestyle and nutrition to the integration of data from markers of multiple OMICS. Caution in the

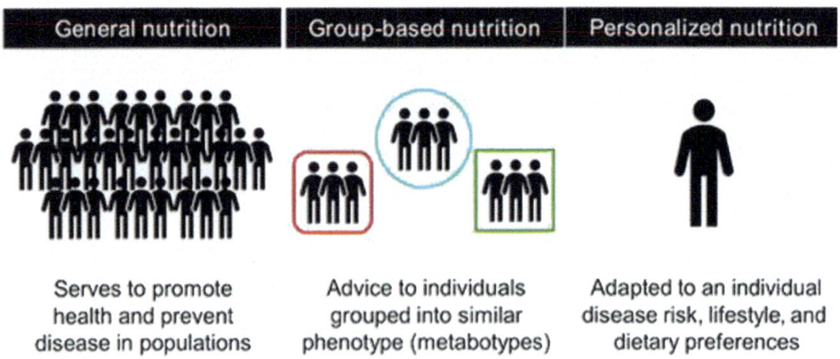

Fig. 5.23 Metabotyping and group-based nutrition in the context of the conventional population-based guidelines and personalized nutrition (Palmnäs et al. 2020)

claim of benefits of personalized nutrition comes from the review that scientific evidence is mostly based on observational studies with a low level of reproducibility. But still, personalization may need sustained change in behavior. (https://doi.org/10.3390/jcm8071065).

5.5.3 Personalized Precision Nutrition and Consumer Aspects

Consumer attitudes toward molecular tests aiming to reveal the risks of a predisposition to various illnesses have already been examined by several papers. Consumer acceptance of nutrigenomics-based personalized nutrition, however, has only been examined by a few. The most important motivator to have a genetic test done was the consumers' own health and the health of their family members. Whereas, early genetic testing, especially direct-to-consumer testing, encountered substantial criticism, modern ways of a personalized precision nutrition have reflected these restrictions and therefore get a much broader approval (Reinders et al. 2020).

Consumer attitude and behavioral change: A systematic review analyzed consumer attitudes toward direct-to-consumer genetic testing (DTC-GT). DTC-GT showed that there is generally low awareness of DTC-GT in the general population even though most study participants had high education levels. Nevertheless, participants in most studies were interested in knowing their disease risk, especially if they were parents (to know their child's risk) or if they had a higher disease risk (e.g. family history). Interest increased when the test results were positive and decreased with price of genetic testing and knowing the risks of DTC-GT (less regulated and accurate results). It was also found that participants preferred genetic testing to be performed by a health professional, instead of it being marketed toward consumers (Covolo 2015).

The impact of genetic testing results on lifestyle changes is fragmentary. Some individuals expressed concern and the intention to change their lifestyle, while others showed indifference to the test results even though they were at a higher risk for disease. Moderate lifestyle changes could be observed though a 3 month follow up showed no impact on user behavior. Furthermore, one-year follow-ups showed no difference in concern compared to individuals who were not tested for disease risk (Ruhl et al. 2019).

Related to genetic tests, considerable concerns were consistently raised about internet privacy, data security, data use, and data destiny; participants articulated their fears about the potential for information to be used by companies for commercial gain or to fall into the hands of insurers, employers, or government agencies. In contrast with the more numerous research studies examining consumer judgements related to genetic tests, only, a few studies examined consumer preferences for genetic-based personalized nutrition. Based on both qualitative and quantitative studies, consumers usually show positive attitudes toward genetic-based personalized nutrition; about one third to a half of respondents would use a service of this kind and would follow a personalized diet although there are significant international differences in this area. Costs and benefits of personalized nutrition turned out to be of primary importance in consumer judgements (Szakály et al. 2021).

A driver of personalized nutrition is the increasing awareness of consumers of their individuality. Personalized nutrition fits well into the current marketing trend of moving the consumer company relationship from the mass model toward a customized model. Consumer goods have become increasingly personalized, particularly during the second half of the twentieth century. Individualized products appeal to today's sophisticated consumers and allow them to feel empowered (van der Lans et al. 2016).

A special ethical question concerns how access to personalized nutrition could be ensured for lower-income classes which could have a greater need for it but lack the means adopt it.

Trust of consumers in advice of scientific bodies is rather commuting depending on upcoming new findings often challenging established thinking, personal priority of values such as animal welfare or unsubstantiated public news. In 2019, the American Society for Nutrition (ASN) commissioned an independent Advisory Committee to look specifically at the public's trust in nutrition science and the factors that can influence it. These areas include (Fig. 5.24):

However, there is also a group of often well-educated and informed consumers who seem to prefer personal studies of literature and personal experience rather than to believe in opinions and advice of scientific bodies which may be influenced by stake holder groups. An extreme form of this development is biohacking which has gone viral in many social groups. Internet followers of biohacking read that Twitter CEO Jack Dorsey experienced the benefits of fasting intermittently and drinking 'salt juice' each morning. Even 'dopamine fasting' received interest. These are all types of biohacking, a broad term for a lifestyle that is growing increasingly popular, and not just in Silicon Valley, where it really took off.

Biohacking—also known as DIY biology—"is a term that can cover a huge range of activities, from performing science experiments on experimental organisms to tracking the personal diet to optimize the own biology" https://daveasprey.com/beginners-guide-to-biohacking-101/.

Biohackers experiment on their own bodies with the hope of boosting their physical and cognitive performance. They form one branch of transhumanism, believing that human beings can and should use technology to augment and evolve performance. https://www.medicaldevice-network.com/analysis/medical-biohacking/.

Hot topics of the biohacking community are the optimization of lifestyle habits, modulation of daily routines, nutrition, and metabolism, fasting, ketogenic diets, cognitive performance, sleep, brain health, nootropics, nutraceutical, etc., in combination with usage of biosensing wearables, genetic analysis, and lab based of information. Biohacking is certainly a tool of a generation that is used to look for information in the Web and in social networks, often neglecting scientific sources of information.

- **Conflict of interest and objectivity**. Consumers are more trusting when they believe scientists are acting independently of financial gain. Any conflicts of interest, as well as, personal bias/beliefs, business associations or personal relationships should be fully disclosed.
- **Standards of scientific rigor and reproducibility**. Consumers want to know that the quality of the research is extremely thorough and accurate, but this can be especially difficult with nutrition science as many individuals hold "unscientific beliefs" about food.
- **Transparency**. Transparency in communicating the scientific process is one of the keys to increasing public trust in science. This requires acknowledgment of all funders, beneficiaries, and opponents of the research and its outcomes. It's also vital to state all potential biases and competing interests that influence the overall research and interpretation of outcomes.
- **Equity**. Typically, in the U.S., inequities in health research center around the lack of women, specific age groups and under-represented income or ethnic groups in clinical research and trials. A lack of equity can undermine trust.
- **Information dissemination (education, communication and marketing)**. The strong and ever-increasing evidence that links food to health is generating a growing interest in nutrition and retail dietitians can advance consumer trust with educational efforts that help link science and decision-making. ASN recommends helping individuals increase their critical thinking and reasoning skills when it comes to scientific information and the ability to critically evaluate the media. In addition, dietitians should continue using multiple educational touch points (i.e.: social media, website and community events), to provide meaningful advice and tools that help consumers make sound dietary decisions based on science.

Fig. 5.24 Are consumers losing trust in science? | Retail dietitians business alliance

5.5.4 Consumer Supporting Organizations in Between Multiple Interests, Discussion

Many consumers certainly rely on government and expert committees approved concepts, guidelines, and products. The general believe in dietary guidelines and nutrition pyramids is high and developments into personalized ways of thinking are picked up only step by step depending on convincing personal advantages. On the other side, many consumers strongly believe that everyone must find out the best way, concept, the best technology, the best product for himself. The adherence in devices from regulatory-based expert committees and guidelines suiting for all seems to be somehow limited in this group. A broader understanding of epigenetic principles by consumers may result in an increased interest and use of epigenetic tests and a better acceptance compared to genetic testing only.

Despite public advertising in their engagement in sustainability and social competence, industry must orient their developments and products on economic success. Whereas, profound risk assessment and management in new developments and products is generally broadly accepted; rigid regulation often result in monopolization and draw backs in health-supporting opportunities. Engagement of industry in preventive, personalized nutrition, and disease prevention will depend on public interest and the believe of consumers into personal advantages of these developments.

Consumer organizations will have to find a delicate balance between serving consumer groups with their rather diverse thinking's as well as the need for research and development. In contrast, basis sciences need to be free of marketing aspects. Therefore, the development of a personalized, precision health care and nutrition needs to be driven by solid science and an improved communication about advantages and needs between consumer groups, research, and industry.

References

Abdelmohsen K, Panda A, Kang MJ, Xu J, Selimyan R, Yoon JH, Martindale JL, De S, Wood WH, Becker KG, Gorospe M (2013) Senescence-associated lncRNAs: senescence-associated long noncoding RNAs. Aging Cell 12(5):890–900. https://doi.org/10.1111/acel.12115

Adams J (2008) No title genetic control of aging and life span. Nat Educ 1(1):130

Adams SH, Anthony JC, Carvajal R, Chae L, Khoo CSH, Latulippe ME, Matusheski NV, McClung HL, Rozga M, Schmid CH, Wopereis S, Yan W (2020) Perspective: guiding principles for the implementation of personalized nutrition approaches that benefit health and function. Adv Nutr 11(1):25–34. http://doi.org/10.1093/ADVANCES/NMZ086

Agostoni C, Bresson J-L, Fairweather-Tait S, Flynn A, Golly I, Korhonen H, Lagiou P, Løvik M, Marchelli R, Martin A, Moseley B, Neuhäuser-Berthold M, Przyrembel H, Salminen S, Sanz Y, Strain S, Strobel S, Tetens I, Tomé D et al (2010) Scientific opinion on principles for deriving and applying dietary reference values. EFSA J 8(3):1458. http://doi.org/10.2903/J.EFSA.2010.1458

Ahadi S, Zhou W, Schüssler-Fiorenza Rose SM, Sailani MR, Contrepois K, Avina M, Ashland M, Brunet A, Snyder M (2020) Personal aging markers and ageotypes revealed by deep longitudinal profiling. Nat Med 26(1):83. https://doi.org/10.1038/S41591-019-0719-5

Allis CD, Jenuwein T (2016) The molecular hallmarks of epigenetic control. Nat Rev Genet 17(8):487–500. http://doi.org/10.1038/nrg.2016.59

Alsulami S, Aji AS, Ariyasra U, Sari SR, Tasrif N, Yani FF, Lovegrove JA, Sudji IR, Lipoeto NI, Vimaleswaran KS (2020) Interaction between the genetic risk score and dietary protein intake on cardiometabolic traits in Southeast Asian. Genes Nutr 15(1). http://doi.org/10.1186/S12263-020-00678-W

Anastasiou D, Krek W (2006) SIRT1: linking adaptive cellular responses to aging-associated changes in organismal physiology. Physiology (Bethesda, Md.) 21(6):404–410. http://doi.org/10.1152/physiol.00031.2006

Anton S, Leeuwenburgh C (2013) Fasting or caloric restriction for healthy aging. Exp Gerontol 48(10):1003–1005. https://doi.org/10.1016/j.exger.2013.04.011

Armenise C, Lefebvre G, Carayol J, Bonnel S, Bolton J, di Cara A, Gheldof N, Descombes P, Langin D, Saris WHM, Astrup A, Hager J, Viguerie N, Valsesia A (2017) Transcriptome profiling from adipose tissue during a low-calorie diet reveals predictors of weight and glycemic outcomes in obese, nondiabetic subjects. Am J Clin Nutr 106(3):736–746. https://doi.org/10.3945/ajcn.117.156216

Asif S, Morrow NM, Mulvihill EE, Kim KH (2020) Understanding dietary intervention-mediated epigenetic modifications in metabolic diseases. Front Genet 11:1242. https://doi.org/10.3389/FGENE.2020.590369/BIBTEX

Baird PA (1990) Genetics and health care: a paradigm shift. Perspect Biol Med 33(2):203–213. https://doi.org/10.1353/PBM.1990.0009

Banjabi AA, Al-Ghafari AB, Kumosani TA, Kannan K, Fallatah SM (2020) Genetic influence of vitamin D receptor gene polymorphisms on osteoporosis risk. Int J Health Sci 14(4):22. /pmc/articles/PMC7346971/

Basavarajappa BS, Subbanna S (2021) Histone methylation regulation in neurodegenerative disorders. Int J Mol Sci 22(9). http://doi.org/10.3390/IJMS22094654

Bautista-García P, González-López L, González-Esparza B, del Castillo-Rosas C (2017) Effect of bioactive nutriments in health and disease: the role of epigenetic modifications. In: Functional food—improve health through adequate food. https://doi.org/10.5772/INTECHOPEN.68789

Berná G, Oliveras-López MJ, Jurado-Ruíz E, Tejedo J, Bedoya F, Soria B, Martín F (2014) Nutrigenetics and nutrigenomics insights into diabetes etiopathogenesis. Nutrients 6(11):5338–5369. https://doi.org/10.3390/NU6115338

Bernadotte A, Mikhelson VM, Spivak IM (2016) Markers of cellular senescence. Telomere shortening as a marker of cellular senescence. Aging 8(1):3–11. http://doi.org/10.18632/aging.100871

Biesiekierski JR, Livingstone KM, Moschonis G (2019) Personalised nutrition: updates, gaps and next steps. Nutrients 11(8). http://doi.org/10.3390/NU11081793

Boqué N, de la Iglesia R, de la Garza AL, Milagro FI, Olivares M, Bañuelos Ó, Soria AC, Rodríguez-Sánchez S, Martínez JA, Campión J (2013) Prevention of diet-induced obesity by apple polyphenols in Wistar rats through regulation of adipocyte gene expression and DNA methylation patterns. Mol Nutr Food Res 57(8):1473–1478. https://doi.org/10.1002/MNFR.201200686

Bowman GD, Poirier MG (2015) Post-translational modifications of histones that influence nucleosome dynamics. Chem Rev 115(6):2274. https://doi.org/10.1021/CR500350X

Braeckman BP, Vanfleteren JR (2007) Genetic control of longevity in C. elegans. Exp Gerontol 42(1–2):90–98. http://doi.org/10.1016/j.exger.2006.04.010

Briceño O, Lissina A, Wanke K, Afonso G, von Braun A, Ragon K, Miquel T, Gostick E, Papagno L, Stiasny K, Price DA, Mallone R, Sauce D, Karrer U, Appay V (2016) Reduced naïve CD8+ T-cell priming efficacy in elderly adults. Aging Cell 15(1):14–21. https://doi.org/10.1111/acel.12384

Bronkhorst AJ, Ungerer V, Holdenrieder S (2019) The emerging role of cell-free DNA as a molecular marker for cancer management. Biomol Detect Quantification 17. https://doi.org/10.1016/J.BDQ.2019.100087

Brunet A, Berger SL (2014a) Epigenetics of aging and aging-related disease. J Gerontol Ser A Biol Sci Med Sci 69(Suppl. 1):17–20. https://doi.org/10.1093/gerona/glu042

Brunet A, Berger SL (2014b) Epigenetics of aging and aging-related disease. J Gerontol Ser A Biol Sci Med Sci 69(Suppl 1):S17–S20. http://doi.org/10.1093/gerona/glu042

Busslinger M, Tarakhovsky A (2014) Epigenetic control of immunity. Cold Spring Harbor Perspect Biol 6(6). http://doi.org/10.1101/cshperspect.a019307

Byrne CS, Chambers ES, Morrison DJ, Frost G (2015) The role of short chain fatty acids in appetite regulation and energy homeostasis. Int J Obes 39(9):1331. http://doi.org/10.1038/IJO.2015.84

Cambier J (2005) Immunosenescence: a problem of lymphopoiesis, homeostasis, microenvironment, and signaling. Immunol Rev 205:5–6. http://doi.org/10.1111/j.0105-2896.2005.00276.x

Campisi J (2013) Aging, cellular senescence, and cancer. Annu Rev Physiol 75:685–705

Casas R, Sacanella E, Estruch R (2016) The immune protective effect of the Mediterranean diet against chronic low-grade inflammatory diseases. Endocr Metab Immune Disord Drug Targets 14(4):245. https://doi.org/10.2174/1871530314666140922153350

Celis-Morales C, Livingstone KM, Marsaux CFM, Macready AL, Fallaize R, O'Donovan CB, Woolhead C, Forster H, Walsh MC, Navas-Carretero S, San-Cristobal R, Tsirigoti L, Lambrinou CP, Mavrogianni C, Moschonis G, Kolossa S, Hallmann J, Godlewska M, Surwiłło A et al (2017) Effect of personalized nutrition on health-related behaviour change: evidence from the Food4Me European randomized controlled trial. Int J Epidemiol 46(2):578–588. http://doi.org/10.1093/IJE/DYW186

Chaleckis R, Murakami I, Takada J, Kondoh H, Yanagida M (2016) Individual variability in human blood metabolites identifies age-related differences. Proc Natl Acad Sci U S A 113(16):4252–4259. https://doi.org/10.1073/PNAS.1603023113/SUPPL_FILE/PNAS.1603023113.SD01.XLSX

Chambers SM, Shaw CA, Gatza C, Fisk CJ, Donehower LA, Goodell MA (2007) Aging hematopoietic stem cells decline in function and exhibit epigenetic dysregulation. PLoS Biol 5(8):1750–1762. https://doi.org/10.1371/journal.pbio.0050201

Colby S, Zhou W, Allison C, Mathews AE, Olfert MD, Morrell JS, Byrd-Bredbenner C, Greene G, Brown O, Kattelmann K, Shelnutt K (2020) Development and validation of the short healthy eating index survey with a college population to assess dietary quality and intake. Nutrients 12(9):1–24. https://doi.org/10.3390/NU12092611

Cool Temperature Alters Human Fat and Metabolism | National Institutes of Health (NIH) (n.d.) Retrieved April 20, 2022, from https://www.nih.gov/news-events/nih-research-matters/cool-temperature-alters-human-fat-metabolism

Covolo L, Rubinelli S, Ceretti E, Gelatti U (2015) Internet-based direct-to-consumer genetic testing: a systematic review. J Med Internet Res 17(12). http://doi.org/10.2196/jmir.4378

Crujeiras AB, Campion J, Díaz-Lagares A, Milagro FI, Goyenechea E, Abete I, Casanueva FF, Martínez JA (2013) Association of weight regain with specific methylation levels in the NPY and POMC promoters in leukocytes of obese men: a translational study. Regul Pept 186:1–6. https://doi.org/10.1016/J.REGPEP.2013.06.012

Cuenca-Micó O, Aceves C (2020) Micronutrients and breast cancer progression: a systematic review. Nutrients 12(12):1–18. https://doi.org/10.3390/NU12123613

D'argenio V (2018) The prenatal microbiome: a new player for human health. High-Throughput 7(4). http://doi.org/10.3390/ht7040038

Dato S, Crocco P, Rambaldi Migliore N, Lescai F (2021) Omics in a digital world: the role of bioinformatics in providing new insights into human aging. Front Genet 12:689824. https://doi.org/10.3389/FGENE.2021.689824/FULL

de Moraes ACF, Fernandes GR, da Silva IT, Almeida-Pititto B, Gomes EP, da Pereira AC, Ferreira SRG (2017) Enterotype may drive the dietary-associated cardiometabolic risk factors. Front Cell Infect Microbiol 7:47. http://doi.org/10.3389/FCIMB.2017.00047/FULL

de Toro-Martín J, Arsenault BJ, Després JP, Vohl MC (2017) Precision nutrition: a review of personalized nutritional approaches for the prevention and management of metabolic syndrome. Nutrients 9(8):913. https://doi.org/10.3390/NU9080913

DeBord DG, Carreón T, Lentz TJ, Middendorf PJ, Hoover MD, Schulte PA (2016) Use of the "exposome" in the practice of epidemiology: a primer on-omic technologies. Am J Epidemiol 184(4):302. https://doi.org/10.1093/AJE/KWV325

Den Besten G, Van Eunen K, Groen AK, Venema K, Reijngoud DJ, Bakker BM (2013) The role of short-chain fatty acids in the interplay between diet, gut microbiota, and host energy metabolism. J Lipid Res 54(9):2325–2340. http://doi.org/10.1194/jlr.R036012

Dhakal R, Bajpai VK, Baek KH (2012) Production of gaba (gamma—aminobutyric acid) by microorganisms: a review. Braz J Microbiol 43(4):1230–1241. https://doi.org/10.1590/s1517-83822012000400001

Dillin A, Kelly JW (2007) Medicine. The yin-yang of sirtuins. Science (New York, N.Y.) 317(5837):461–462. http://doi.org/10.1126/science.1146585

Drew L (2022) Turning back time with epigenetic clocks. Nature 601(7893):S20–S22. https://doi.org/10.1038/d41586-022-00077-8. PMID: 35046594

Duranti S, Ruiz L, Lugli GA, Tames H, Milani C, Mancabelli L, Mancino W, Longhi G, Carnevali L, Sgoifo A, Margolles A, Ventura M, Ruas-Madiedo P, Turroni F (2020) Bifidobacterium adolescentis as a key member of the human gut microbiota in the production of GABA. Sci Rep 10(1):1–13. http://doi.org/10.1038/s41598-020-70986-z

Eckburg PB, Bik EM, Bernstein CN, Purdom E, Dethlefsen L, Sargent M, Gill SR, Nelson KE, Relman DA (2005) Microbiology: diversity of the human intestinal microbial flora. Science 308(5728):1635–1638. https://doi.org/10.1126/science.1110591

Fan C, Zhang L, Fu H, Liu C, Li W, Cheng Q, Zhang H, Jia S, Zhang Y (2020) Enterotypes of the gut microbial community and their response to plant secondary compounds in plateau pikas. Microorganisms 8(9):1–16. https://doi.org/10.3390/MICROORGANISMS8091311

Faragher RGA (2015) Should we treat aging as a disease? The consequences and dangers of miscategorisation. Front Genet 6:171. http://doi.org/10.3389/FGENE.2015.00171/BIBTEX

Fenley AT, Anandakrishnan R, Kidane YH, Onufriev AV (2018) Modulation of nucleosomal DNA accessibility via charge-altering post-translational modifications in histone core. Epigenetics Chromatin 11(1):1–19. https://doi.org/10.1186/S13072-018-0181-5/METRICS

Ferguson LR, de Caterina R, Görman U, Allayee H, Kohlmeier M, Prasad C, Choi MS, Curi R, de Luis DA, Gil Á, Kang JX, Martin RL, Milagro FI, Nicoletti CF, Nonino CB, Ordovas JM, Parslow VR, Portillo MP, Santos JL et al (2016) Guide and position of the international society of nutrigenetics/nutrigenomics on personalised nutrition: part 1—fields of precision nutrition. Lifestyle Genomics 9(1):12–27. http://doi.org/10.1159/000445350

Feser J, Tyler J (2011) Chromatin structure as a mediator of aging. FEBS Lett 585(13):2041–2048. http://doi.org/10.1016/j.febslet.2010.11.016

Finkel T, Deng CX, Mostoslavsky R (2009) Recent progress in the biology and physiology of sirtuins. Nature 460(7255):587–591. http://doi.org/10.1038/nature08197

Flint HJ, Scott KP, Duncan SH, Louis P, Forano E (2012) Microbial degradation of complex carbohydrates in the gut. Gut Microbes 3(4):289. https://doi.org/10.4161/GMIC.19897

Fluitman KS, Wijdeveld M, Nieuwdorp M, IJzerman RG (2018) Potential of butyrate to influence food intake in mice and men. Gut 67(7):1203–1204. https://doi.org/10.1136/GUTJNL-2017-315543

Frayling TM (2014) Genome-wide association studies: the good, the bad and the ugly. Clin Med 14(4):428. https://doi.org/10.7861/CLINMEDICINE.14-4-428

Freitas RDS, Campos MM (2019) Protective effects of omega-3 fatty acids in cancer-related complications. Nutrients 11(5). http://doi.org/10.3390/NU11050945

Frommherz L, Bub A, Hummel E, Rist MJ, Roth A, Watzl B, Kulling SE (2016) Age-related changes of plasma bile acid concentrations in healthy adults-results from the cross-sectional Karmen study. PLoS ONE 11(4). http://doi.org/10.1371/JOURNAL.PONE.0153959

Fuhler GM (2020) The immune system and microbiome in pregnancy. Best Pract Res Clin Gastroenterol 44–45:101671. https://doi.org/10.1016/J.BPG.2020.101671

Fymat AL (2017) Genetics, epigenetics and cancer. Cancer Ther Oncol Int J 4(2). http://doi.org/10.19080/ctoij.2017.04.555634

Gabanyi I, Lepousez G, Wheeler R, Vieites-Prado A, Nissant A, Wagner S, Moigneu C, Dulauroy S, Hicham S, Polomack B, Verny F, Rosenstiel P, Renier N, Boneca IG, Eberl G, Lledo P-M (2022) Bacterial sensing via neuronal Nod2 regulates appetite and body temperature. Science 376(6590). http://doi.org/10.1126/SCIENCE.ABJ3986

García-Giménez JL, Seco-Cervera M, Tollefsbol TO, Romá-Mateo C, Peiró-Chova L, Lapunzina P, Pallardó FV (2017) Epigenetic biomarkers: current strategies and future challenges for their use in the clinical laboratory. Crit Rev Clin Lab Sci 54(7–8):529. https://doi.org/10.1080/10408363.2017.1410520

Garcia-Perez I, Posma JM, Chambers ES, Mathers JC, Draper J, Beckmann M, Nicholson JK, Holmes E, Frost G (2020) Dietary metabotype modelling predicts individual responses to dietary interventions. Nat Food 1(6):355–364. https://doi.org/10.1038/S43016-020-0092-Z

Garmany A, Yamada S, Terzic A (2021) Longevity leap: mind the healthspan gap. npj Regenerative Med 6(1):1–7. http://doi.org/10.1038/s41536-021-00169-5

Gelato KA, Fischle W (2008) Role of histone modifications in defining chromatin structure and function. Biol Chem 389(4):353–363. https://doi.org/10.1515/BC.2008.048
Gems D (2011) Aging: to treat or not to treat? Am Sci 99(4):278–280. https://doi.org/10.1511/2011.91.278
Gomez-Cabrero D, Walter S, Abugessaisa I, Miñambres-Herraiz R, Palomares LB, Butcher L, Erusalimsky JD, Garcia-Garcia FJ, Carnicero J, Hardman TC, Mischak H, Zürbig P, Hackl M, Grillari J, Fiorillo E, Cucca F, Cesari M, Carrie I, Colpo M et al (2021) A robust machine learning framework to identify signatures for frailty: a nested case-control study in four aging European cohorts. GeroScience 43(3):1317–1329. http://doi.org/10.1007/S11357-021-00334-0
Goodall EF, Leach V, Wang C, Cooper-Knock J, Heath PR, Baker D, Drew DR, Jill Saffrey M, Simpson JE, Romero IA, Wharton SB (2019) Age-associated mRNA and miRNA expression changes in the blood-brain barrier. Int J Mol Sci 20(12). http://doi.org/10.3390/ijms20123097
Grabowski M, Barski J, Liśkiewicz D (2020) [Metabolic phenotyping in the development personalized nutrition]. Postepy Biochemii 66(2). http://doi.org/10.18388/PB.2020_329
Gruendler R, Hippe B, Sendula Jengic V, Peterlin B, Haslberger AG (2020) Nutraceutical approaches of autophagy and neuroinflammation in Alzheimer's disease: a systematic review. Molecules (Basel, Switzerland) 25(24). http://doi.org/10.3390/molecules25246018
Guarente L (2011) Sirtuins, aging, and medicine. N Engl J Med 364(23):2235–2244. http://doi.org/10.1056/NEJMra1100831
Guarente L, Guarente L (2007) Sirtuins in aging and disease. Cold Spring Harb Symp Quant Biol 72:483–488. http://doi.org/10.1101/sqb.2007.72.024
Guasch-Ferre M, Bhupathiraju SN, Hu FB (2018) Use of metabolomics in improving assessment of dietary intake. Clin Chem 64(1):82. https://doi.org/10.1373/CLINCHEM.2017.272344
Haigis MC, Sinclair DA (2010) Mammalian Sirtuins: biological insights and disease relevance. Annu Rev Pathol 5(1):253–295. https://doi.org/10.1146/annurev.pathol.4.110807.092250
Haithcock E, Dayani Y, Neufeld E, Zahand AJ, Feinstein N, Mattout A, Gruenbaum Y, Liu J (2005) Age-related changes of nuclear architecture in Caenorhabditis elegans. Proc Natl Acad Sci U S A 102(46):16690–16695. https://doi.org/10.1073/pnas.0506955102
Hall JA, Dominy JE, Lee Y, Puigserver P (2013) The Sirtuin family's role in aging and age-associated pathologies. J Clin Investig 123(3):973–979. https://doi.org/10.1172/JCI64094
Hannum G, Guinney J, Zhao L, Zhang L, Hughes G, Sadda SV, Klotzle B, Bibikova M, Fan JB, Gao Y, Deconde R, Chen M, Rajapakse I, Friend S, Ideker T, Zhang K (2013) Genome-wide methylation profiles reveal quantitative views of human aging rates. Mol Cell 49(2):359–367. https://doi.org/10.1016/J.MOLCEL.2012.10.016
Haslberger A (2009) Epigenetics and human health: linking hereditary, environmental and nutritional aspects. Wiley-VCH
Hassan FU, Rehman MSU, Khan MS, Ali MA, Javed A, Nawaz A, Yang C (2019) Curcumin as an alternative epigenetic modulator: mechanism of action and potential effects. Front Genet 10. http://doi.org/10.3389/FGENE.2019.00514
Henry NL, Hayes DF (2012) Cancer biomarkers. Mol Oncol 6(2):140. https://doi.org/10.1016/J.MOLONC.2012.01.010
Herforth A, Arimond M, Álvarez-Sánchez C, Coates J, Christianson K, Muehlhoff E (2019) A global review of food-based dietary guidelines. Adv Nutr (Bethesda, Md.) 10(4):590–605. http://doi.org/10.1093/ADVANCES/NMY130
Hernández MAG, Canfora EE, Jocken JWE, Blaak EE (2019) The short-chain fatty acid acetate in body weight control and insulin sensitivity. Nutrients 11(8). http://doi.org/10.3390/NU11081943
Herskind AM, McGue M, Holm NV, Sørensen TIA, Harvald B, Vaupel JW (1996) The heritability of human longevity: a population-based study of 2872 Danish twin pairs born 1870–1900. Hum Genet 97(3):319–323. https://doi.org/10.1007/BF02185763
Hillesheim E, Brennan L (2020) Metabotyping and its role in nutrition research. Nutr Res Rev 33(1):33–42. https://doi.org/10.1017/S0954422419000179

Hillesheim E, Ryan MF, Gibney E, Roche HM, Brennan L (2020) Optimisation of a metabotype approach to deliver targeted dietary advice. Nutr Metab 17(1):1–12. https://doi.org/10.1186/S12986-020-00499-Z/FIGURES/2

Hippe B, Zwielehner J, Liszt K, Lassl C, Unger F, Haslberger AG (2011) Quantification of butyryl CoA: acetate CoA-transferase genes reveals different butyrate production capacity in individuals according to diet and age. FEMS Microbiol Lett 316(2):130–135. https://doi.org/10.1111/j.1574-6968.2010.02197.x

Holscher HD (2017) Dietary fiber and prebiotics and the gastrointestinal microbiota. Gut Microbes 8(2):172. https://doi.org/10.1080/19490976.2017.1290756

Horne JR, Nielsen DE, Madill J, Robitaille J, Vohl MC, Mutch DM (2022). Guiding global best practice in personalized nutrition based on genetics: the development of a nutrigenomics care map. J Acad Nutr Diet 122:259–268, 269. https://doi.org/10.1016/j.jand.2021.02.008

Horvath S (2013) DNA methylation age of human tissues and cell types. Genome Biol 14(10):1–20. https://doi.org/10.1186/GB-2013-14-10-R115/COMMENTS

Hosseini-Esfahani F, Mirmiran P, Daneshpour MS, Mehrabi Y, Hedayati M, Soheilian-Khorzoghi M, Azizi F (2015) Dietary patterns interact with APOA1/APOC3 polymorphisms to alter the risk of the metabolic syndrome: the Tehran lipid and glucose study. Br J Nutr 113(4):644–653. https://doi.org/10.1017/S0007114514003687

How are biomarkers used to treat cancer? | MD Anderson Cancer Center (n.d.). Retrieved April 19, 2022, from https://www.mdanderson.org/cancerwise/how-are-biomarkers-used-in-cancer-treatment.h00-159460056.html

Huan T, Chen G, Liu C, Bhattacharya A, Rong J, Chen BH, Seshadri S, Tanriverdi K, Freedman JE, Larson MG, Murabito JM, Levy D (2018) Age-associated microRNA expression in human peripheral blood is associated with all-cause mortality and age-related traits. Aging Cell 17(1). http://doi.org/10.1111/acel.12687

Huang ZB, Zhang HT, Yu B, Yu DH (2021) Cell-free DNA as a liquid biopsy for early detection of gastric cancer. Oncol Lett 21(1). http://doi.org/10.3892/OL.2020.12264

Huarte M (2015) The emerging role of lncRNAs in cancer. Nat Med 21(11):1253–1261. http://doi.org/10.1038/nm.3981

Hughes RL, Kable ME, Marco M, Keim NL (2019) The role of the gut microbiome in predicting response to diet and the development of precision nutrition models. Part II: results. Adv Nutr 10(6):979. https://doi.org/10.1093/ADVANCES/NMZ049

Hwang JY, Aromolaran KA, Zukin RS (2017) The emerging field of epigenetics in neurodegeneration and neuroprotection. Nat Rev Neurosci 18(6):347. https://doi.org/10.1038/NRN.2017.46

Jardon KM, Canfora EE, Goossens GH, Blaak EE (2022) Dietary macronutrients and the gut microbiome: a precision nutrition approach to improve cardiometabolic health. Gut. gutjnl-2020-323715. http://doi.org/10.1136/GUTJNL-2020-323715

Jiang-hua Q, De-chuang J, Zhen-duo L, Shu-de C, Zhenzhen L (2014) Association of methylenetetrahydrofolate reductase and methionine synthase polymorphisms with breast cancer risk and interaction with folate, vitamin B6, and vitamin B12 intakes. Tumour Biol J Int Soc Oncodevelopmental Biol Med 35(12):11895–11901. https://doi.org/10.1007/S13277-014-2456-1

Jin H, Zhang Y, Ding Q, Wang SS, Rastogi P, Dai DF, Lu D, Purvis M, Cao C, Wang A, Liu D, Ren C, Elhadi S, Hu MC, Chai Y, Zepeda-Orozco D, Campisi J, Attanasio M (2019) Epithelial innate immunity mediates tubular cell senescence after kidney injury. JCI Insight 4(2). http://doi.org/10.1172/JCI.INSIGHT.125490

Juengst ET (2000) Concepts of disease after the human genome project. In: Ethical issues in health care on the frontiers of the twenty-first century, pp 127–154. https://doi.org/10.1007/0-306-46879-4_7

Kasper N, Mandell C, Ball S, Miller AL, Lumeng J, Peterson KE (2016) The healthy meal index: a tool for measuring the healthfulness of meals served to children. Appetite 103:54. https://doi.org/10.1016/J.APPET.2016.02.160

Kastenmüller G, Raffler J, Gieger C, Suhre K (2015) Genetics of human metabolism: an update. Hum Mol Genet 24(R1):R93–R101. https://doi.org/10.1093/HMG/DDV263

Keenan CR, Allan RS (2019) Epigenomic drivers of immune dysfunction in aging. Aging Cell 18(1). http://doi.org/10.1111/ACEL.12878

Kenyon C, Chang J, Gensch E, Rudner A, Tabtiang R (1993) A C. elegans mutant that lives twice as long as wild type. Nature 366(6454):461–464. http://doi.org/10.1038/366461a0

Khalilisamani N, Thomson PC, Raadsma HW, Khatkar MS (2022) Estimating heritability using family-pooled phenotypic and genotypic data: a simulation study applied to aquaculture. Heredity 128(3):178–186. http://doi.org/10.1038/s41437-022-00502-8

Kinser HE, Pincus Z (2020) MicroRNAs as modulators of longevity and the aging process. Hum Genet 139(3):291–308. Springer. http://doi.org/10.1007/s00439-019-02046-0

Klutstein M, Reizel Y, Galow A-M, Peleg S (2022) How to slow down the ticking clock: age-associated epigenetic alterations and related interventions to extend life span. Cells 11(3):468. http://doi.org/10.3390/CELLS11030468

Kolda TG, Bader BW (2009) Tensor decompositions and applications. Soc Ind Appl Math 51(3):455–500. https://doi.org/10.1137/07070111X

Kolodziejczyk AA, Zheng D, Elinav E (n.d.) Diet-microbiota interactions and personalized nutrition. Nat Rev Microbiol. http://doi.org/10.1038/s41579-019-0256-8

Kolodziejczyk AA, Zheng D, Elinav E (2019) Diet–microbiota interactions and personalized nutrition. Nat Rev Microbiol 17(12):742–753. http://doi.org/10.1038/s41579-019-0256-8

Koren O, Goodrich JK, Cullender TC, Spor A, Laitinen K, Kling Bäckhed H, Gonzalez A, Werner JJ, Angenent LT, Knight R, Bäckhed F, Isolauri E, Salminen S, Ley RE (2012) Host remodeling of the gut microbiome and metabolic changes during pregnancy. Cell 150(3):470–480. https://doi.org/10.1016/j.cell.2012.07.008

Lacal I, Ventura R (2018) Epigenetic Inheritance: concepts, mechanisms and perspectives. Front Mol Neurosci 11. https://doi.org/10.3389/FNMOL.2018.00292

Lazarus A, Banerjee KK, Kolthur-Seetharam U (2013) Small changes, big effects: chromatin goes aging. Subcell Biochem 61:151–176. https://doi.org/10.1007/978-94-007-4525-4_8

Lee B, Zhang S, Poleksic A, Xie L (2020) Heterogeneous multi-layered network model for omics data integration and analysis. Front Genet 10:1381. https://doi.org/10.3389/FGENE.2019.01381/BIBTEX

Leeming ER, Johnson AJ, Spector TD, Roy CIL (2019) Effect of diet on the gut microbiota: rethinking intervention duration. Nutrients 11(12). http://doi.org/10.3390/NU11122862

Legube G, Trouche D (2003) Regulating histone acetyltransferases and deacetylases. EMBO Rep 4(10):944. https://doi.org/10.1038/SJ.EMBOR.EMBOR941

Li X, Watanabe K, Kimura I (2017) Gut microbiota dysbiosis drives and implies novel therapeutic strategies for diabetes mellitus and related metabolic diseases. Front Immunol 8. http://doi.org/10.3389/FIMMU.2017.01882

Li Y, Song Y, Wang Z, Zhang Z, Lu M, Wang Y (2019) Long non-coding RNA LINC01787 drives breast cancer progression via disrupting miR-125b generation. Front Oncol 9. https://doi.org/10.3389/fonc.2019.01140

Longo VD, Kennedy BK (2006) Sirtuins in aging and age-related disease. Cell 126(2):257–268. http://doi.org/10.1016/j.cell.2006.07.002

López-Cortegano E, Caballero A (2019) Inferring the nature of missing heritability in human traits using data from the GWAS catalog. Genetics 212(3):891–904. https://doi.org/10.1534/GENETICS.119.302077

López-Otín C, Blasco MA, Partridge L, Serrano M, Kroemer G (2013) The hallmarks of aging. Cell 153(6):1194. http://doi.org/10.1016/j.cell.2013a.05.039

Lozupone CA, Stombaugh JI, Gordon JI, Jansson JK, Knight R (2012) Diversity, stability and resilience of the human gut microbiota. Nature 489(7415):220. https://doi.org/10.1038/NATURE11550

Lu L, Zhang J, Xie Y, Gao F, Xu S, Wu X, Ye Z (2020) Wearable health devices in health care: narrative systematic review. JMIR MHealth UHealth 8(11). http://doi.org/10.2196/18907

Mahmoud AM (2022) An overview of epigenetics in obesity: the role of lifestyle and therapeutic interventions. Int J Mol Sci 23(3). http://doi.org/10.3390/IJMS23031341

Manolio TA, Collins FS, Cox NJ, Goldstein DB, Hindorff LA, Hunter DJ, McCarthy MI, Ramos EM, Cardon LR, Chakravarti A, Cho JH, Guttmacher AE, Kong A, Kruglyak L, Mardis E, Rotimi CN, Slatkin M, Valle D, Whittemore AS et al (2009) Finding the missing heritability of complex diseases. Nature 461(7265):747–753. http://doi.org/10.1038/nature08494

Márquez-Quiñones A, Mutch DM, Debard C, Wang P, Combes M, Roussel B, Holst C, Martinez JA, Handjieva-Darlenska T, Kalouskova P, Jebb S, Babalis D, Pfeiffer AFH, Larsen TM, Astrup A, Saris WHM, Mariman E, Clément K, Vidal H et al (2010) Adipose tissue transcriptome reflects variations between subjects with continued weight loss and subjects regaining weight 6 mo after caloric restriction independent of energy intake. Am J Clin Nutr 92(4):975–984. http://doi.org/10.3945/AJCN.2010.29808

Martin GM (2005) Epigenetic drift in aging identical twins. Proc Natl Acad Sci U S A 102(30):10413–10414. https://doi.org/10.1073/pnas.0504743102

Martin AM, Sun EW, Rogers GB, Keating DJ (2019) The influence of the gut microbiome on host metabolism through the regulation of gut hormone release. Front Physiol 10:428. http://doi.org/10.3389/FPHYS.2019.00428/BIBTEX

Mazzoli R, Pessione E (2016) The neuro-endocrinological role of microbial glutamate and GABA signaling. Front Microbiol 7:1934. http://doi.org/10.3389/FMICB.2016.01934/BIBTEX

McGregor RA, Seo DY (2016) miRNAs as nutritional targets in aging. In: Molecular basis of nutrition and aging: a volume in the molecular nutrition series. Elsevier Inc., pp 277–291. http://doi.org/10.1016/B978-0-12-801816-3.00021-2

McOrist AL, Miller RB, Bird AR, Keogh JB, Noakes M, Topping DL, Conlon MA (2011) Fecal butyrate levels vary widely among individuals but are usually increased by a diet high in resistant starch. J Nutr 141(5):883–889. https://doi.org/10.3945/JN.110.128504

Melzer D, Pilling LC, Ferrucci L (2020) The genetics of human ageing. Nat Rev Genet 21(2):88–101. https://doi.org/10.1038/S41576-019-0183-6

Menzel PT (2012) Justice and fairness: a critical element in U.S. health system reform. J Law Med Ethics 40(3):582–597. http://doi.org/10.1111/J.1748-720X.2012.00691.X

Ming DK, Sangkaew S, Chanh HQ, Nhat PTH, Yacoub S, Georgiou P, Holmes AH (2020) Continuous physiological monitoring using wearable technology to inform individual management of infectious diseases, public health and outbreak responses. Int J Infect Dis 96:648. https://doi.org/10.1016/J.IJID.2020.05.086

Modrackova N, Copova I, Stovicek A, Makovska M, Schierova D, Mrazek J, Sabolova M, Vlkova E, Hradsky O, Bronsky J, Nevoral J, Neuzil-Bunesova V (2021) Microbial shifts of faecal microbiota using enteral nutrition in vitro. J Funct Foods 77:104330. https://doi.org/10.1016/J.JFF.2020.104330

Molina-Serrano D, Kyriakou D, Kirmizis A (2019) Histone modifications as an intersection between diet and longevity. Front Genet 10. http://doi.org/10.3389/FGENE.2019a.00192

Moradi Sarabi M, Zahedi SA, Pajouhi N, Khosravi P, Bagheri S, Ahmadvand H, Shahryarhesami S (2018) The effects of dietary polyunsaturated fatty acids on miR-126 promoter DNA methylation status and VEGF protein expression in the colorectal cancer cells. Genes Nutr 13(1). http://doi.org/10.1186/S12263-018-0623-5

Morris BJ (2013) Seven sirtuins for seven deadly diseases of aging. Free Radical Biol Med 56:133–171. https://doi.org/10.1016/j.freeradbiomed.2012.10.525

Morris BJ, Willcox BJ, Donlon TA (2019) Genetic and epigenetic regulation of human aging and longevity. Biochim Biophys Acta (BBA) Mol Basis Dis 1865(7):1718–1744. http://doi.org/10.1016/J.BBADIS.2018.08.039

Muñoz-Najar U, Sedivy JM (2011) Epigenetic control of aging. In: Antioxidants and redox signaling, vol 14, Issue 2, pp 241–259. http://doi.org/10.1089/ars.2010.3250

Mutch DM, Temanni MR, Henegar C, Combes F, Pelloux V, Holst C, Sørensen TIA, Astrup A, Martinez JA, Saris WHM, Viguerie N, Langin D, Zucker JD, Clément K (2007) Adipose gene expression prior to weight loss can differentiate and weakly predict dietary responders. PLoS ONE 2(12). http://doi.org/10.1371/JOURNAL.PONE.0001344

Mutch DM, Pers TH, Temanni MR, Pelloux V, Marquez-Quinõnes A, Holst C, Martinez JA, Babalis D, van Baak MA, Handjieva-Darlenska T, Walker CG, Astrup A, Saris WHM, Langin D, Viguerie N, Zucker JD, Clement K (2011) A distinct adipose tissue gene expression response to caloric restriction predicts 6-mo weight maintenance in obese subjects. Am J Clin Nutr 94(6):1399–1409. https://doi.org/10.3945/AJCN.110.006858

Nebbioso A, Tambaro FP, Dell'Aversana C, Altucci L (2018) Cancer epigenetics: moving forward. PLOS GEnet 14(6):e1007362. https://doi.org/10.1371/JOURNAL.PGEN.1007362

Nicholson JK, Holmes E, Kinross J, Burcelin R, Gibson G, Jia W, Pettersson S (2012) Host-gut microbiota metabolic interactions. Science (New York, N.Y.) 336(6086):1262–1267. http://doi.org/10.1126/SCIENCE.1223813

Olafuyi O, Parekh N, Wright J, Koenig J (2021) Inter-ethnic differences in pharmacokinetics—is there more that unites than divides? Pharmacol Res Perspect 9(6):e00890. https://doi.org/10.1002/PRP2.890

Olshansky SJ (2018) From lifespan to healthspan. JAMA 320(13):1323–1324. https://doi.org/10.1001/jama.2018.12621

Ordovas JM, Ferguson LR, Tai ES, Mathers JC (2018) Personalised nutrition and health. BMJ 361. https://doi.org/10.1136/BMJ.K2173

O'Sullivan RJ, Karlseder J (2012) The great unravelling: chromatin as a modulator of the aging process. Trends Biochem Sci 37(11):466–476. http://doi.org/10.1016/j.tibs.2012.08.001

Padin AC, Hébert JR, Woody A, Wilson SJ, Shivappa N, Belury MA, Malarkey WB, Sheridan JF, Kiecolt-Glaser JK (2019) A proinflammatory diet is associated with inflammatory gene expression among healthy, non-obese adults: can social ties protect against the risks? Brain Behav Immun 82:36. https://doi.org/10.1016/J.BBI.2019.07.031

Pal S, Tyler JK (2016b) Epigenetics and aging. Sci Adv 2(7). http://doi.org/10.1126/sciadv.1600584

Palacín-aliana I, García-romero N, Asensi-puig A, Carrión-navarro J, González-rumayor V, Ayuso-sacido Á (2021) Clinical utility of liquid biopsy-based actionable mutations detected via ddPCR. Biomedicines 9(8). http://doi.org/10.3390/BIOMEDICINES9080906

Palmnäs M, Brunius C, Shi L, Rostgaard-Hansen A, Torres NE, González-Domínguez R, Zamora-Ros R, Ye YL, Halkjær J, Tjønneland A, Riccardi G, Giacco R, Costabile G, Vetrani C, Nielsen J, Andres-Lacueva C, Landberg R (2020) Perspective: metabotyping—a potential personalized nutrition strategy for precision prevention of cardiometabolic disease. Adv Nutr 11(3):524–532. https://doi.org/10.1093/ADVANCES/NMZ121

Panchal SK, Brown L (2020) DNA methylation in adipose tissue and metabolic syndrome. J Clin Med 9(9):1–4. https://doi.org/10.3390/JCM9092699

Panzeri I, Pospisilik JA (2018) Epigenetic control of variation and stochasticity in metabolic disease. Mol Metab 14:26–38. https://doi.org/10.1016/J.MOLMET.2018.05.010

Pokorska-Bocci A, Stewart A, Sagoo GS, Hall A, Kroese M, Burton H (2014) "Personalized medicine": what's in a name? Pers Med 11(2):197–210. https://doi.org/10.2217/PME.13.107/ASSET/IMAGES/LARGE/FIGURE2.JPEG

Poulsen P, Esteller M, Vaag A, Fraga MF (2007) The epigenetic basis of twin discordance in age-related diseases. Pediatr Res 61:38R–42R. Retrieved March 25, 2020, from https://www.google.com/search?q=P.+Poulsen%25252C+M.+Esteller%25252C+A.+Vaag%25252C+M.+F.+Fraga%25252C+The+epigenetic+basis+of+twin+discordance+in+age-related+diseases.+Pediatr.+Res.+61%25252C+38R-42R+(2007).&rlz=1C1CHBF_deAT866AT867&oq=P.+Poulsen%25252C+M.+Esteller%25

Randhawa V, Kumar M (2021) Analysis of aging-related protein interactome and cross-network module comparisons across tissues provide new insights into aging. Comput Biol Chem 92. https://doi.org/10.1016/J.COMPBIOLCHEM.2021.107506

Reinders MJ, Bouwman EP, van den Puttelaar J, Verain MCD (2020) Consumer acceptance of personalised nutrition: the role of ambivalent feelings and eating context. PLoS ONE 15(4). http://doi.org/10.1371/JOURNAL.PONE.0231342

Ren C, An G, Zhao C, Ouyang Z, Bo X, Shu W (2018) Lnc2Catlas: an atlas of long noncoding RNAs associated with risk of cancers. Sci Rep 8(1):1–8. https://doi.org/10.1038/s41598-018-20232-4

Riedl A, Wawro N, Gieger C, Meisinger C, Peters A, Rathmann W, Koenig W, Strauch K, Quante AS, Thorand B, Huth C, Daniel H, Hauner H, Linseisen J (2020) Modifying effect of metabotype on diet-diabetes associations. Eur J Nutr 59(4):1357–1369. https://doi.org/10.1007/S00394-019-01988-5

Riggs AD, Porter TN (1996) Overview of epigenetic mechanisms. Cold Spring Harb Monogr Arch 32:29–45. https://doi.org/10.1101/0.29-45

Rist MJ, Roth A, Frommherz L, Weinert CH, Krüger R, Merz B, Bunzel D, Mack C, Egert B, Bub A, Görling B, Tzvetkova P, Luy B, Hoffmann I, Kulling SE, Watzl B (2017) Metabolite patterns predicting sex and age in participants of the Karlsruhe Metabolomics and Nutrition (KarMeN) study. PLoS One 12(8). http://doi.org/10.1371/JOURNAL.PONE.0183228

Ruhl GL, Hazel JW, Clayton EW, Malin BA (2019) Public attitudes toward direct to consumer genetic testing. In: AMIA annual symposium proceedings, p 774. /pmc/articles/PMC7153088/

Salonen A (2014) Impact of diet and individual variation on intestinal microbiota composition and fermentation products in obese men. ISME J 8:2218–2230

Samblas M, Milagro FI, Gómez-Abellán P, Martínez JA, Garaulet M (2016) Methylation on the circadian gene BMAL1 is associated with the effects of a weight loss intervention on serum lipid levels. J Biol Rhythms 31(3):308–317. https://doi.org/10.1177/0748730416629247

Samblas M, Milagro FI, Martínez A (2019) DNA methylation markers in obesity, metabolic syndrome, and weight loss. Epigenetics 14(5):421. https://doi.org/10.1080/15592294.2019.1595297

Sargent M (2010) Why twins age differently. Nature 464(7292):1130–1131. https://doi.org/10.1038/4641130a

Sato Y, Nagasaki M, Nakai N, Fushimi T (2003) Physical exercise improves glucose metabolism in lifestyle-related diseases. Exp Biol Med 228(10):1208–1212. https://doi.org/10.1177/153537020322801017

She J, Wong CC, Yu J (2021) Targeted prebiotics alter the obese gut microbiome in humans. Signal Transduct Targeted Ther 6(1):1–2. http://doi.org/10.1038/s41392-021-00758-2

Shigeyasu K, Toden S, Zumwalt TJ, Okugawa Y, Goel A (2017) Emerging role of microRNAs as liquid biopsy biomarkers in gastrointestinal cancers. https://doi.org/10.1158/1078-0432.CCR-16-1676

Shin J, Noh JR, Choe D, Lee N, Song Y, Cho S, Kang EJ, Go MJ, Ha SK, Chang DH, Kim JH, Kim YH, Kim KS, Jung H, Kim MH, Sung BH, Lee SG, Lee DH, Kim BC et al (2021) Ageing and rejuvenation models reveal changes in key microbial communities associated with healthy ageing. Microbiome 9(1):1–19. http://doi.org/10.1186/S40168-021-01189-5/FIGURES/6

Shulpekova Y, Shirokova E, Zharkova M, Tkachenko P, Tikhonov I, Stepanov A, Sinitsyna A, Izotov A, Butkova T, Shulpekova N, Nechaev V, Damulin I, Okhlobystin A, Ivashkin V (2022) A recent ten-year perspective: bile acid metabolism and signaling. Molecules 27(6):1983. http://doi.org/10.3390/MOLECULES27061983

Sidler C, Kovalchuk O, Kovalchuk I (2017) Epigenetic regulation of cellular senescence and aging. Front Genet 8. http://doi.org/10.3389/FGENE.2017.00138

Silva YP, Bernardi A, Frozza RL (2020) The role of short-chain fatty acids from gut microbiota in gut-brain communication. Front Endocrinol 11:25. https://doi.org/10.3389/FENDO.2020.00025

Simpson HL, Campbell BJ (2015) Review article: dietary fibre–microbiota interactions. Aliment Pharmacol Ther 42(2):158. https://doi.org/10.1111/APT.13248

Skinner MK (2015) Environmental epigenetics and a unified theory of the molecular aspects of evolution: a neo-Lamarckian concept that facilitates neo-Darwinian evolution. Genome Biol Evol 7(5):1296–1302. https://doi.org/10.1093/GBE/EVV073

Snow A, Chen D, Lang JE (2019) The current status of the clinical utility of liquid biopsies in cancer. Expert Rev Mol Diagn 19(11):1031. https://doi.org/10.1080/14737159.2019.1664290

Soriano-Tárraga C, Jiménez-Conde J, Roquer J (2019) Epigenetics and aging. In: Handbook of nutrition, diet, and epigenetics, vol 2, pp 1413–1433. https://doi.org/10.1007/978-3-319-55530-0_123

Stromsnes K, Correas AG, Lehmann J, Gambini J, Olaso-gonzalez G (2021) Anti-inflammatory properties of diet: role in healthy aging. Biomedicines 9(8):922. http://doi.org/10.3390/BIOMEDICINES9080922

Szafranski K, Abraham KJ, Mekhail K (2015) Non-coding RNA in neural function, disease, and aging. Front Genet 6. http://doi.org/10.3389/fgene.2015.00087

Szakály Z, Kovács B, Szakály M, T Nagy-Pető D, Popovics P, Kiss M (2021) Consumer acceptance of genetic-based personalized nutrition in Hungary. Genes Nutr 16(1). http://doi.org/10.1186/S12263-021-00683-7

Szczyrek M, Bitkowska P, Chunowski P, Czuchryta P, Krawczyk P, Milanowski J (2021) Diet, Microbiome, and cancer immunotherapy—a comprehensive review. Nutrients 13(7):2217. http://doi.org/10.3390/NU13072217

Tiffon C (2018) The impact of nutrition and environmental epigenetics on human health and disease. Int J Mol Sci 19(11). http://doi.org/10.3390/IJMS19113425

Tollefsbol TO (2014) Dietary epigenetics in cancer and aging. Cancer Treat Res 159:257–267. https://doi.org/10.1007/978-3-642-38007-5_15

Toomey S, Eustace AJ, Pritzker LB, Pritzker KPH, Fay J, O'Grady A, Cummins R, Grogan L, Kennedy J, O'Connor D, Young L, Kay EW, O'Donovan N, Gallagher WM, Kalachand R, Crown J, Hennessy BT (2016) RE: RNA disruption assay as a biomarker of pathological complete response in Neoadjuvant Trastuzumab-treated human epidermal growth factor receptor 2-positive breast cancer. J Natl Cancer Inst 108(8):7–8. https://doi.org/10.1093/jnci/djw111

Topart C, Werner E, Arimondo PB (2020) Wandering along the epigenetic timeline. Clin Epigenetics 12(1):1–13. https://doi.org/10.1186/S13148-020-00893-7/TABLES/1

van der Lans IA et al (2016) Willingness to pay for personalised nutrition across Europe.http://doi.org/10.1093/eurpub/ckw045

Vijayan V, Connolly J, Condell J, McKelvey N, Gardiner P (2021) Review of wearable devices and data collection considerations for connected health. Sensors 21(16):5589. http://doi.org/10.3390/S21165589

Vinke PC, el Aidy S, van Dijk G (2017) The role of supplemental complex dietary carbohydrates and gut microbiota in promoting cardiometabolic and immunological health in obesity: lessons from healthy non-obese individuals. Front Nutr 4:34. https://doi.org/10.3389/FNUT.2017.00034

Wagner W (2019) The link between epigenetic clocks for aging and senescence. Front Genet 10:303. http://doi.org/10.3389/fgene.2019.00303

Wang DD, Hu FB (2018) Precision nutrition for prevention and management of type 2 diabetes. Lancet Diab Endocrinol 6(5):416–426. https://doi.org/10.1016/S2213-8587(18)30037-8

Wang J, Guleria S, Koffas MAG, Yan Y (2016) Microbial production of value-added nutraceuticals. Curr Opin Biotechnol 37:97–104. http://doi.org/10.1016/j.copbio.2015.11.003

West CE, Renz H, Jenmalm MC, Kozyrskyj AL, Allen KJ, Vuillermin P, Prescott SL (2015) The gut microbiota and inflammatory noncommunicable diseases: associations and potentials for gut microbiota therapies. J Allergy Clin Immunol 135(1):3–13. https://doi.org/10.1016/J.JACI.2014.11.012

Williams J, Smith F, Kumar S, Vijayan M, Reddy PH (2017) Are microRNAs true sensors of ageing and cellular senescence? Ageing Res Rev 35:350–363. http://doi.org/10.1016/j.arr.2016.11.008

Wilmanski T, Diener C, Rappaport N, Patwardhan S, Wiedrick J, Lapidus J, Earls JC, Zimmer A, Glusman G, Robinson M, Yurkovich JT, Kado DM, Cauley JA, Zmuda J, Lane NE, Magis AT, Lovejoy JC, Hood L, Gibbons SM et al (2021) Gut microbiome pattern reflects healthy aging and predicts survival in humans. Nat Metab 3(2):274. http://doi.org/10.1038/S42255-021-00348-0

Wilson KL (2005) Integrity matters: linking nuclear architecture to lifespan. Proc Natl Acad Sci U S A 102(52):18767–18768. https://doi.org/10.1073/pnas.0509224102

Wolever TM, Fernandes J, Rao AV (1996) Serum acetate: propionate ratio is related to serum cholesterol in men but not women. J Nutr 126:2790–2797. https://academic.oup.com/jn/article/126/11/2790/4724669

Wolters M, Ahrens J, Romaní-Pérez M, Watkins C, Sanz Y, Benítez-Páez A, Stanton C, Günther K (2019) Dietary fat, the gut microbiota, and metabolic health—a systematic review conducted within the MyNewGut project. Clin Nutr (edinburgh, Scotland) 38(6):2504–2520. https://doi.org/10.1016/J.CLNU.2018.12.024

Woodcox A (2018) Aristotle's theory of aging. Cahiers Des Études Anciennes, LV

Xiao FH, Wang HT, Kong QP (2019) Dynamic DNA methylation during aging: a "prophet" of age-related outcomes. Front Genet 10. http://doi.org/10.3389/fgene.2019.00107

Yeoh YK, Chen Z, Hui M, Wong MCS, Ho WCS, Chin ML, Ng SC, Chan FKL, Chan PKS (2019) Impact of inter- and intra-individual variation, sample storage and sampling fraction on human stool microbial community profiles. PeerJ (1). http://doi.org/10.7717/PEERJ.6172/SUPP-6

Yin J, Ren W, Chen S, Li Y, Han H, Gao J, Liu G, Wu X, Li T, Woo Kim S, Yin Y (2018) Metabolic regulation of methionine restriction in diabetes. Mol Nutr Food Res 62(10):1700951. https://doi.org/10.1002/MNFR.201700951

Youngid AI (2019) Solving the missing heritability problem. https://doi.org/10.1371/journal.pgen.1008222

Zeevi D, Korem T, Zmora N, Israeli D, Rothschild D, Weinberger A, Ben-Yacov O, Lador D, Avnit-Sagi T, Lotan-Pompan M, Suez J, Mahdi JA, Matot E, Malka G, Kosower N, Rein M, Zilberman-Schapira G, Dohnalová L, Pevsner-Fischer M, Bikovsky R, Halpern Z, Elinav E, Segal E (2015) Personalized nutrition by prediction of glycemic responses. Cell 163(5):1079–1094. https://doi.org/10.1016/j.cell.2015.11.001. PMID: 26590418

Zhang F, Gigout S, Liu Y, Wang Y, Hao H, Buckley NJ, Zhang H, Wood IC, Gamper N (2019) Repressor element 1-silencing transcription factor drives the development of chronic pain states. Pain 160(10):2398–2408. https://doi.org/10.1097/J.PAIN.0000000000001633

Zuk O, Hechter E, Sunyaev SR, Lander ES (2012) The mystery of missing heritability: genetic interactions create phantom heritability. Proc Natl Acad Sci U S A 109(4):1193–1198. https://doi.org/10.1073/PNAS.1119675109/SUPPL_FILE/SAPP.PDF

Precise Nutrition and Metabolic Syndrome, Remodeling the Microbiome with Polyphenols, Probiotics, and Postbiotics

6

Djordjevic I. Brizita and Ivanovic Dj. Nevena

Contents

Abstract

Metabolic syndrome (MetS) has become a global epidemic that continues to grow despite numerous efforts to determine the cause of its occurrence and create effective dietary interventions. The unsatisfactory long-term efficacy of otherwise non-personalized dietary recommendations can be explained by significant variations in inter-individual responses to diet and lifestyle intervention. Precise nutrition considers the factors responsible for variations in response to diet to generate a personalized dietary intervention. The gut microbiome, a mediator between diet and the pathogenesis of MetS, is considered an important source of variation that modulates food responses. Given that food is considered a key determinant in microbiome remodeling, it is not surprising that the microbiota is seen as a target for dietary intervention to prevent diseases associated

D. I. Brizita · I. Dj. Nevena (✉)
Department of Bromatology, Faculty of Pharmacy, University of Belgrade, Vojvode Stepe 450, Belgrade, Serbia
e-mail: nevena.ivanovic@pharmacy.bg.ac.rs

A. G. Haslberger (ed.), *Advances in Precision Nutrition, Personalization and Healthy Aging*, https://doi.org/10.1007/978-3-031-10153-3_6

with dysbiosis, such as MetS. This section summarizes the latest findings on the effects of probiotics, prebiotics, and postbiotics as critical components of individually oriented dietary strategies to optimize symbiosis between microbiome and host to prevent or treat MetS.

Abbreviations

BMI	Body mass index;
BP	Blood pressure;
FPG	Fasting plasma glucose concentration;
HDLC	High-density lipoprotein cholesterol;
IDF	International diabetes federation;
IR	Insulin resistance;
MetS	Metabolic syndrome;
NCEP:ATP III	National cholesterol education program expert panel on detection, evaluation, and treatment of high blood cholesterol in adults (adult treatment panel III);
T2D	Type 2 diabetes;
TG	Triglycerides;
WC	Waist circumference;
WHO	World Health Organization.

6.1 Introduction

Metabolic syndrome (MetS) has grown to epidemic proportions and has become one of the major health challenges of the twenty-first century. MetS refers to a set of metabolic disorders that represent the most significant risk factor for developing cardiovascular disease and type 2 diabetes (T2D) and thus the leading cause of increased mortality and morbidity worldwide (McCracken et al. 2018). Due to the high costs of treating Mets complications, emphasis is placed on developing and implementing preventive strategies to reduce the prevalence of MetS. Therefore, many studies have focused on determining the components and dietary patterns to constitute a healthy and beneficial diet. However, current dietary recommendations are based on the average population and often do not consider inter-individual variations in response to diet and lifestyle intervention. As the incidence of metabolic disorders rises, it has become clear that the one-size-fits-all approach is not practical enough (Berry et al. 2020). The development of numerous advanced techniques has provided insight into the causes of these variations, pointing to the role of the gut microbiome, in addition to the human genome, as a potential causative agent responsible for the individual dietary response (Bashiardes et al. 2018). A personalized diet considers the factors that are the source of variation in response to food to generate personalized dietary recommendations (Iizuka and Yabe 2020). The gut microbiota consists of a community of over 100 trillion microorganisms with a mutualistic and symbiotic relationship in the host; it participates in nutrient metabolism, immune system development, intestinal barriers, and host metabolism. Numerous studies have demonstrated an association between perturbation in microbiomes and the development of obesity and MetS (De Filippis et al. 2020; Mills et al. 2019a). Given that food is considered a key determinant in microbiome remodeling, it is not surprising that the microbiota is seen as a target for dietary intervention to prevent diseases associated with dysbiosis, such as MetS (Kviatcovsky et al. 2021). In light of this, probiotics, prebiotics, symbiotic, and postbiotics may represent critical components of individually oriented dietary strategies to alter the microbiome toward a state of homeostasis to prevent and treat metabolic diseases.

6.2 Metabolic Syndrome—Definition, Prevalence, and Pathophysiology

MetS has become one of the significant public health challenges worldwide. The global prevalence of MetS differs depending on the diagnostic criteria used and the geographic and sociodemographic factors, ranging between 20 and 45% (Engin 2017; McCracken et al. 2018). MetS is a complex disorder commonly defined as a cluster of several interrelated metabolic disorders, including abdominal (central) obesity, insulin resistance (IR)/impaired glucose tolerance, hypertension, and dyslipidemia (Alberti et al. 2009; Grundy 2016). MetS are associated with many other clinical conditions, such as pro-inflammatory and pro-thrombotic state, oxidative stress, non-alcoholic fatty liver disease, polycystic ovary syndrome, obstructive sleep apnea, vascular dementia, and several types of cancer (Cornier et al. 2008). The presence of MetS is associated with a twofold increase in the risk of atherosclerotic cardiovascular disease and a fivefold increase in the risk of T2D (Despres et al. 2008; Grundy 2016; Mameli et al. 2017). Accordingly, MetS is a significant risk factor for all-cause mortality (Engin 2017; Yu et al. 2019). The World Health Organization (WHO) first defined MetS in 1998 (Alberti and Zimmet 1998). Subsequently, various definitions with diagnostic criteria have been suggested by different organizations: European Group for the Study of Insulin Resistance (EGIR) (Ferrannini et al. 1997), the National Cholesterol Education Program's Adult Treatment Panel III (NCEP: ATP III) (Expert Panel on Detection 2001), the American Heart Association/National Heart, Lung, and Blood Institute (AHA/NHLBI) (Grundy et al. 2005), and the International Diabetes Federation (IDA) (Alberti et al. 2005). The various diagnostic MetS criteria are listed in Table 6.1.

Although there are subtle differences in the diagnostic criteria given by these organizations regarding focus on different metabolic alterations, all of them describe the same clinical condition. Specifically, the definition reported by WHO

Table 6.1 Different diagnostic criteria for Mets diagnostic

Clinical and biochemical features	WHO 1998	NCEP ATP III 2004	IDF 2005	Consensus (AHA/NHLBI + IDF) 2009
Diagnosis of MetS	IR plus any two of the following risk factors	Presence 3 of 5 risk factors	Obesity plus any two of the following risk factors	IR plus any two of the following risk factors
Obesity	Abdominal obesity (BMI > 30 kg/m^2 or waist-to-hip ratio >0.9 in men; waist-to-hip ratio >0.85 in women,	WC > 102 cm in men; WC > 88 cm in women	BMI > 30 kg/m^2 or WC with ethnicity-specific values[a]	Raised WC (population- and country-specific definitions)
Glucose homeostasis	T2D, or impaired fasting glucose, or impaired glucose tolerance	FPG > 6,1 mmol/l (includes the presence of T2D)	FPG > 5,6 mmol/l (includes the presence of T2D)	FPG > 5,6 mmol/l (includes the presence of T2D)
Hypertension	BP ≥ 140/90 mm Hg	BP ≥ 130/85 mm Hg	≥ 130/85 mm Hg or on antihypertensive medication	≥ 130/85 mm Hg or on antihypertensive medication
Dyslipidemia	TG ≥ 150 mg/dl HDLC < 40 mg/dl in men and HDLC < 50 mg/dl in women	TG ≥ 150 mg/dl HDLC < 40 mg/dl in men and HDLC < 50 mg/dl in women	TG ≥ 150 mg/dl HDLC < 40 mg/dl in men and HDLC < 50 mg/dl in women, or on treatment	TG ≥ 150 mg/dl HDLC < 40 mg/dl in men and HDLC < 50 mg/dl in women, or on treatment
Others	Microalbuminuria			

BMI body mass index; *BP* blood pressure; *FPG* fasting plasma glucose concentration; *HDLC* high-density lipoprotein cholesterol; *IDF* International Diabetes Federation; *IR* insulin resistance; *NCEP ATP III* National cholesterol education program expert panel on detection, evaluation, and treatment of high blood cholesterol in adults (adult treatment panel III); *T2D* type 2 diabetes mellitus; *TG* triglycerides; *WC* waist circumference; *WHO* World Health Organization

[a]Waist circumference: for Europids, >94 cm in men and >80 cm in women; for South Asians, Chinese, and Japanese, >90 cm in men and >80 cm in women; for ethnic South and Central Americans, use South Asian data; for sub-Saharan Africans and Eastern Mediterranean and Middle East (Arab) populations, use European data

emphasizes the significance of IR (identified either as DT-2, impaired glucose tolerance, or impaired fasting glycemia) in MetS diagnosis. Contrarily, NCEP:ATP III criteria do not focus on the central pathology but the co-presence of multiple risk factors. According to these criteria, MetS is present if at least three of the following five metabolic risk factors are met: elevated waist circumference (WC), high blood pressure, lowered HDL-cholesterol (HDLc), elevated triglycerides, and increased fasting glycemia. At last, the focus of IDF criteria is on abdominal obesity, as a central component of MetS, determined by WC, while diagnosis is set if additional two metabolic risk factors are present. In 2009, however, a harmonized consensus definition was made by several organizations where it was agreed that WC might be considered as a useful screening tool for MetS, instead of insisting on compulsory components (Alberti et al. 2009).

The pathogenic mechanism of MetS is complex and multifactorial, with the interplay of genetic susceptibility, epigenetics, metagenomics, and environmental factors, including dietary and lifestyle (Pigeyre et al. 2016). The underlying etiology that could explain the clustering of metabolic disorders remains to be completely elucidated. Insulin resistance, adipocyte dysfunction, and chronic low-grade inflammation together with visceral adipose tissue are supposed to have a significant role in the progression of the syndrome by impairing glucose and lipid homeostasis in all insulin-sensitive tissues, such as the liver, muscle, and adipocytes (Zafar et al. 2018; Zimmet et al. 2019). Adipose tissue is considered to play a central role in the pathophysiology of MetS. It has been recognized as a biologically active endocrine and paracrine organ responsible for producing inflammatory cytokines and non-esterified fatty acids, which link central obesity, IR, inflammation, and atherogenesis (McCracken et al. 2018; Zafar et al. 2018). Although the MetS are closely associated with obesity, about 10–30% of obese individuals, primarily ones with normal WC, present no metabolic abnormalities, which further confirms the role of abdominal adipose tissue in the development of MetS (Latifi et al. 2017; Zeng et al. 2021). However, in multiple human microbiome studies, a close relationship has been demonstrated between alteration in the gut microbiota composition or diversity, known as dysbiosis, and the development of MetS. These accumulated research evidence have focused on the gut microbiome as a pre-eminent target for potential MetS amelioration.

6.3 The Microbiome—Composition, Establishment, and Functions

A microbiota is a complex and dynamic ecosystem composed of symbiotic, commensal, and pathogenic microorganisms (Haque and Haque 2017). It encompasses 10^{13} to 10^{14} resident microorganisms, including bacteria, viruses, archaea, fungi, and protozoans, whose number exceeds ten times the total number of host cells (Gilbert et al. 2018; Singh et al. 2017). Moreover, the genetic content of the human gut microbiota, commonly referred to as the gut microbiome (>100 times larger than the host genome), expands the host's biochemical and metabolic capabilities substantially, thus considered to be host "second genome" (Bäckhed et al. 2005; Norman et al. 2015). Advances in DNA sequencing technologies, coupled with advances in bioinformatics tools, have revealed that the composition of human gut microbiota at species and strains level, including their number, has varied dramatically in an individual, thus being unique for every individual. There are no two individuals who share the same microbiome, including identical twins, although some bacterial species are present in most individuals. It is estimated that the number of species that inhabit the human gut is greater than 1000 (Lozupone et al. 2012). Given high inter-individual variation, defining the "healthy" gut microbiota composition in any age population has been virtually impossible. However, most of these species belong to a limited number of phyla: *Firmicutes* (with belonging Clostridium clusters and members of *Eubacterium*, *Faecalibacterium*, *Roseburia*, and *Ruminococcus*), Bacteriodetes (with belonging genus *Bacteroides* and *Prevotella*), and in a lesser extent to *Actinobacteria* (with belonging *Bifidobacterium* species), *Proteobacteria* (mainly *Escherichia coli*), *Fusobacteria*, and *Verrumicrobia* (includes the genus *Akkermansia*) (Yadav et al. 2018). Furthermore, in 2011 the "enterotype" concept was introduced when metagenomic analysis of fecal microbiota from American, European, and Japanese populations has revealed that microbiota can effectively be subdivided into different enterotypes, each enriched by particular bacterial genera. Three enterotypes have been proposed based on both the relative number of certain bacterial genera isolated from feces and the presence of certain metabolic pathways: *Bacteroides* (enterotype 1), *Prevotella* (enterotype 2), and *Ruminococcus* (enterotype 3). These enterotypes are independent regardless of age, gender, or geography (Arumugam et al. 2011; Clemente et al. 2012). Further analysis has resulted in the identification of only two enterotypes, one dominated by genera *Prevotella* (P-type) and the other by genera *Bacteroides* (B-type). Recently, these two enterotypes have been interpreted as biomarkers of diet since *Prevotella* has been linked to long-term carbohydrate-rich diets and *Bacteroides* to fat- and protein-rich diet (western diet) (Gorvitovskaia et al. 2016; Precup and Vodnar 2019).

The establishment of the human intestinal microbiota begins during or shortly after birth, during exposure of newborns to maternal and environmental microbes. After initial establishment, the microbial population develops rapidly until 2–3 years of age, when microbiota gains in complexity and stability, which are characteristics of adult microbiota. Studies have demonstrated that microbial colonization of the infant's gut is influenced by several factors: delivery mode (vaginal vs. caesarian), sanitary conditions, antibiotic exposure, feeding regime (breastfeeding vs. infant formula), and host genetic. The factor that significantly affects the gut microbiome of a newborn is the delivery mode. Additionally, it has been found that the greatest diversity of the microbiota at this stage is in spontaneously vaginally delivered infants compared to infants delivered by cesarean section (Hill et al. 2017).

Moreover, the cesarean section is responsible for infant microbiota perturbations, which have been associated with immune and metabolic disorders in the adult stage (Kim et al. 2020). Diet also plays an essential role in the development of the infant gut microbiome. For example, human breast milk contains oligosaccharides (e.g., human milk oligosaccharides) which nourish the gut microbiome and lead to a significant increase in the relative abundance of bifidobacteria and lactic acid bacteria, including genera *Lactobacillus* and *Streptococcus*, with more than 70% of strains belonging to bifidobacteria species (Moore and Townsend 2019; Thomson et al. 2018).

Within the first year of life, the microbiota is typically characterized by low species diversity and high instability, dominated mainly by Actinobacteria phylum members. Maturation into adult-like microbiota is driven by cessation of breast-feeding rather than the introduction of solid food drives the (Bäckhed et al. 2015). A stable adult-like microbiota, established between 2–3 years of age, is dominated by *Firmicutes* and *Bacteroidetes* phylum. The composition and activities of healthy adult microbiota are stable over long periods although can be influenced by many factors, including those derived from the host (e.g., genetics, immune, and metabolic regulations) and environmental factors (e.g., diet, stress, physical activity, geographical location, antibiotic usage) (Leeming et al. 2019; Tang et al. 2021). Diet and host genetics are the most significant contributors to variability in microbiota composition and functionality among individuals. Metagenomics and analysis of twins' data have revealed that diet and household cohabitation greatly outweigh the contribution of heritable genetic (Rothschild et al. 2018). According to the data analysis of more than 1000 twins, genetics role in shaping the gut microbiota is on average around 8.8% (Goodrich et al. 2016).

A large amount of data has contributed to a deeper understanding of the symbiotic relationship between the gut microbiota and the host. This relationship is regulated by complex metabolic, immunological, and neuroendocrine interactions that are involved in regulating numerous physiological processes such as energy homeostasis, nutrient metabolism, immune system maturation and function, regulation of structural and morphological maturation of the gastrointestinal tract, resistance to infection, intestinal barrier function, and brain development, function, and behavior (Bäckhed 2012; Frame et al. 2020).

The gut microbiome supplies the host metabolism with enzymes not encoded by the human genome, enabling digestion and energy harvest from otherwise indigestible polysaccharides (i.e., fiber). The major fermented end products of dietary polysaccharides are short-chain fatty acids (SCFAs) and gases. The three most abundant SCFAs, primarily produced in the proximal colon, are acetate, propionate, and butyrate, which account for >95% of SCFA content in the gut (Hu et al. 2018). These SCFAs are involved in several regulatory and cellular processes. SCFAs act as energy substrates, accounting for as much as 10% of host energy requirements. Moreover, SCFAs have been shown to enhance the absorption of dietary minerals such as calcium and water and to serve locally as crucial nourishment. Specifically, butyrate is a vital energy source for colon epithelial cells, involving 70% of their total energy consumption, thus allowing colonic cells to proliferate. Additionally, butyrate contributes to maintaining gut barrier function (Cani 2018; Donohoe et al. 2011). Butyrate also plays an important role in brain function and may prevent carcinogenesis and inflammation in colonocytes (Frame et al. 2020). All three SCFAs are involved in energy homeostasis, energy regulation, and the regulation of immune response (Frame et al. 2020; Mills et al. 2019b). In addition, the gut microbiota significantly enriches the metabolism of glycans, xenobiotics, methanogenesis, and the biosynthesis of isoprenoids (Santos-Marcos et al. 2019). Moreover, the gut microbiota is highly efficient in degrading proteins, peptides, and amino acids. End products of these proteins' fermentations are organic acids, branched-chain fatty acids, and trace amounts of phenols, indoles, ammonia, and amines. Some of these fermentation by-products could have detrimental effects on health (Yadav et al. 2018).

Besides SCFAs, the gut microbiota produces a variety of other nutrients, including B group vitamins and vitamin K, which are essential for their metabolism and the host, maintaining host physiology (Rowland et al. 2018; Yadav et al. 2018). The metabolic function of gut microbiota also includes the bile acids metabolism involved in the digestion and absorption of dietary fats and lipid-soluble nutrients. A small fraction of bile acids that escape enterohepatic circulation reaches the colon, where they are converted into secondary bile acids through the action of bile salt hydrolases secreted by several microbiota bacterial species. These secondary bile acids regulate lipid and cholesterol metabolism by regulating gene expression in the liver and intestines. Moreover, secondary bile acids possess antimicrobial characteristics providing a host mechanism to control bacterial population and protect from infection pathogens (Kho and Lal 2018; Rowland et al. 2018).

Gastrointestinal microbiota profoundly affects the development and functionality of host gut physiological processes, immune system maturation, and immune homeostasis (Parker et al. 2018; Wang et al. 2019). These critical roles of microbiota for the host are seen through the consequence of its absence in germ-free animals. Concerning conventional mice, the gut anatomy of these animals was changed and less effective in nutrient absorption. Moreover, germ-free mice have shown immunodeficiency in terms of immune cells number, production of the immune response mediators (e.g., decreased immunoglobin A and antimicrobial peptides), and deficit in local and systemic local structures. In addition, there was a decrease in the number of specialized epithelial cells that secret mucus resulting in a thinner mucus layer and consequently to the impaired protective barrier (Dieterich et al. 2018; Parker et al. 2018; Zheng et al. 2020).

In addition, it is essential to mention the role of the microbiota as one of the key regulators of the gut–brain axis. Recent findings have indicated that metabolic products of microbiota can act as neurochemicals and affect the peripheral enteric and central nervous systems. One of these products is gamma-aminobutyric acid (GABA), a major inhibitory neurotransmitter in the brain, produced by various strains in the human intestine (Cryan et al. 2019). Furthermore, SCFAs, end products of bacterial fermentation of carbohydrates, exert significant hormone-like activity and interact with the autonomic nervous system. Moreover, SCFAs are involved in behavior modulation (Appleton 2018).

Human microbiota has a vital role in protecting the host from exogenous pathogens and preventing overgrowth of potentially pathogenic microbiota members (pathobionts), referred to as colonization resistance (Kho and Lal 2018; Mills et al. 2019b). This protection of host by microbiota is enabled through an effective biological mechanism which could be direct or indirect. Direct mechanisms include competition of human microbiota and pathogen for shared niches and nutrients and inhibition of pathogens through the production of antimicrobial substances. Indirect mechanisms include modulation of the luminal environment, including decreasing pH during the production of SCFAs, modulation of host immune system, and via host commensal interactions involving epithelium barrier function (Mills et al. 2019b).

6.4 Role of Microbiome in Development of Metabolic Syndrome

Human microbiome studies have linked MetS development to the alternations in the gut microbiota composition or diversity, known as dysbiosis. Given that gut microbiota has a crucial role in regulating various host physiological processes, including energy and immune homeostasis, as well as metabolic function, a dysbiosis could be seen as a trigger for the development of metabolic disorders, including obesity, T2D, dyslipidemia, and non-alcoholic fatty liver disease (NAFLD). In most of these conditions, the mechanisms leading to disease development involve an altered interaction between the gut microbiome, their metabolites, and the host immune system (Green et al. 2020; Ivanovic et al. 2015).

The link between gut microbiota and obesity was initially documented in a germ-free mouse protected against the development of obesity after consuming a high-fat, sugar-rich diet (Western diet) (Bäckhed et al. 2007). Colonization of germ-free mice with a normal microbiota from conventionally raised animals resulted in a 60% increase in body fat and IR without additional food intake or observed differences in energy expenditure (Bäckhed et al. 2004). In terms of microbiota composition, significant differences were found in the relative abundance of two dominant bacterial phyla between genetically obese and lean mice. Compared to their lean counterparts, obese mice have a reduction in 50% in *Bacteroidetes* and a proportional increase in *Firmicutes* abundance, and thus an increase in *Firmicutes*/*Bacteroidetes* (F/B) ratio (Ley et al. 2005). Moreover, such an "obese microbiota" is more efficient in harvesting energy from the diet indicating a causal link between this alteration in the intestinal microbiota and the development of obesity (Turnbaugh et al. 2006). A decrease in the relative proportion of *Bacteroidetes* was also reported in obese people in comparison with lean people, and this proportion was found to increase with weight loss (Armougom et al. 2009; Ley et al. 2006). However, not all studies in humans confirmed the involvement of the F/B ratio in obesity (Finucane et al. 2014; Ley 2010), reporting even contradictory results (Ley 2010). Nevertheless, a recent study has shown that in obese individuals with MetS, dysbiosis was characterized by an increase in F/B ratio compared with obese individuals without MetS, suggesting that the F/B ratio could be more related to the presence or absence of metabolic disturbances rather than the presence of obesity itself (Haro et al. 2017). These inconsistencies between studies could be related to the inter-individual variability in the gut microbiome which is influence by differences in dietary habits, environmental factors, and host genetics (Zeng et al. 2021).

In various studies, several bacterial groups at deeper taxonomic levels (e.g., family, genus, and even species), and individual bacterial species, were directly or inversely associated with metabolic disorders. However, literature data have suggested that loss in microbial richness and bacterial genes and metabolic pathways, including those involved in nutrient digestion, is closely associated with increased calories harvesting and thus with the development of obesity and other metabolic disorders (Santos-Marcos et al. 2019; Vallianou et al. 2019).

Recent evidence has shown that many different microbial metabolites influence host metabolisms, which could cause the development of MetS, including SCFAs, lipopolysaccharide (LPS) of Gram-negative bacteria, trimethylamine *N*-oxide (TMAO), indoxyl sulfate, and p-cresol sulfate (Croci et al. 2021).

Even though several studies reported higher fecal concentrations of total or individual SCFA in obese individuals, their role in obesity development is still unclear (Bäckhed et al. 2004; Fernandes et al. 2014; Teixeira et al. 2013). However, there is agreement from many studies, both in animals and humans, that in obesity, different SCFAs have been presented compared with the non-obese phenotype (Petraroli et al. 2021). Additionally, the gut microbiota of MetS patients is characterized by a reduction in the abundance of several bacterial species with important saccharolytic activity. The different end products of digestion, such as acetate, are used by many gut bacteria to produce propionate and butyrate in a growth-promoting cross-feeding process (Kumar et al. 2020). A reduction in butyrate-producing bacteria may affect the secretion of peripheral hormones, such as insulin, leptin, and ghrelin, influencing appetite control, thus leading to the development of obesity and MetS (Santos-Marcos et al. 2019). Butyrate, besides its role as a signal molecule, is responsible for maintaining microbial homeostasis. The lack of butyrate leads to the inhibition of mitochondrial beta-oxidation in colonocytes, resulting in more accessible oxygen for pathogenic facultative anaerobes such as *E. coli* (Hills et al. 2019). According to metagenomic analysis, the gut microbiota of individuals with T2D has been characterized by moderate gut microbial dysbiosis, a decrease in the abundance of some universal butyrate-producing bacteria, and an increase in various opportunistic pathogens has also been detected (Qin et al. 2012). Furthermore, findings from two population-based cohort studies, which included more than 1600 adult individuals, have shown an association between markers of IR, poor blood glucose levels, and systematic inflammation with lower diversity of the gut microbiome and distinct structure of the microbial community. These results proved that lower gut microbial diversity is responsible for impaired metabolic control (Zouiouich et al. 2021).

As previously mentioned, butyrate has a vital role in maintaining healthy gut barrier function, which deterioration causes leaking of various harmful substances, mainly lipopolysaccharides (LPS), resulting in chronic inflammation (Vrieze et al. 2010). Specifically, a vast array of research has been reported an association between changes in the gut microbiota composition and impaired gut barrier functions, increased gut permeability, and increased plasma lipopolysaccharide concentrations (i.e., metabolic endotoxemia), which causes low-grade inflammation that triggers the development of obesity, IR, and MetS (Festi et al. 2014; Tseng and Wu 2019). The link between obesity and chronic inflammation was observed in a study conducted by Cani et al. in which it has been found that a high-fat diet leads to an increase in LPS-containing bacteria abundance and LPS-induced inflammation, a process defined as metabolic endotoxemia (Cani et al. 2007). Metabolic endotoxemia results from the interaction between the host's immune system and luminal bacteria (de La Serre et al. 2010). Specifically, metabolic endotoxemia is promoted by increased intestinal permeability caused by disorganized tight junction proteins in colonocytes or thinning the mucus layer and translocation of bacteria toxins and antigens from the gastrointestinal lumen into the bloodstream. LPS triggers downstream inflammation through LPS receptor CD14 and thus enhances the transcription of several pro-inflammatory cytokines involved in the pathogenesis of obesity and other metabolic diseases (Cani et al. 2007). Additionally, in human studies, exposure to LPS has been shown to promote systemic IR and adipose tissue-related inflammation (Mehta et al. 2012).

Trimethylamine-*N*-oxide (TMAO) is one of the microbial metabolites which present in circulation is associated with the development of MetS and a higher risk of major cardiovascular events (Croci et al. 2021; Croyal et al. 2020). In the presence of specific microbes, primarily members of the Enterobacteriaceae family, dietary choline, and its derivate (L-carnitine, betaine, lecithin) are metabolized into trimethylamine (TMA). Choline is a metabolite of phosphatidylcholine, a dietary lipid found in high quantities in egg yolk, liver, and other high-fat animal products. After entering the circulation, TMA is metabolized in the liver to TMAO via the enzyme flavin-containing monooxygenase-3 (FMO_3), a nitrosamine precursor with carcinogenic activity (Croci et al. 2021; Hills et al. 2019). Increased plasma TMAO concentrations have been correlated with the accumulation of adipose depots in the liver and blood vessels, leading to visceral obesity and atherosclerosis (Croci et al. 2021). The results of a recent meta-analysis involving 19 prospective studies indicated that elevated plasma TMAO levels and their precursors could be risk biomarkers for major adverse cardiovascular events, including death events. Moreover, it has been determined that TMAO as a risk factor is independent of other traditional risk factors (Heianza et al. 2017). Higher plasma TMAO levels are also associated with diabetes (Dambrova et al. 2016). Nevertheless, TMAO level could be affected by both intra-individual and inter-individual variations over time which may be an obstacle for using TMAO as a risk marker in long-term epidemiological studies (Kuhn et al. 2017). Interestingly, circulating TMAO generated from L-carnitine by gut microbes has been detected in humans, suggesting the mechanism for the connection between atherosclerosis and consumption of red meat (Koeth et al. 2013).

Recently, mounting evidence suggests a causal link between the biosynthesis of branched-chain amino acids (BCAAs) and IR. BCAAs (leucine, isoleucine, and valine) represent three essential amino acids found in the diet that are the building blocks for protein synthesis. However, members of the gut microbiota can also regulate the biosynthesis, transport, and metabolism of BCAAs (Zeng et al. 2020). A meta-analysis of 8 prospective studies involving 8000 subjects found a statistically significant association between plasma concentrations of BCAAs and aromatic amino acids with T2D (Guasch-Ferre et al. 2016). Furthermore, a systematic review of 23 studies involving 20,091 subjects indicated that circulating

BCAAs may be a valuable biomarker for detecting IR and thus diabetic risk later on (Zhao et al. 2016).

6.5 Precision Nutrition-Gut Microbiota as a Target for Metabolic Syndrome Treatment

An effective strategy is required to prevent and control diet-related non-communicable diseases, including Mets. Given the pandemic rise in metabolic diseases, conventional recommendations and dietary guidelines focused on population averages are not compelling enough. The low effectiveness of dietary interventions in treating obesity and its complications can be explained with inter-individual variabilities in response to food, weight-loss diets (Zeevi et al. 2015) and lifestyle interventions (Delgado-Floody et al. 2019). Factors proposed as a source of inter-individual variations in response to dietary and lifestyle interventions are the genetic, epigenetic, microbiome and behavioral/psychological features (Gonzalez-Muniesa and Martinez 2019). Based on this knowledge, the concept of precise nutrition was established. Precise nutrition refers to an individual diet or intervention that is designed to prevent or treat various diseases. In planning a precise diet, the interplay between metabolic, genetic, social, and environmental factors should be considered (Iizuka and Yabe 2020).

Unlike genetic factors whose contribution to disease risk is well known, the role of the microbiota as a significant source of individualized responses to food has been for a long time neglected. As already have mentioned in the previous section, there are significant inter-individual differences in microbiome composition and diversity, which is linked to differences in the biological functions it performs for the host. On the other hand, the composition and functions of the microbiota depend on many factors, primarily on diet and genetics (Rothschild et al. 2018). In addition, the diversity and function of the microbiota present in healthy individuals can be significantly altered in certain diseases and conditions.

Numerous studies have observed a certain percentage of participants defined as "non-responders" to applied dietary intervention. It has been shown that baseline microbiome signatures can influence an individual's response to diet (Mills et al. 2019b). Moreover, inter-individual response to obese-related dietary interventions has also been associated with the presence of specific bacterial species at the baseline (Korpela et al. 2014). Accordingly, the microbiome can potentially serve as a target organ of dietary intervention (e.g., precision microbiomics) and serve as a biomarker for predicting responsiveness to diets and interventions, enabling greater opportunities in health promotion and disease prevention (Hughes et al. 2019; Mills et al. 2019b). Precision microbiomics advanced when certain studies have shown that the glycemic responses to diets could be predicted by including the gut microbiota markers. For example, applying appropriate mathematical algorithms that integrate microbiome composition data alone, or combine them with other clinical blood parameters, an individual postprandial glycemic response to a particular food (Korem et al. 2017), or complete meal (Zeevi et al. 2015) could be accurately predicted. Postprandial hyperglycemia is a risk factor for developing cardiovascular disease and mortality, even in people with normal fasting glucose (Berry et al. 2020). Furthermore, the metabolic improvements induced by specific diets or food may also be closely related to the presence of a specific microbiota species (Kovatcheva-Datchary et al. 2015) or correlate with the change in its presence before and after the dietary intervention (Martinez et al. 2013).

Numerous recent studies have found evidence that stratification of obese individuals according to Prevotella to *Bacteroides* ratio may help predict responses on weight-loss intervention. This enterotype-based approach could be used in personalized nutrition in weight-loss strategy and obesity management. This approach could be justified in differences between two enterotypes in digestive functions with a preference for specific dietary substrates (Costea et al. 2018). According to these findings, a fiber-rich diet will result in effective weight loss among P-enterotype subjects but not among B-enterotypes subjects (Hjorth et al. 2019, 2018). Contrary, in terms of metabolic parameters improvements, B-enterotypes subjects could benefit from bifidobacteria-increasing interventions (Christensen et al. 2018).

In addition, the pre-treatment abundance of certain fecal bacterial species, predominantly derived from *Firmicutes* phylum, may serve as a predictor of response predictors to weight-loss dietary interventions in obese men. A more precise prediction of the host responses to various diets could be achieved by identifying the specific bacterial taxa and functions that are closely related to the host responses. For example, in overweight/obese individuals with a higher baseline abundance of *Akkermansia muciniphila*, a restrictive caloric diet led to a healthier metabolic status and a better clinical outcome, suggesting a predictive role of *A. muciniphila* in assessing response to dietary interventions (Dao et al. 2016). Furthermore, the results of a recently published longitudinal study involving 1089 deeply phenotyped subjects have shown that the presence of only two bacterial species, Prevotella copri and Blastocystis spp., was a reliable microbial predictor of favorable postprandial glycemic responses to diet (Asnicar et al. 2021).

The relevance of the gastrointestinal microbiota in precise nutrition may also be seen in its role in the relationship between red meat consumption and the development of atherosclerosis and CVD, which is associated with elevated plasma levels of TMAO. According to some authors, the general recommendations for reducing red meat intake should be relevant primarily to individuals whose microbiota composition is more prone to metabolize choline and L-carnitine to proatherogenic metabolites (Zmora et al. 2016). Those could be individuals with significantly higher F/B ratio and significantly less microbiota diversity (Cho et al. 2017) or individuals classified within Prevotella enterotype (Koeth et al. 2013). Furthermore, another general recommendation regarding replacing sugar with artificial sweeteners may be harmful to a particular population if the approach is based on the microbiota composition. Suez et al. have reported that increased intake of artificial sweeteners leads to glucose intolerance in a subgroup of individuals with sensitive microbiota (Suez et al. 2014). Considering that high doses of saccharin were used and that the number of subjects was only seven, the results are still controversial.

Given the complexity and individual uniqueness of the microbiota, designing personalized microbiome-based nutrition remains a challenge (Kolodziejczyk et al. 2019). A strategy based on an individual approach could lead the microbiota from dysbiosis to eubiosis or improve the response to a particular diet. Although the primary determinant of the microbiota composition is food, probiotics, prebiotics, and recently postbiotics offer opportunities for the "normalization of the microbiota". Therefore, they may represent a new potential for achieving an effective dietary responses in weight-loss intervention and obesity management.

6.5.1 Probiotics

One of the strategies aimed at modulating the intestinal microbiota to prevent obesity and MetS is the use of probiotics to restore balance or maintain the diversity and functionality of the microbiome when its homeostasis is expected to be disturbed, as is the case in metabolic diseases. In addition, numerous pieces of evidence suggest that certain probiotic strains may modulate the inflammatory response, which may also reduce the risk of developing MetS (Ivanovic et al. 2015; Xavier-Santos et al. 2020).

According to the International Association for Probiotics and Prebiotics (ISAPP) consensus document, the term probiotic means "live microorganisms that, when administered in adequate amounts, confer a health benefit on the host" (Hill et al. 2014). Among the most widely used commercial probiotics, which have also been the most studied in animal and human studies, are strains belonging to the genus *Lactobacillus* and *Bifidobacterium*. Intensive studies of their beneficial effects in humans have led to several recommendations for prophylactic and therapeutic use in both children and adults, including prevention or treatment of acute, antibiotic, and *Clostridium difficile*-induced diarrhea, treatment of functional constipation, treatment of irritable bowel syndrome, and inflammatory bowel disease, eradication of *Helicobacter pillory*, prevention of infantile colic (Guarner et al. 2017; Dimitrijevic et al. 2014). Studies indicating the efficacy of probiotics in preventing and treating other diseases can also be found in the literature. Some of the proposed underlying mechanisms of probiotic beneficial effects are given in Table 6.2.

The link between microbiota modification and functional effects on MetS is increasingly attracting attention, which is why the number of intervention studies examining the effectiveness of probiotics on obesity and associated metabolic pathophysiology is growing. The results of systematic reviews of randomized control studies and meta-analyses that evaluated the efficacy of probiotics in obese/overweight adults demonstrated their efficacy in lowering body weight, BMI, body fat mass, and WC (John et al. 2018; Koutnikova et al. 2019). Moreover, specific *Lactobacillus* and *Bifidobacterium* strains have shown constant anti-obesity activity in many animal and human studies (*Bifidobacterium*

Table 6.2 Mechanism of probiotic and host interaction (Guarner et al. 2017)

Immunological benefits	• Stimulation of IgA production • Modulation of cytokine production • Induction of tolerance to food antigens
Nonimmunological benefits	• Production of bacteriocins to inhibit pathogens • Digestion of food and competition for nutrients with pathogens • Alteration of local pH to create an unfavorable local environment for pathogens • Stimulation of epithelium mucin production • Enhance intestinal barrier function • Competition for adhesion with pathogens • Modification of pathogen-derived toxins • Scavenge of superoxide radicals

breve, *B. longum*, *B. infantis*, *Lacticaseibacillus casei*, *L. rhamnosus*, *L. gasseri*, *L. plantarum*) (Ejtahed et al. 2019; Ivanovic et al. 2015). Some of the proposed mechanisms involved in the amelioration of MetS are modulation of host energy metabolism (SCFA production, modulation of satiety signaling pathways, bile acid deconjugation), strengthening of intestinal mucosal barrier (modulation of tight junction, stimulation of mucus secretion), interaction with host immune system, and interaction with microbiota members (production of bacteriocins and nutrients used by other bacteria) (Abenavoli et al. 2019; Le Barz et al. 2015; Ivanovic et al. 2015). However, inconsistent results can be found in the literature showing that administration of probiotics did not significantly affect the body weight or the effects obtained were not statistically significant compared to the placebo group (Borgeraas et al. 2018; Park and Bae 2015). One of the reasons for the inconsistency of the results between the conducted studies is the large number of microorganisms whose effects were examined in the studies. In particular, each of these genera includes several species, subspecies, and strains that may have the same but also opposite effects. Thus, the results of studies show that the benefits of probiotics that contribute to the host are highly specific for the species and strain, but also depend on the dose, duration of probiotic administration, as well as the basal characteristics of the host and microbiome (Green et al. 2020; Ivanovic et al. 2015, 2016; Mills et al. 2019a; Suez et al. 2019). For example, contrary to most *Lactobacillus* species tested, which mainly lead to weight loss, *L. acidophilus* has been shown to increase weight in humans and animals. On the other hand, while *L. gasseri* BNR17 reduces body weight, *L. gasseri* L66-5 promotes it (Mills et al. 2019a).

Furthermore, to achieve a health benefit for the host, as the definition of probiotics requires, probiotics must be administered in an adequate dose. This dose varies greatly depending on the strain used and ranges between 100 million and 10 billion CFU/day. In a recently published review of studies examining the probiotic dose–response relationship in humans, authors concluded that dose effects on probiotic efficacy were inconsistent and that number of studies are insufficient for drawing a conclusion. However, it has been found that studies examining the impact of probiotics on fecal recovery and antibiotic-induced diarrhea have shown a clear dose–response indicating that higher doses are necessary for biological effects (Ouwehand 2017).

In a recently published paper, Zmora et al. observed individual differences concerning intestinal mucosal colonization by the applied 11-strain probiotic mix in humans. Based on the degree of their colonization, the authors categorized the volunteers into two groups: probiotics "permissives" and probiotics "resisters." In the permissive individuals, some of the strains from the probiotic mixture were detected a few weeks after administration, and the effects on microbiome composition, function, diversity, and bacterial load were significantly more pronounced than in resistant individuals. Moreover, this individual susceptibility to probiotic colonization could be predicted by combining the basal characteristics of the host and the microbiome (Zmora et al. 2018). For example, stable colonization of the probiotic strain *B. longum* AH1206 was found in only 30% of individuals, associated with low basal levels of resident *B. longum* and underrepresented microbiome-related genes for carbohydrate utilization (Maldonado-Gomez et al. 2016). The role of the microbiome in the mechanism of probiotic resistance was confirmed during transplantation of fecal microbiome from persistent or resistant individuals into germ-free mice, which led, after the application of probiotics, to the recapitulation of the resistance to colonization from donors (Suez et al. 2020). Also, Zhang et al. have observed a more significant impact of probiotic supplementation on fecal microbiota diversity in individuals colonized with the probiotic strain from a given supplement than individuals resistant to colonization (Zhang et al. 2016).

Interestingly, Suez et al. observed that antibiotics may enhance probiotic colonization in the human intestine, most likely due to depletion of resident bacteria. On the contrary, post-antibiotic application of probiotics delayed the reconstruction of the pre-antibiotic intestinal microbiota composition, diversity, and functionality compared to spontaneous recovery and autologous fecal transplantation (Suez et al. 2018). In addition, Ferrario et al. reported that consumption of *L. paracasei* DG led to individually specific changes in microbiota composition and SCFAs production, indicating the role of the initial microbiota composition in exhibiting the effects of the probiotics (Ferrario et al. 2014).

These studies indicate that probiotics may have a limited effect on microbiota composition and, thus, its functionality; their effects depend on the characteristics of the host, primarily on the initial microbiota composition. The above suggests the importance of developing personalized probiotics to represent a more precise therapeutic approach than empirical universal probiotics whose effects could show inter-individual variations.

Until the development of personalized probiotics, one way to avoid the risks posed by the use of exogenous strains (transmission of antibiotic resistance genes and the risk of bacteremia and fungemia in immunocompromised individuals) is the use of next-generation probiotics, i.e., human intestinal bacteria such as *A. muciniphila* and *Faecalibacterium prausnitzii*. Human studies have shown a negative correlation between the prevalence of *A. muciniphila* and obesity, T2D, hypertension, and dyslipidemia. The presence of this bacterium in the microbiome is considered a feature of a healthy metabolic status. Moreover, supplementation of *A. muciniphila* at a dose of 10 billion CFU/g has been shown to improve metabolic parameters (IR, total cholesterol, and inflammatory biomarkers) while being safe and well-tolerated (Cuevas-Sierra et al. 2019; Vallianou et al. 2019, 2020). However, some obstacles need to be overcome for commercial application of this strain, such as its high sensitivity to oxygen and mucus-based medium requirements (Vallianou et al. 2019).

6.5.2 Postbiotics

The results of numerous studies have demonstrated that probiotics can effectively treat various diseases, especially in infections of the gastrointestinal tract and inflammatory bowel diseases, and in extra-intestinal diseases. However, the use of probiotics may also be associated with some health risks, such as the possibility of developing bacteremia or fungemia as a consequence of translocation from the gut to the systemic circulation, primarily in immunocompromised individuals (Rannikko et al. 2021; Yelin et al. 2019), as well as with the risk of antibiotic resistance gene transfer (Montassier et al. 2021). Contrarily, there is a vast number of evidence suggesting that viability is not necessary for the manifestation of the probiotic effect, i.e., that non-viable microorganisms, their components, and their metabolites can also exhibit bioactivity, similar or something different from their living counter partners (Cuevas-González et al. 2020; Pique et al. 2019). These observations have led to increased interest in using non-viable microorganisms called "postbiotics" (also known as "paraprobiotics," "metabiotics," "ghost probiotics," "tyndallized probiotics," "bacterial lysates"). According to the recently published ISAPP consensus paper, postbiotics are defined as "a preparation of inanimate microorganisms and/or their components that confers a health benefit the host" where the term "inanimate" is intended to emphasize that the microorganism was previously alive but did not lose physiological benefit to the host with loss of viability (Salminen et al. 2021). The main characteristics of postbiotics are its advantages over probiotics, such as safety of use in sensitive categories (infants, people in intensive care units). Since the postbiotics are not viable cells, their effectiveness is not compromised by reducing the number of bacteria at the end of use, and they are more stable during industrial processes and storage (Cabello-Olmo et al. 2021; Salminen et al. 2021) (Fig. 6.1).

Different technological processes (heat, sonication, irradiation, and high pressure) used for the inactivation of probiotics may have different effects on the structural components of the cell and the metabolite profile, and thus on the characteristics of postbiotics and their biological activity (Deshpande et al. 2018). There are a large number of probiotics classes such as cell wall components (peptidoglycan and lipoteichoic acids as major cell wall components of Gram-positive bacteria), exopolysaccharides and cell surface proteins (S-layer), cell-free supernatants, polysaccharide fermentation products (SCFAs and succinate), metabolites

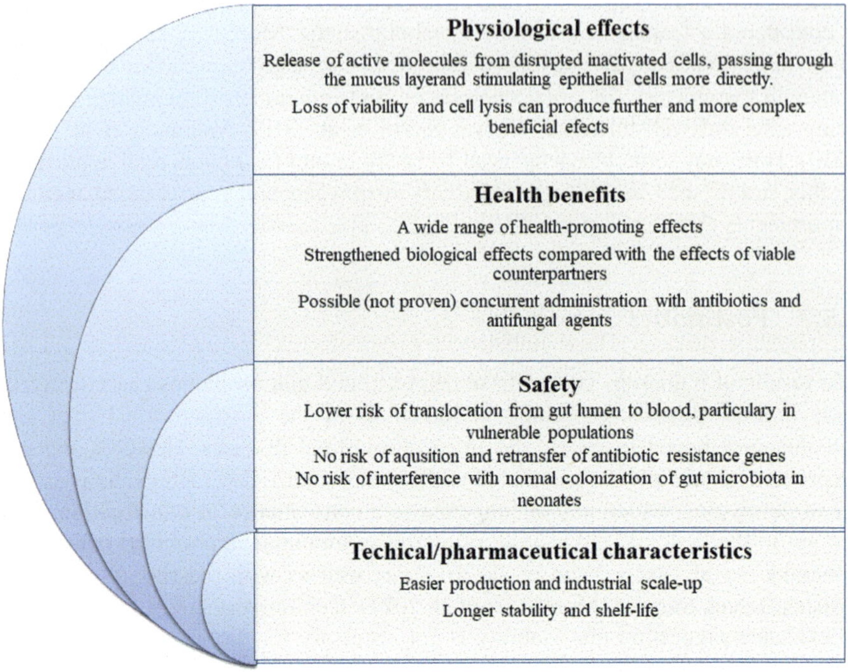

Fig. 6.1 General characteristics of postbiotics

(vitamins, aromatic amino acids, antimicrobial peptides), bacterial lysates, or enzymes (Aguilar-Toalá et al. 2018; Hernández-Granados and Franco-Robles 2020; Pique et al. 2019; Zolkiewicz et al. 2020). Although the mechanisms involved in the health effects of postbiotics have not yet been fully elucidated, scientific evidence suggests that postbiotics exhibit health effects through several different mechanisms, which may act independently or in combination. These mechanisms may be similar to those known for probiotics and include the following: (i) modulation of the resident microbiota, (ii) maintenance of epithelial barrier function, (iii) modulation of the local immune response, (iv) modulation of the systemic metabolism, and (v) signaling via the nervous system (Salminen et al. 2021).

In recent years, many in vitro and in vivo studies have been conducted to determine the health effects of various postbiotics. In most cases, postbiotics derived from *Lactobacillus* and *Bifidobacterium* strains were used, although other bacterial species have also been reported as probiotics sources (Aguilar-Toalá et al. 2018; Cuevas-González et al. 2020). Although data from human studies are limited, they have shown efficacy for orally administered, inactivated lactic acid bacteria (primarily *Lactobacillus* species) in reducing symptoms in patients with inflammatory bowel disease and chronic diarrhea, eradicating H. pylori infection (Salminen et al. 2021). These studies have also shown that postbiotics may be a safer alternative to probiotics, although an assessment of safety for postbiotics is needed before use (Salminen et al. 2021; Wegh et al. 2019).

Literature data indicate that postbiotics are promising candidates for treating metabolic complications related to obesity and promote cardiometabolic benefits by activating the innate immune response (Anhê et al. 2019b, 2021). Among postbiotics, SCFAs are essential modulators of host metabolism (Salminen et al. 2021). Numerous strong evidence from animal studies has indicated that SCFAs play an essential role in treating and preventing obesity-induced IR. Moreover, there is an increasing number of evidence from human studies on the beneficial effects of SCFAs on body weight control, glucose and lipid homeostasis, and inflammatory status (Canfora et al. 2019). SCFAs, as already mentioned in the previous sections, affect appetite and energy intake through various mechanisms, such as stimulation of satiety hormones production (peptide YY and glucagon-like peptide 1) from enteroendocrine cells via G receptors and secretion of the adipocyte-derived satiety hormone leptin. One of the mechanisms by which SCFA can have a suppressive effect on appetite and food intake is the central nervous system, i.e., the gut–brain axis. Animal studies have provided evidence that SCFA supplementation may play an essential role in the prevention of HFD-induced obesity through SCFA-induced upregulation of genes related to thermogenesis and lipid oxidation (Den Besten et al. 2015; Reynés et al. 2019) or through improved hepatic metabolic conditions without altering gut microbial composition (Shimizu et al. 2019). Furthermore, colonic administration of SCFA mixtures increased fasting lipid oxidation and resting energy expenditure and decreased lipolysis in overweight/obese normoglycemic volunteers (Canfora et al. 2017). The same effects have been shown in healthy volunteers after acute oral propionate administration where it was observed that these beneficial effects of propionate resulted from induced PYY and GLP-1 secretion in colonic cells (Chambers et al. 2018).

Findings confirming the usefulness of postbiotics in the treatment of MetS also result from a human study that has demonstrated that pasteurized *A. muciniphila* has superior effects on inflammation and several features of MetS metabolic syndrome associated with fatty liver disease and cardiometabolic risk compared to live bacterium (Depommier et al. 2019). Moreover, a component isolated from the cell wall of this bacterium (Amuc_1100) has been shown to exhibit the same favorable metabolic benefits in a mouse model of diet-induced obesity as it was reported after administration of pasteurized *A. muciniphila* (Plovier et al. 2017). Another example demonstrating the role of postbiotics in MetS treatment includes muramyl dipeptide, a component of the Gram + bacterial cell wall that has been shown to reduce adipose inflammation and glucose intolerance in a mouse model of obesity without affecting body weight or microbiota composition (Cavallari et al. 2017).

6.5.3 Polyphenols

Polyphenols are non-nutritive phytochemicals commonly found in human diets. They represent a very different class of secondary metabolites synthesized by a plant to perform various functions, such as protection against UV radiation, microbial infections, or mechanical damage (Cueva et al. 2020). Polyphenols comprise a large heterogeneous group of compounds with over 10,000 structural variants, characterized by aromatic rings and ligand groups; these structures can vary from monomers to complex polymers of high molecular weight (De Filippis et al. 2020; Singla et al. 2019). These compounds are usually classified into two main groups: flavonoids and nonflavonoids. Each class can be subdivided into many subclasses depending on their chemical structure (e.g., number of phenol units, substituted groups or bonds between phenolic units). Subclasses of flavonoids are flavanols, flavones and isoflavones, anthocyanidins, and anthocyanins, while nonflavonoids include phenolic acids, stilbenes, and lignans (Singla et al. 2019). Classification of polyphenol components and their primary dietary sources are shown in Table 6.3.

Due to their chemical structure, only 5–10% of ingested polyphenols are absorbed in the small intestine (Ray and Mukherjee 2021). Approximately 90% of ingested polyphenols reach the colon, where they interact with the gut microbiota (Rodriguez-Daza et al. 2021). There is a two-way interaction between polyphenols and the microbiota, which is determinant in the effects of polyphenols (Cortés-Martín et al. 2020). Polyphenols modulate the gut microbiota composition through their antimicrobial and prebiotic effects, which can per se affect the host's metabolism. Also, members of the gut microbiota catabolize ingested polyphenols

Table 6.3 Polyphenolic classes and their primary dietary sources

Class	Sub-class	Compounds	Dietary sources
Flavonoids	Anthocyanins	Cyanidin, delphinidin, peonidin, malvidin	Berries, cherries, red grapes, currants, beans
	Flavan-3-ols	Catechin, epicatechin, gallocatechin, epigallocatechin, tannins	Green tea, cocoa, grapes, berries
	Flavones	Apigenin, luteolin	Olives, apple, cabbage
	Flavanols	Kaempferol, quercetin,	Green and black tea, various fruits, vegetables, nuts
	Flavanones	Naringenin, hesperetin, neohesperidin	Citrus fruits, tomato, berries
	Isoflavones	Daidzein, genistein	Soybeans and soy products
Non-flavonoids	Phenolic acids	Benzoic acids, cinnamic acids, benzoic acids	Berries, grapes, nuts
	Stilbenes	Resveratrol	Red wines, grapes, peanuts, plums, pine nuts
	Lignans	Enterodiol, enterolactone	Flaxseed, beans, berries, sesame

into various low molecular weight metabolites that can be easily absorbed and thus again exhibit a potential effect on host health (Alves-Santos et al. 2020; Cueva et al. 2020). Based on recent evidence that polyphenols can be selectively used by the gut microbiota and thus exert beneficial effects on host health, polyphenols can also be considered prebiotics according to the revised definition, previously limited to carbohydrates only: "A prebiotic is a non-digestible compound that, through its metabolism by microorganisms in the gut modulates the composition and (or) activity of the gut microbiota, thus conferring a beneficial physiological effect on the host" (Gibson et al. 2017). However, the effects of polyphenols on the gut microbiome ecology are mainly based on the combined antimicrobial and grow-stimulating (prebiotic) effects (Rodriguez-Daza et al. 2021).

Accumulating literature have demonstrated that regular consumption of food rich in polyphenols such as fruits, vegetables, tea, coffee, wine, and grains is associated with a reduced risk of cardiometabolic diseases and cancer (Čakar et al. 2018; Anhê et al. 2019a; Van Hul and Cani 2019). Also, a wide range of biological activity of different polyphenols has been described in numerous in vitro and in vivo studies, as it is shown in Fig. 6.2.

Numerous in vivo studies, both in animals and humans, indicate that certain classes of polyphenols such as anthocyanins, proanthocyanidins, flavanones, flavanols, ellagitannins, and stilbenoid resveratrol may alleviate some of the features of MetS. Proposed mechanisms of action include reduction of inflammation, modulation of glucose homeostasis, repair of endothelial dysfunction, suppression of adipogenesis and lipid synthesis, an increase of energy expenditure through thermogenesis, stimulation of fat oxidation, and reduction of nutrient intake by interaction with digestive enzymes (De Filippis et al. 2020; Van Hul and Cani 2019) Moreover, the findings from these studies indicate that their effects depend primarily on the interaction with gut microbiome ecology, i.e., that gut microbiota is a crucial mediator of the health effects of polyphenols (Anhê et al. 2019a, b). A recently published review involving 44 animal studies have reported that polyphenols are effective in improving metabolic derangements (weight gain, visceral obesity, plasma TAG, glucose intolerance) and that these effects were the result of polyphenol ability to improve gut dysbiosis, induced by HFD (Moorthy et al. 2021). Additionally, polyphenols can play an essential role in maintaining the microbial richness and thus better metabolic fitness (Anhê et al. 2019a, b). Analysis of the microbiota of 1135 subjects showed that the frequency of fruits, vegetables, red wine, coffee, and tea intake, which are the major sources of polyphenols in diet, is one of the most critical factors that positively correlate with microbial richness. Moreover, it has been shown that the intake of these foods was among the main variables that firmly explain the inter-individual differences in the gut microbiota composition (Zhernakova et al. 2016).

Quercetin is one of the polyphenols for which there is strong evidence, primarily from animal studies, that it can be highly effective in preventing various risk factors for MetS. Quercetin has been shown alone or in combination with other compounds such as resveratrol or green tea extract to correct HFD-induced dysbiosis in mice and thus prevent metabolic changes associated with obesity. The

Fig. 6.2 Biological activities of dietary polyphenols

observed effects of quercetin alone or in combination with these compounds on the microbiota composition were, among others, an increase in the prevalence of *A. muciniphila* and a decrease in F/B ratio, one of the hallmarks of metabolic fitness (Koch 2019; Tan et al. 2021; Zhao et al. 2017). In addition, there is much evidence demonstrating the antidiabetic activity of quercetin, and one of the proposed mechanisms is that quercetin activates 5′ adenosine monophosphate-activated protein kinase (AMPK) in the liver (Bule et al. 2019). This mechanism is analogous to the action of metformin, the most commonly used antidiabetic drug. A group of researchers proposed the so-called "AMPK hypothesis" that polyphenols can activate AMPK by phosphorylation and thus regulate energy metabolism (Yang et al. 2016). Given that AMPK is the central regulator of energy homeostasis and thus an important target for the treatment of MetS, the role and potential of polyphenols in the treatment of MetS can then be explained.

Furthermore, certain polyphenols can benefit cardiometabolic health via their prebiotic effect, such as increasing the abundance of *A. muciniphila.* The exact mechanism is not established, but it has been shown that polyphenols intake is also associated with an increase in mucus synthesis, which is the primary energy source for this mucin-degrading bacteria (Anhê et al. 2019a, b; Rodriguez-Daza et al. 2021; Van Hul and Cani 2019).

Regarding the results from human studies, a meta-analysis that included 117 studies (published between 1997–2015) revealed a significant beneficial effect of flavanols (from tea, cocoa, and apples) on BMI, WC, and serum lipids levels. Although these effects were modest, they could be compared quantitatively with those obtained, for example, by a change in lifestyle (González-Sarrías et al. 2017). Furthermore, a meta-analysis of 11 randomized controlled trials demonstrated that resveratrol could significantly improve glucose homeostasis in people with diabetes (Liu et al. 2014). However, the results from numerous clinically randomized studies are contradictory. Thus, the final consensus on the beneficial effects of polyphenols on health remains elusive. The reason lies in the fact that human subjects have different responses to dietary intervention, and thus, it is more evidence that there is no universal (one-size-fit-all) diet and that the effects are significantly determined by the characteristics of the host and the gut microbiota. Therefore, the possibility of polyphenols exhibiting favorable health effects depends on many factors, including the dietary source in which they are found, the amounts ingested, i.e., nutritionally relevant doses, and on inter-individual differences in genetics, physiological status, lifestyle.

Additionally, the set of microbial polyphenolic metabolites, and the effect of polyphenols, depend primarily on the microbiota's characteristics, which is the primary intermediate between polyphenols and health (Cortés-Martín et al. 2020; Pushpass et al. 2021). In this matter, the results of numerous studies have indicated the need to cluster human subjects into metabotypes, which refers to a differential gut microbial metabolism of polyphenols, to explain the inter-individual variability of effects observed after consumption of dietary polyphenols. The introduction of metabotypes aims to design a customized polyphenol-rich diet for specific individuals to prevent or treat certain diseases (Cortés-Martín et al. 2020); this achieves the concept of a personalized diet focused on the microbiota as the target organ of intervention.

Until now, only two polyphenols, ellagic acid and isoflavone daidzein, are known to show inter-individual differences in their metabolism related to specific intestinal microbial ecology, otherwise responsible for the formation of different metabolites (postbiotics). As for daidzein, soy isoflavones, it has been found that only 30% of the Western population possesses certain bacterial species in the microbiota capable of metabolizing daidzein to S-equol. The existence of two metabotypes that differ in their ability to synthesize S-equol may be relevant in achieving the beneficial effects of soy isoflavone on glucose and lipid homeostasis, which is the necessary presence of this metabolite (Usui et al. 2013). In addition to the equol-producer metabotype, O-desmethylangolensin (ODMA)-producer metabotype was also identified; for this, two metabotypes have been reported to be independent of each other. Regarding the microbial conversion of ellagic acid to urolithins, three metabotypes have been proposed depending on the final postbiotic produced: metabotype A (UM-A) produces urolithin A, metabotype B (UM-B) produces urolithin B and isourolithin A in addition to urolithin A, and metabotype O (UM-0) is unable to produce urolithins from ellagic acid (Romo-Vaquero et al. 2019). The importance of clustering human subjects into metabotypes can be seen in a study in which pomegranate extract was given to obese subjects stratified as UM-A or UM-B metabotypes. Although UM-A individuals had a better basal lipid profile, administration of pomegranate extract led to a greater improvement in lipid profile in UM-B individuals (Gonzalez-Sarrias et al. 2017). Furthermore, consumption of walnuts over three days led to UM-depending modulation of the gut microbiota. The short-term intervention with walnuts led to an increase in *Blautia*, *Bifidobacterium*, and members of the *Coriobacteriaceae* family only in UM-B individuals, while in UM-A individuals, only a decrease in members of the Lachnospiraceae family was observed (García-Mantrana et al. 2019).

Romo Vaquero et al. reported a difference in diversity and richness between UM metabotypes, which was not found when individuals were stratified on three enterotypes (*Bacteroides*, *Prevotella*, and *Ruminococcus*). UM-O metabotype was characterized by lower diversity and richness than UM-A and UM-B, while more than half of the discriminant genera between UM-A and UM-B belonged to the *Coriobacteriaceae* family. The presence of *Coriobacteriaceae* was increased in UM-B compared with UM-A and UM-O metabotypes, and it was positively associated with BMI, total, and LDL cholesterol (Romo-Vaquero et al. 2019). So far, no clear link has been reported between UM metabotypes and dietary pattern, gender, or ethnicity. However, preliminary observations show that UM-B metabotype is associated with higher BMI and cardiometabolic disorders caused by dysbiosis, while UM-A is marked as a "protective" metabotype (Cortés-Martín et al. 2020). Recent cross-sectional studies have reported significantly different content and functioning of the microbial gut community between equol producers and non-producers. Further, lower prevalences of dyslipidemia were found within equol producers suggesting that S-equol might play an essential role in lipid metabolism by gut microbiota (Zheng et al. 2019).

6.6 Conclusion

The link between diet, intestinal microbiota, and MetS has been intensively investigated in numerous studies published in recent years. Current evidence suggests that obesity and associated metabolic disorders are closely related to a perturbation in the composition and function of the microbiota dictated by several factors, of which nutrition plays the most crucial role. Numerous interventional studies have indicated the role of the microbiome as a source of inter-individual variation in response to diet. Therefore, microbiome-focused dietary interventions can be an essential tool for achieving a personalized approach for preventing and treating cardiometabolic diseases. Probiotics, prebiotics, postbiotics, and polyphenols could be critical components of personalized dietary intervention to alter the gut microbiome to a more beneficial configuration for preventing and treating MetS.

However, the use of the microbiome as a target in creating a precise diet has been accompanied by many unknowns. For example, many data are still missing, such as the association of certain enterotypes, diversity, richness, and specific taxa, with different responses in different dietary contexts. Moreover, the limited data from human studies suggest that more profound large-scale clinical randomized studies are needed to evaluate the conditions for their personalized use, including the adequate dose, duration of supplementation, the durability of their beneficial effects, and in the case of probiotics, determination of microbiota and host factors influencing colonization of probiotics, and thus their efficiency.

Literatures

Abenavoli L, Scarpellini E, Colica C, Boccuto L, Salehi B, Sharifi-Rad J et al (2019) Gut microbiota and obesity: a role for probiotics. Nutrients 11(11):2690

Aguilar-Toalá J, Garcia-Varela R, Garcia H, Mata-Haro V, González-Córdova A, Vallejo-Cordoba B, Hernández-Mendoza A (2018) Postbiotics: an evolving term within the functional foods field. Trends Food Sci Technol 75:105–114

Alberti KG, Zimmet PZ (1998) Definition, diagnosis and classification of diabetes mellitus and its complications. Part 1: diagnosis and classification of diabetes mellitus provisional report of a WHO consultation. Diabet Med 15(7):539–553. https://doi.org/10.1002/(SICI)1096-9136(199807)15:7<539::AID-DIA668>3.0.CO;2-S

Alberti KGM, Zimmet P, Shaw J (2005) The metabolic syndrome—a new worldwide definition. Lancet 366(9491):1059–1062

Alberti KG, Eckel RH, Grundy SM, Zimmet PZ, Cleeman JI, Donato KA et al, International Association for the Study of, O (2009) Harmonizing the metabolic syndrome: a joint interim statement of the International Diabetes Federation Task Force on Epidemiology and Prevention; National Heart, Lung, and Blood Institute; American Heart Association; World Heart Federation; International Atherosclerosis Society; and International Association for the Study of Obesity. Circulation 120(16):1640–1645. https://doi.org/10.1161/CIRCULATIONAHA.109.192644

Alves-Santos AM, Sugizaki CSA, Lima GC, Naves MMV (2020) Prebiotic effect of dietary polyphenols: a systematic review. J Funct Foods 74:104169

Anhê FF, Choi B, Dyck J, Schertzer J, Marette A (2019a) Host–microbe interplay in the cardiometabolic benefits of dietary polyphenols. Trends Endocrinol Metab 30(6):384–395

Anhê FF, Schertzer JD, Marette A (2019b) Bacteria to alleviate metabolic syndrome. Nat Med 25(7):1031–1033

Anhê FF, Jensen BA, Perazza LR, Tchernof A, Schertzer JD, Marette A (2021) Bacterial postbiotics as promising tools to mitigate cardiometabolic diseases. J Lipid Atherosclerosis 10(2):123

Appleton J (2018) The gut-brain axis: influence of microbiota on mood and mental health. Integr Med Clin J 17(4):28

Armougom F, Henry M, Vialettes B, Raccah D, Raoult D (2009) Monitoring bacterial community of human gut microbiota reveals an increase in Lactobacillus in obese patients and Methanogens in anorexic patients. PLoS ONE 4(9):e7125. https://doi.org/10.1371/journal.pone.0007125

Arumugam M, Raes J, Pelletier E, Le Paslier D, Yamada T, Mende DR et al (2011) Enterotypes of the human gut microbiome. Nature 473(7346):174–180

Asnicar F, Berry SE, Valdes AM, Nguyen LH, Piccinno G, Drew DA et al (2021) Microbiome connections with host metabolism and habitual diet from 1,098 deeply phenotyped individuals. Nat Med 27(2):321–332. https://doi.org/10.1038/s41591-020-01183-8

Bäckhed F (2012) Host responses to the human microbiome. Nutr Rev 70(suppl_1):S14–S17

Bäckhed F, Ding H, Wang T, Hooper LV, Koh GY, Nagy A et al (2004) The gut microbiota as an environmental factor that regulates fat storage. Proc Natl Acad Sci 101(44):15718–15723

Bäckhed F, Ley RE, Sonnenburg JL, Peterson DA, Gordon JI (2005) Host-bacterial mutualism in the human intestine. Science 307(5717):1915–1920

Bäckhed F, Manchester JK, Semenkovich CF, Gordon JI (2007) Mechanisms underlying the resistance to diet-induced obesity in germ-free mice. Proc Natl Acad Sci 104(3):979–984

Bäckhed F, Roswall J, Peng Y, Feng Q, Jia H, Kovatcheva-Datchary P et al (2015) Dynamics and stabilization of the human gut microbiome during the first year of life. Cell Host Microbe 17(5):690–703

Bashiardes S, Godneva A, Elinav E, Segal E (2018) Towards utilization of the human genome and microbiome for personalized nutrition. Curr Opin Biotechnol 51:57–63. https://doi.org/10.1016/j.copbio.2017.11.013

Berry SE, Valdes AM, Drew DA, Asnicar F, Mazidi M, Wolf J et al (2020) Human postprandial responses to food and potential for precision nutrition. Nat Med 26(6):964–973. https://doi.org/10.1038/s41591-020-0934-0

Borgeraas H, Johnson LK, Skattebu J, Hertel JK, Hjelmesaeth J (2018) Effects of probiotics on body weight, body mass index, fat mass and fat percentage in subjects with overweight or obesity: a systematic review and meta-analysis of randomized controlled trials. Obes Rev 19(2):219–232. https://doi.org/10.1111/obr.12626

Bule M, Abdurahman A, Nikfar S, Abdollahi M, Amini M (2019) Antidiabetic effect of quercetin: a systematic review and meta-analysis of animal studies. Food Chem Toxicol 125:494–502

Cabello-Olmo M, Araña M, Urtasun R, Encio IJ, Barajas M (2021) Role of postbiotics in diabetes mellitus: current knowledge and future perspectives. Foods 10(7):1590

Čakar U, Grozdanić N, Pejin B, Vasić V, Čakar M, Petrović A, Djordjević B (2018) Impact of vinification procedure on fruit wine inhibitory activity against α-glucosidase. Food Biosci 25:1–7

Canfora EE, van der Beek CM, Jocken JWE, Goossens GH, Holst JJ, Olde Damink SWM et al (2017) Colonic infusions of short-chain fatty acid mixtures promote energy metabolism in overweight/obese men: a randomized crossover trial. Sci Rep 7(1):2360. https://doi.org/10.1038/s41598-017-02546-x

Canfora EE, Meex RCR, Venema K, Blaak EE (2019) Gut microbial metabolites in obesity, NAFLD and T2DM. Nat Rev Endocrinol 15(5):261–273. https://doi.org/10.1038/s41574-019-0156-z

Cani PD (2018) Human gut microbiome: hopes, threats and promises. Gut 67(9):1716–1725. https://doi.org/10.1136/gutjnl-2018-316723

Cani PD, Amar J, Iglesias MA, Poggi M, Knauf C, Bastelica D et al (2007) Metabolic endotoxemia initiates obesity and insulin resistance. Diabetes 56(7):1761–1772. https://doi.org/10.2337/db06-1491

Cavallari JF, Fullerton MD, Duggan BM, Foley KP, Denou E, Smith BK et al (2017) Muramyl dipeptide-based postbiotics mitigate obesity-induced insulin resistance via IRF4. Cell Metab 25(5):1063–1074, e1063

Chambers ES, Byrne CS, Aspey K, Chen Y, Khan S, Morrison DJ, Frost G (2018) Acute oral sodium propionate supplementation raises resting energy expenditure and lipid oxidation in fasted humans. Diabetes Obes Metab 20(4):1034–1039

Cho CE, Taesuwan S, Malysheva OV, Bender E, Tulchinsky NF, Yan J et al (2017) Trimethylamine-*N*-oxide (TMAO) response to animal source foods varies among healthy young men and is influenced by their gut microbiota composition: a randomized controlled trial. Mol Nutr Food Res 61(1). https://doi.org/10.1002/mnfr.201600324

Christensen L, Roager HM, Astrup A, Hjorth MF (2018) Microbial enterotypes in personalized nutrition and obesity management. Am J Clin Nutr 108(4):645–651. https://doi.org/10.1093/ajcn/nqy175

Clemente JC, Ursell LK, Parfrey LW, Knight R (2012) The impact of the gut microbiota on human health: an integrative view. Cell 148(6):1258–1270

Cornier MA, Dabelea D, Hernandez TL, Lindstrom RC, Steig AJ, Stob NR et al (2008) The metabolic syndrome. Endocr Rev 29(7):777–822. https://doi.org/10.1210/er.2008-0024

Cortés-Martín A, Selma MV, Tomás-Barberán FA, González-Sarrías A, Espín JC (2020) Where to look into the puzzle of polyphenols and health? The postbiotics and gut microbiota associated with human metabotypes. Mol Nutr Food Res 64(9):1900952

Costea PI, Hildebrand F, Arumugam M, Bäckhed F, Blaser MJ, Bushman FD et al (2018) Enterotypes in the landscape of gut microbial community composition. Nat Microbiol 3(1):8–16

Croci S, D'Apolito LI, Gasperi V, Catani MV, Savini I (2021) Dietary strategies for management of metabolic syndrome: role of gut microbiota metabolites. Nutrients 13(5):1389. https://doi.org/10.3390/nu13051389

Croyal M, Saulnier PJ, Aguesse A, Gand E, Ragot S, Roussel R et al (2020) Plasma trimethylamine *N*-oxide and risk of cardiovascular events in patients with type 2 diabetes. J Clin Endocrinol Metab 105(7). https://doi.org/10.1210/clinem/dgaa188

Cryan JF, O'Riordan KJ, Cowan CS, Sandhu KV, Bastiaanssen TF, Boehme M et al (2019) The microbiota-gut-brain axis. Physiol Rev

Cueva C, Silva M, Pinillos I, Bartolomé B, Moreno-Arribas M (2020) Interplay between dietary polyphenols and oral and gut microbiota in the development of colorectal cancer. Nutrients 12(3):625

Cuevas-González P, Liceaga A, Aguilar-Toalá J (2020) Postbiotics and paraprobiotics: from concepts to applications. Food Res Int 109502

Cuevas-Sierra A, Ramos-Lopez O, Riezu-Boj JI, Milagro FI, Martinez JA (2019) Diet, gut microbiota, and obesity: links with host genetics and epigenetics and potential applications. Adv Nutr 10(suppl_1):S17–S30. https://doi.org/10.1093/advances/nmy078

Dambrova M, Latkovskis G, Kuka J, Strele I, Konrade I, Grinberga S et al (2016) Diabetes is associated with higher trimethylamine *N*-oxide plasma levels. Exp Clin Endocrinol Diab 124(4):251–256. https://doi.org/10.1055/s-0035-1569330

Dao MC, Everard A, Aron-Wisnewsky J, Sokolovska N, Prifti E, Verger EO et al (2016) Akkermansia muciniphila and improved metabolic health during a dietary intervention in obesity: relationship with gut microbiome richness and ecology. Gut 65(3):426–436. https://doi.org/10.1136/gutjnl-2014-308778

de La Serre CB, Ellis CL, Lee J, Hartman AL, Rutledge JC, Raybould HE (2010) Propensity to high-fat diet-induced obesity in rats is associated with changes in the gut microbiota and gut inflammation. Am J Physiol Gastrointest Liver Physiol 299(2):G440-448. https://doi.org/10.1152/ajpgi.00098.2010

Delgado-Floody P, Álvarez C, Cadore EL, Flores-Opazo M, Caamaño-Navarrete F, Izquierdo M (2019) Preventing metabolic syndrome in morbid obesity with resistance training: reporting interindividual variability. Nutr Metab Cardiovasc Dis 29(12):1368–1381

Den Besten G, Bleeker A, Gerding A, van Eunen K, Havinga R, van Dijk TH et al (2015) Short-chain fatty acids protect against high-fat diet–induced obesity via a PPARγ-dependent switch from lipogenesis to fat oxidation. Diabetes 64(7):2398–2408

Depommier C, Everard A, Druart C, Plovier H, Van Hul M, Vieira-Silva S et al (2019) Supplementation with *Akkermansia muciniphila* in overweight and obese human volunteers: a proof-of-concept exploratory study. Nat Med 25(7):1096–1103. https://doi.org/10.1038/s41591-019-0495-2

Deshpande G, Athalye-Jape G, Patole S (2018) Para-probiotics for preterm neonates—the next frontier. Nutrients 10(7):871

Despres JP, Lemieux I, Bergeron J, Pibarot P, Mathieu P, Larose E et al (2008) Abdominal obesity and the metabolic syndrome: contribution to global cardiometabolic risk. Arterioscler Thromb Vasc Biol 28(6):1039–1049. https://doi.org/10.1161/ATVBAHA.107.159228

De Filippis A, Ullah H, Baldi A, Dacrema M, Esposito C, Garzarella EU et al (2020) Gastrointestinal disorders and metabolic syndrome: dysbiosis as a key link and common bioactive dietary components useful for their treatment. Int J Mol Sci 21(14). https://doi.org/10.3390/ijms21144929

Dieterich W, Schink M, Zopf Y (2018) Microbiota in the gastrointestinal tract. Medical Sciences 6(4):116

Dimitrijevic R, Ivanovic N, Mathiesen G, Petrusic V, Zivkovic I, Djordjevic B, Dimitrijevic L (2014) Effects of Lactobacillus rhamnosus LA68 on the immune system of C57BL/6 mice upon oral administration. J Dairy Res 81(2):202–207

Donohoe DR, Garge N, Zhang X, Sun W, O'Connell TM, Bunger MK, Bultman SJ (2011) The microbiome and butyrate regulate energy metabolism and autophagy in the mammalian colon. Cell Metab 13(5):517–526. https://doi.org/10.1016/j.cmet.2011.02.018

Ejtahed H-S, Angoorani P, Soroush A-R, Atlasi R, Hasani-Ranjbar S, Mortazavian AM, Larijani B (2019) Probiotics supplementation for the obesity management; a systematic review of animal studies and clinical trials. J Funct Foods 52:228–242

Engin A (2017) The definition and prevalence of obesity and metabolic syndrome. Adv Exp Med Biol 960:1–17. https://doi.org/10.1007/978-3-319-48382-5_1

Expert Panel on Detection E (2001) Executive summary of the third report of the National Cholesterol Education Program (NCEP) expert panel on detection, evaluation, and treatment of high blood cholesterol in adults (adult treatment panel III). Jama 285(19):2486–2497

Fernandes J, Su W, Rahat-Rozenbloom S, Wolever TM, Comelli EM (2014) Adiposity, gut microbiota and faecal short chain fatty acids are linked in adult humans. Nutr Diabetes 4:e121. https://doi.org/10.1038/nutd.2014.23

Ferrannini E, Natali A, Capaldo B, Lehtovirta M, Jacob S, Yki-Jarvinen H (1997) Insulin resistance, hyperinsulinemia, and blood pressure: role of age and obesity. European Group for the Study of Insulin Resistance (EGIR). Hypertension 30(5):1144–1149. https://doi.org/10.1161/01.hyp.30.5.1144

Ferrario C, Taverniti V, Milani C, Fiore W, Laureati M, De Noni I et al (2014) Modulation of fecal Clostridiales bacteria and butyrate by probiotic intervention with *Lactobacillus paracasei* DG varies among healthy adults. J Nutr 144(11):1787–1796. https://doi.org/10.3945/jn.114.197723

Festi D, Schiumerini R, Eusebi LH, Marasco G, Taddia M, Colecchia A (2014) Gut microbiota and metabolic syndrome. World J Gastroenterol 20(43):16079–16094. https://doi.org/10.3748/wjg.v20.i43.16079

Finucane MM, Sharpton TJ, Laurent TJ, Pollard KS (2014) A taxonomic signature of obesity in the microbiome? Getting to the guts of the matter. PLoS ONE 9(1):e84689

Frame LA, Costa E, Jackson SA (2020) Current explorations of nutrition and the gut microbiome: a comprehensive evaluation of the review literature. Nutr Rev 78(10):798–812. https://doi.org/10.1093/nutrit/nuz106

García-Mantrana I, Calatayud M, Romo-Vaquero M, Espín JC, Selma MV, Collado MC (2019) Urolithin metabotypes can determine the modulation of gut microbiota in healthy individuals by tracking walnuts consumption over three days. Nutrients 11(10):2483

Gibson GR, Hutkins R, Sanders ME, Prescott SL, Reimer RA, Salminen SJ et al (2017) Expert consensus document: The International Scientific Association for Probiotics and Prebiotics (ISAPP) consensus statement on the definition and scope of prebiotics. Nat Rev Gastroenterol Hepatol 14(8):491–502

Gilbert JA, Blaser MJ, Caporaso JG, Jansson JK, Lynch SV, Knight R (2018) Current understanding of the human microbiome. Nat Med 24(4):392–400

González-Sarrías A, Combet E, Pinto P, Mena P, Dall'Asta M, Garcia-Aloy M et al (2017) A systematic review and meta-analysis of the effects of flavanol-containing tea, cocoa and apple products on body composition and blood lipids: exploring the factors responsible for variability in their efficacy. Nutrients 9(7):746

González-Muniesa P, Martinez JA (2019) Precision nutrition and metabolic syndrome management. Nutrients 11(10). https://doi.org/10.3390/nu11102411

Gonzalez-Sarrias A, Garcia-Villalba R, Romo-Vaquero M, Alasalvar C, Orem A, Zafrilla P et al (2017) Clustering according to urolithin metabotype explains the interindividual variability in the improvement of cardiovascular risk biomarkers in overweight-obese individuals consuming pomegranate: a randomized clinical trial. Mol Nutr Food Res 61(5). https://doi.org/10.1002/mnfr.201600830

Goodrich JK, Davenport ER, Beaumont M, Jackson MA, Knight R, Ober C et al (2016) Genetic determinants of the gut microbiome in UK twins. Cell Host Microbe 19(5):731–743

Gorvitovskaia A, Holmes SP, Huse SM (2016) Interpreting prevotella and bacteroides as biomarkers of diet and lifestyle. Microbiome 4(1):1–12

Green M, Arora K, Prakash S (2020) Microbial medicine: prebiotic and probiotic functional foods to target obesity and metabolic syndrome. Int J Mol Sci 21(8):2890

Grundy SM (2016) Metabolic syndrome update. Trends Cardiovasc Med 26(4):364–373. https://doi.org/10.1016/j.tcm.2015.10.004

Grundy SM, Cleeman JI, Daniels SR, Donato KA, Eckel RH, Franklin BA et al (2005) Diagnosis and management of the metabolic syndrome: an American Heart Association/National Heart, Lung, and Blood Institute scientific statement. Circulation 112(17):2735–2752

Guarner F, Sanders M, Eliakim R, Fedorak R, Gangl A, Garisch J (2017) Probiotics and prebiotics. World Gastroenterology Organisation Global Guidelines. World Gastroenterol Organ Milwaukee WI USA 46:468–481

Guasch-Ferre M, Hruby A, Toledo E, Clish CB, Martinez-Gonzalez MA, Salas-Salvado J, Hu FB (2016) Metabolomics in prediabetes and diabetes: a systematic review and meta-analysis. Diab Care 39(5):833–846. https://doi.org/10.2337/dc15-2251

Haque SZ, Haque M (2017) The ecological community of commensal, symbiotic, and pathogenic gastrointestinal microorganisms—an appraisal. Clin Exp Gastroenterol 10:91

Haro C, Garcia-Carpintero S, Rangel-Zuniga OA, Alcala-Diaz JF, Landa BB, Clemente JC et al (2017) Consumption of two healthy dietary patterns restored microbiota dysbiosis in obese patients with metabolic dysfunction. Mol Nutr Food Res 61(12). https://doi.org/10.1002/mnfr.201700300

Heianza Y, Ma W, Manson JE, Rexrode KM, Qi L (2017) Gut microbiota metabolites and risk of major adverse cardiovascular disease events and death: a systematic review and meta-analysis of prospective studies. J Am Heart Assoc 6(7). https://doi.org/10.1161/JAHA.116.004947

Hernández-Granados MJ, Franco-Robles E (2020) Postbiotics in human health: Possible new functional ingredients? Food Res Int 137:109660

Hill C, Guarner F, Reid G, Gibson GR, Merenstein DJ, Pot B et al (2014) Expert consensus document: The International Scientific Association for Probiotics and Prebiotics consensus statement on the scope and appropriate use of the term probiotic. Nat Rev Gastroenterol Hepatol

Hill CJ, Lynch DB, Murphy K, Ulaszewska M, Jeffery IB, O'Shea CA et al (2017) Evolution of gut microbiota composition from birth to 24 weeks in the INFANTMET Cohort. Microbiome 5(1):1–18

Hills RD Jr, Pontefract BA, Mishcon HR, Black CA, Sutton SC, Theberge CR (2019) Gut microbiome: profound implications for diet and disease. Nutrients 11(7). https://doi.org/10.3390/nu11071613

Hjorth MF, Roager HM, Larsen TM, Poulsen SK, Licht TR, Bahl MI et al (2018) Pre-treatment microbial prevotella-to-bacteroides ratio, determines body fat loss success during a 6-month randomized controlled diet intervention. Int J Obes (lond) 42(3):580–583. https://doi.org/10.1038/ijo.2017.220

Hjorth MF, Blaedel T, Bendtsen LQ, Lorenzen JK, Holm JB, Kiilerich P et al (2019) Prevotella-to-bacteroides ratio predicts body weight and fat loss success on 24-week diets varying in macronutrient composition and dietary fiber: results from a post-hoc analysis. Int J Obes (lond) 43(1):149–157. https://doi.org/10.1038/s41366-018-0093-2

Hu J, Lin S, Zheng B, Cheung PC (2018) Short-chain fatty acids in control of energy metabolism. Crit Rev Food Sci Nutr 58(8):1243–1249
Hughes RL, Kable ME, Marco M, Keim NL (2019) The role of the gut microbiome in predicting response to diet and the development of precision nutrition models. Part II: results. Adv Nutr 10(6):979–998. https://doi.org/10.1093/advances/nmz049
Iizuka K, Yabe D (2020) The role of metagenomics in precision nutrition. Nutrients 12(6). https://doi.org/10.3390/nu12061668
Ivanovic N, Minic R, Dimitrijevic L, Skodric SR, Zivkovic I, Djordjevic B (2015) Lactobacillus rhamnosus LA68 and Lactobacillus plantarum WCFS1 differently influence metabolic and immunological parameters in high fat diet-induced hypercholesterolemia and hepatic steatosis. Food Funct 6(2):558–565
Ivanovic N, Minic R, Djuricic I, Skodric SR, Zivkovic I, Sobajic S, Djordjevic B (2016) Active Lactobacillus rhamnosus LA68 or Lactobacillus plantarum WCFS1 administration positively influences liver fatty acid composition in mice on a HFD regime. Food Funct 7(6):2840–2848
John GK, Wang L, Nanavati J, Twose C, Singh R, Mullin G (2018) Dietary alteration of the gut microbiome and its impact on weight and fat mass: a systematic review and meta-analysis. Genes (Basel) 9(3). https://doi.org/10.3390/genes9030167
Kho ZY, Lal SK (2018) The human gut microbiome—a potential controller of wellness and disease. Front Microbiol 9(1835). https://doi.org/10.3389/fmicb.2018.01835
Kim G, Bae J, Kim MJ, Kwon H, Park G, Kim S-J et al (2020) Delayed establishment of gut microbiota in infants delivered by cesarean section. Front Microbiol 11:2099
Koch W (2019) Dietary polyphenols—important non-nutrients in the prevention of chronic noncommunicable diseases. A systematic review. Nutrients 11(5):1039
Koeth RA, Wang Z, Levison BS, Buffa JA, Org E, Sheehy BT et al (2013) Intestinal microbiota metabolism of L-carnitine, a nutrient in red meat, promotes atherosclerosis. Nat Med 19(5):576–585. https://doi.org/10.1038/nm.3145
Kolodziejczyk AA, Zheng D, Elinav E (2019) Diet–microbiota interactions and personalized nutrition. Nat Rev Microbiol 17(12):742–753
Korem T, Zeevi D, Zmora N, Weissbrod O, Bar N, Lotan-Pompan M et al (2017) Bread affects clinical parameters and induces gut microbiome-associated personal glycemic responses. Cell Metab 25(6):1243–1253, e1245. https://doi.org/10.1016/j.cmet.2017.05.002
Korpela K, Flint HJ, Johnstone AM, Lappi J, Poutanen K, Dewulf E et al (2014) Gut microbiota signatures predict host and microbiota responses to dietary interventions in obese individuals. PLoS ONE 9(6):e90702. https://doi.org/10.1371/journal.pone.0090702
Koutnikova H, Genser B, Monteiro-Sepulveda M, Faurie JM, Rizkalla S, Schrezenmeir J, Clement K (2019) Impact of bacterial probiotics on obesity, diabetes and non-alcoholic fatty liver disease related variables: a systematic review and meta-analysis of randomised controlled trials. BMJ Open 9(3):e017995. https://doi.org/10.1136/bmjopen-2017-017995
Kovatcheva-Datchary P, Nilsson A, Akrami R, Lee YS, De Vadder F, Arora T et al (2015) Dietary fiber-induced improvement in glucose metabolism is associated with increased abundance of prevotella. Cell Metab 22(6):971–982. https://doi.org/10.1016/j.cmet.2015.10.001
Kuhn T, Rohrmann S, Sookthai D, Johnson T, Katzke V, Kaaks R et al (2017) Intra-individual variation of plasma trimethylamine-*N*-oxide (TMAO), betaine and choline over 1 year. Clin Chem Lab Med 55(2):261–268. https://doi.org/10.1515/cclm-2016-0374
Kumar J, Rani K, Datt C (2020) Molecular link between dietary fibre, gut microbiota and health. Mol Biol Rep 1–9
Kviatcovsky D, Zheng D, Elinav E (2021) Gut microbiome and its potential link to personalized nutrition. Curr Opin Physiol
Latifi SM, Karandish M, Shahbazian H, Cheraghian B, Moradi M (2017) Prevalence of metabolically healthy obesity (MHO) and its relation with incidence of metabolic syndrome, hypertension and type 2 diabetes amongst individuals aged over 20 years in Ahvaz: a 5 year cohort study (2009–2014). Diab Metab Syndr 11:S1037–S1040
Le Barz M, Anhê FF, Varin TV, Desjardins Y, Levy E, Roy D et al (2015) Probiotics as complementary treatment for metabolic disorders. Diab Metab J 39(4):291–303

Leeming ER, Johnson AJ, Spector TD, Le Roy CI (2019) Effect of diet on the gut microbiota: rethinking intervention duration. Nutrients 11(12):2862

Ley RE (2010) Obesity and the human microbiome. Curr Opin Gastroenterol 26(1):5–11

Ley RE, Backhed F, Turnbaugh P, Lozupone CA, Knight RD, Gordon JI (2005) Obesity alters gut microbial ecology. Proc Natl Acad Sci U S A 102(31):11070–11075. https://doi.org/10.1073/pnas.0504978102

Ley RE, Turnbaugh PJ, Klein S, Gordon JI (2006) Microbial ecology: human gut microbes associated with obesity. Nature 444(7122):1022–1023. https://doi.org/10.1038/4441022a

Liu K, Zhou R, Wang B, Mi M-T (2014) Effect of resveratrol on glucose control and insulin sensitivity: a meta-analysis of 11 randomized controlled trials. Am J Clin Nutr 99(6):1510–1519

Lozupone CA, Stombaugh JI, Gordon JI, Jansson JK, Knight R (2012) Diversity, stability and resilience of the human gut microbiota. Nature 489(7415):220–230

Maldonado-Gomez MX, Martinez I, Bottacini F, O'Callaghan A, Ventura M, van Sinderen D et al (2016) Stable engraftment of *Bifidobacterium longum* ah1206 in the human gut depends on individualized features of the resident microbiome. Cell Host Microbe 20(4):515–526. https://doi.org/10.1016/j.chom.2016.09.001

Mameli C, Zuccotti GV, Carnovale C, Galli E, Nannini P, Cervia D, Perrotta C (2017) An update on the assessment and management of metabolic syndrome, a growing medical emergency in paediatric populations. Pharmacol Res 119:99–117. https://doi.org/10.1016/j.phrs.2017.01.017

Martinez I, Lattimer JM, Hubach KL, Case JA, Yang J, Weber CG et al (2013) Gut microbiome composition is linked to whole grain-induced immunological improvements. ISME J 7(2):269–280. https://doi.org/10.1038/ismej.2012.104

McCracken E, Monaghan M, Sreenivasan S (2018) Pathophysiology of the metabolic syndrome. Clin Dermatol 36(1):14–20. https://doi.org/10.1016/j.clindermatol.2017.09.004

Mehta NN, Heffron SP, Patel PN, Ferguson J, Shah RD, Hinkle CC et al (2012) A human model of inflammatory cardio-metabolic dysfunction; a double blind placebo-controlled crossover trial. J Transl Med 10:124. https://doi.org/10.1186/1479-5876-10-124

Mills S, Lane JA, Smith GJ, Grimaldi KA, Ross RP, Stanton C (2019a) Precision nutrition and the microbiome part II: potential opportunities and pathways to commercialisation. Nutrients 11(7). https://doi.org/10.3390/nu11071468

Mills S, Stanton C, Lane JA, Smith GJ, Ross RP (2019b) Precision nutrition and the microbiome, part I: current state of the science. Nutrients 11(4). https://doi.org/10.3390/nu11040923

Montassier E, Valdés-Mas R, Batard E, Zmora N, Dori-Bachash M, Suez J, Elinav E (2021) Probiotics impact the antibiotic resistance gene reservoir along the human GI tract in a person-specific and antibiotic-dependent manner. Nat Microbiol 6(8):1043–1054

Moore RE, Townsend SD (2019) Temporal development of the infant gut microbiome. Open Biol 9(9):190128

Moorthy M, Sundralingam U, Palanisamy UD (2021) Polyphenols as prebiotics in the management of high-fat diet-induced obesity: a systematic review of animal studies. Foods 10(2). https://doi.org/10.3390/foods10020299

Norman JM, Handley SA, Baldridge MT, Droit L, Liu CY, Keller BC et al (2015) Disease-specific alterations in the enteric virome in inflammatory bowel disease. Cell 160(3):447–460. https://doi.org/10.1016/j.cell.2015.01.002

Ouwehand A (2017) A review of dose-responses of probiotics in human studies. Beneficial Microbes 8(2):143–151

Park S, Bae JH (2015) Probiotics for weight loss: a systematic review and meta-analysis. Nutr Res 35(7):566–575. https://doi.org/10.1016/j.nutres.2015.05.008

Parker A, Lawson MA, Vaux L, Pin C (2018) Host-microbe interaction in the gastrointestinal tract. Environ Microbiol 20(7):2337–2353

Petraroli M, Castellone E, Patianna V, Esposito S (2021) Gut microbiota and obesity in adults and children: the state of the art. Front Pediatr 9:657020. https://doi.org/10.3389/fped.2021.657020

Pigeyre M, Yazdi FT, Kaur Y, Meyre D (2016) Recent progress in genetics, epigenetics and metagenomics unveils the pathophysiology of human obesity. Clin Sci 130(12):943–986

Pique N, Berlanga M, Minana-Galbis D (2019) Health benefits of heat-killed (tyndallized) probiotics: an overview. Int J Mol Sci 20(10). https://doi.org/10.3390/ijms20102534

Plovier H, Everard A, Druart C, Depommier C, Van Hul M, Geurts L et al (2017) A purified membrane protein from *Akkermansia muciniphila* or the pasteurized bacterium improves metabolism in obese and diabetic mice. Nat Med 23(1):107–113. https://doi.org/10.1038/nm.4236

Precup G, Vodnar D-C (2019) Gut Prevotella as a possible biomarker of diet and its eubiotic versus dysbiotic roles: a comprehensive literature review. Br J Nutr 122(2):131–140

Pushpass RG, Alzoufairi S, Jackson KG, Lovegrove JA (2021) Circulating bile acids as a link between the gut microbiota and cardiovascular health: impact of prebiotics, probiotics and polyphenol-rich foods. Nutr Res Rev, 1–20.https://doi.org/10.1017/S0954422421000081

Qin J, Li Y, Cai Z, Li S, Zhu J, Zhang F et al (2012) A metagenome-wide association study of gut microbiota in type 2 diabetes. Nature 490(7418):55–60

Rannikko J, Holmberg V, Karppelin M, Arvola P, Huttunen R, Mattila E et al (2021) Fungemia and other fungal infections associated with use of *Saccharomyces boulardii* probiotic supplements. Emerg Infect Dis 27(8). https://doi.org/10.3201/eid2708.210018

Ray SK, Mukherjee S (2021) Evolving Interplay between dietary polyphenols and gut microbiota—an emerging importance in healthcare. Front Nutr 8

Reynés B, Palou M, Rodríguez AM, Palou A (2019) Regulation of adaptive thermogenesis and browning by prebiotics and postbiotics. Front Physiol 9:1908

Rodriguez-Daza MC, Pulido-Mateos EC, Lupien-Meilleur J, Guyonnet D, Desjardins Y, Roy D (2021) Polyphenol-mediated gut microbiota modulation: toward prebiotics and further. Front Nutr 8:689456. https://doi.org/10.3389/fnut.2021.689456

Romo-Vaquero M, Cortés-Martín A, Loria-Kohen V, Ramírez-de-Molina A, García-Mantrana I, Collado MC et al (2019) Deciphering the human gut microbiome of urolithin metabotypes: association with enterotypes and potential cardiometabolic health implications. Mol Nutr Food Res 63(4):1800958

Rothschild D, Weissbrod O, Barkan E, Kurilshikov A, Korem T, Zeevi D et al (2018) Environment dominates over host genetics in shaping human gut microbiota. Nature 555(7695):210–215

Rowland I, Gibson G, Heinken A, Scott K, Swann J, Thiele I, Tuohy K (2018) Gut microbiota functions: metabolism of nutrients and other food components. Eur J Nutr 57(1):1–24

Salminen S, Collado MC, Endo A, Hill C, Lebeer S, Quigley EM et al (2021) The International Scientific Association of Probiotics and Prebiotics (ISAPP) consensus statement on the definition and scope of postbiotics. Nat Rev Gastroenterol Hepatol 1–19

Santos-Marcos JA, Perez-Jimenez F, Camargo A (2019) The role of diet and intestinal microbiota in the development of metabolic syndrome. J Nutr Biochem 70:1–27. https://doi.org/10.1016/j.jnutbio.2019.03.017

Shimizu H, Masujima Y, Ushiroda C, Mizushima R, Taira S, Ohue-Kitano R, Kimura I (2019) Dietary short-chain fatty acid intake improves the hepatic metabolic condition via FFAR3. Sci Rep 9(1):1–10

Singh RK, Chang H-W, Yan D, Lee KM, Ucmak D, Wong K et al (2017) Influence of diet on the gut microbiome and implications for human health. J Transl Med 15(1):1–17

Singla RK, Dubey AK, Garg A, Sharma RK, Fiorino M, Ameen SM et al (2019) Natural polyphenols: chemical classification, definition of classes, subcategories, and structures. Oxford University Press

Suez J, Korem T, Zeevi D, Zilberman-Schapira G, Thaiss CA, Maza O et al (2014) Artificial sweeteners induce glucose intolerance by altering the gut microbiota. Nature 514(7521):181–186. https://doi.org/10.1038/nature13793

Suez J, Zmora N, Zilberman-Schapira G, Mor U, Dori-Bachash M, Bashiardes S et al (2018) Post-antibiotic gut mucosal microbiome reconstitution is impaired by probiotics and improved by autologous FMT. Cell 174(6):1406–1423, e1416

Suez J, Zmora N, Segal E, Elinav E (2019) The pros, cons, and many unknowns of probiotics. Nat Med 25(5):716–729

Suez J, Zmora N, Elinav E (2020) Probiotics in the next-generation sequencing era. Gut Microbes 11(1):77–93
Tan Y, Tam CC, Rolston M, Alves P, Chen L, Meng S et al (2021) Quercetin ameliorates insulin resistance and restores gut microbiome in mice on high-fat diets. Antioxidants 10(8):1251
Tang J, Wu X, Mou M, Wang C, Wang L, Li F et al (2021) GIMICA: host genetic and immune factors shaping human microbiota. Nucleic Acids Res 49(D1):D715–D722
Teixeira TF, Grzeskowiak L, Franceschini SC, Bressan J, Ferreira CL, Peluzio MC (2013) Higher level of faecal SCFA in women correlates with metabolic syndrome risk factors. Br J Nutr 109(5):914–919. https://doi.org/10.1017/S0007114512002723
Thomson P, Medina DA, Garrido D (2018) Human milk oligosaccharides and infant gut bifidobacteria: molecular strategies for their utilization. Food Microbiol 75:37–46
Tseng CH, Wu CY (2019) The gut microbiome in obesity. J Formos Med Assoc 118(Suppl 1):S3–S9. https://doi.org/10.1016/j.jfma.2018.07.009
Turnbaugh PJ, Ley RE, Mahowald MA, Magrini V, Mardis ER, Gordon JI (2006) An obesity-associated gut microbiome with increased capacity for energy harvest. Nature 444(7122):1027–1031. https://doi.org/10.1038/nature05414
Usui T, Tochiya M, Sasaki Y, Muranaka K, Yamakage H, Himeno A et al (2013) Effects of natural S-equol supplements on overweight or obesity and metabolic syndrome in the Japanese, based on sex and equol status. Clin Endocrinol 78(3):365–372
Vallianou N, Stratigou T, Christodoulatos GS, Dalamaga M (2019) Understanding the role of the gut microbiome and microbial metabolites in obesity and obesity-associated metabolic disorders: current evidence and perspectives. Curr Obes Rep 8(3):317–332. https://doi.org/10.1007/s13679-019-00352-2
Vallianou N, Stratigou T, Christodoulatos GS, Tsigalou C, Dalamaga M (2020) Probiotics, prebiotics, synbiotics, postbiotics, and obesity: current evidence, controversies, and perspectives. Curr Obes Rep 9(3):179–192. https://doi.org/10.1007/s13679-020-00379-w
Van Hul M, Cani PD (2019) Targeting carbohydrates and polyphenols for a healthy microbiome and healthy weight. Curr Nutr Rep 8(4):307–316. https://doi.org/10.1007/s13668-019-00281-5
Vrieze A, Holleman F, Zoetendal E, De Vos W, Hoekstra J, Nieuwdorp M (2010) The environment within: how gut microbiota may influence metabolism and body composition. Diabetologia 53(4):606–613
Wang G, Huang S, Wang Y, Cai S, Yu H, Liu H et al (2019) Bridging intestinal immunity and gut microbiota by metabolites. Cell Mol Life Sci 76(20):3917–3937
Wegh CAM, Geerlings SY, Knol J, Roeselers G, Belzer C (2019) Postbiotics and their potential applications in early life nutrition and beyond. Int J Mol Sci 20(19). https://doi.org/10.3390/ijms20194673
Xavier-Santos D, Bedani R, Lima ED, Saad SMI (2020) Impact of probiotics and prebiotics targeting metabolic syndrome. J Funct Foods 64:103666
Yadav M, Verma MK, Chauhan NS (2018) A review of metabolic potential of human gut microbiome in human nutrition. Arch Microbiol 200(2):203–217
Yang CS, Zhang J, Zhang L, Huang J, Wang Y (2016) Mechanisms of body weight reduction and metabolic syndrome alleviation by tea. Mol Nutr Food Res 60(1):160–174. https://doi.org/10.1002/mnfr.201500428
Yelin I, Flett KB, Merakou C, Mehrotra P, Stam J, Snesrud E et al (2019) Genomic and epidemiological evidence of bacterial transmission from probiotic capsule to blood in ICU patients. Nat Med 25(11):1728–1732
Yu WW, Randhawa AK, Blair SN, Sui X, Kuk JL (2019) Age-and sex-specific all-cause mortality risk greatest in metabolic syndrome combinations with elevated blood pressure from 7 US cohorts. PLoS ONE 14(6):e0218307
Zafar U, Khaliq S, Ahmad HU, Manzoor S, Lone KP (2018) Metabolic syndrome: an update on diagnostic criteria, pathogenesis, and genetic links. Hormones (athens) 17(3):299–313. https://doi.org/10.1007/s42000-018-0051-3

Zeevi D, Korem T, Zmora N, Israeli D, Rothschild D, Weinberger A et al (2015) Personalized Nutrition by Prediction of Glycemic Responses. Cell 163(5):1079–1094. https://doi.org/10.1016/j.cell.2015.11.001

Zeng S-L, Li S-Z, Xiao P-T, Cai Y-Y, Chu C, Chen B-Z et al (2020) Citrus polymethoxyflavones attenuate metabolic syndrome by regulating gut microbiome and amino acid metabolism. Sci Adv 6(1):eaax6208

Zeng Q, Yang Z, Wang F, Li D, Liu Y, Wang D et al (2021) Association between metabolic status and gut microbiome in obese populations. Microb Genomics 7(8):000639. https://doi.org/10.1099/mgen.0.000639

Zhang C, Derrien M, Levenez F, Brazeilles R, Ballal SA, Kim J et al (2016) Ecological robustness of the gut microbiota in response to ingestion of transient food-borne microbes. ISME J 10(9):2235–2245

Zhao X, Han Q, Liu Y, Sun C, Gang X, Wang G (2016) The relationship between branched-chain amino acid related metabolomic signature and insulin resistance: a systematic review. J Diab Res 2016

Zhao L, Zhang Q, Ma W, Tian F, Shen H, Zhou M (2017) A combination of quercetin and resveratrol reduces obesity in high-fat diet-fed rats by modulation of gut microbiota. Food Funct 8(12):4644–4656

Zheng W, Ma Y, Zhao A, He T, Lyu N, Pan Z et al (2019) Compositional and functional differences in human gut microbiome with respect to equol production and its association with blood lipid level: a cross-sectional study. Gut Pathogens 11(1):20

Zheng D, Liwinski T, Elinav E (2020) Interaction between microbiota and immunity in health and disease. Cell Res 30(6):492–506

Zhernakova A, Kurilshikov A, Bonder MJ, Tigchelaar EF, Schirmer M, Vatanen T et al (2016) Population-based metagenomics analysis reveals markers for gut microbiome composition and diversity. Science 352(6285):565–569

Zimmet P, Alberti K, Stern N, Bilu C, El-Osta A, Einat H, Kronfeld-Schor N (2019) The Circadian syndrome: is the metabolic syndrome and much more! J Intern Med 286(2):181–191. https://doi.org/10.1111/joim.12924

Zmora N, Zeevi D, Korem T, Segal E, Elinav E (2016) Taking it personally: personalized utilization of the human microbiome in health and disease. Cell Host Microbe 19(1):12–20. https://doi.org/10.1016/j.chom.2015.12.016

Zmora N, Zilberman-Schapira G, Suez J, Mor U, Dori-Bachash M, Bashiardes S et al (2018) Personalized gut mucosal colonization resistance to empiric probiotics is associated with unique host and microbiome features. Cell 174(6):1388–1405, e1321

Zolkiewicz J, Marzec A, Ruszczynski M, Feleszko W (2020) Postbiotics-a step beyond pre- and probiotics. Nutrients 12(8). https://doi.org/10.3390/nu12082189

Zouiouich S, Loftfield E, Huybrechts I, Viallon V, Louca P, Vogtmann E et al (2021) Markers of metabolic health and gut microbiome diversity: findings from two population-based cohort studies. Diabetologia 64(8):1749–1759. https://doi.org/10.1007/s00125-021-05464-w

Precision Nutrition and Metabolomics, a Model of Alzheimer's Disease

7

Stefan Ledinger, Carmen Ludwig-Papst, and Matthias Scheffler

Contents

Abstract

Metabolomics offers unique insights into the interface between environment, host, and microbiome in the pathophysiology of disease. New metabolomics findings can enhance the understanding obtained through classical molecular biology research and lead to actionable targets not just for diagnosis but also for prevention and therapeutic strategies. In this chapter, we discuss how the typical Western diet contributes to a variety of chronic diseases, through both

S. Ledinger · C. Ludwig-Papst (✉) · M. Scheffler
Biocrates Life Sciences AG, Eduard-Bodem-Gasse 8, 6020 Innsbruck, Austria
e-mail: carmenludwig@me.com; carmen.ludwig@biocrates.com

S. Ledinger
e-mail: stefan.ledinger@biocrates.com

A. G. Haslberger (ed.), *Advances in Precision Nutrition, Personalization and Healthy Aging*, https://doi.org/10.1007/978-3-031-10153-3_7

systemic effects and alterations of the microbiome. Alzheimer's disease (AD) is presented as a case study for exploring how nutritional interventions can support efforts to prevent and manage disease. We discuss how nutritional approaches affect systemic and local biochemistry, potentially reducing the risk for AD and slowing disease progression. Finally, we establish a metabolic model of AD pathogenesis by describing the molecular pathways and pathophysiological that are affected by a Western-style diet, and the changes in the gut microbiome that accompany it. For this purpose, we draw from recent findings of metabolomics studies in epidemiological, clinical, and basic research settings. In conclusion, we present how metabolomics can provide a roadmap for further research in chronic disease, including AD and other neurodegenerative diseases, as well as provide a basis for new interventions based on a thorough understanding of the biochemical processes involved in the respective pathologies.

7.1 Introduction

The role of nutrition and lifestyle in complex, chronic, and age-related diseases is undeniable. There is growing interest in the microbiome's involvement in the pathogenesis of these diseases. Exactly how nutrition affects the bioenergetic state and fate of cells is not yet fully understood.

Metabolomics is a promising tool to help disentangle these complex interactions and processes. In this chapter, we will summarize current knowledge, focusing on the development of neuropsychiatric diseases such as Alzheimer's disease.

7.2 Metabolomics and the -Omics Cascade

Before we discuss the association between nutrition, Alzheimer's disease, and metabolism, the concept of metabolomics shall briefly be introduced.

Metabolomics is an omics technology that aims at the characterization of intermediates of metabolism with low molecular weight, i.e., metabolites. Although there are feedback loops through which metabolites determine upstream—omics levels, metabolites are typically considered the end point of biochemical processes. Consequently, metabolic profiles are considered to represent the biochemical phenotype of an individual or sample.

The recent surge of the use of metabolomics is probably fueled by the high information content inherent to metabolic profiling. While genes are easy to measure, consisting of only four nucleobases, they do not necessarily inform about the functional importance in a specific disease context at a certain time. Metabolites, on the other hand, are very diverse with assumed tens of thousands of endogenous intermediates of metabolism believed to exist. In addition, they show

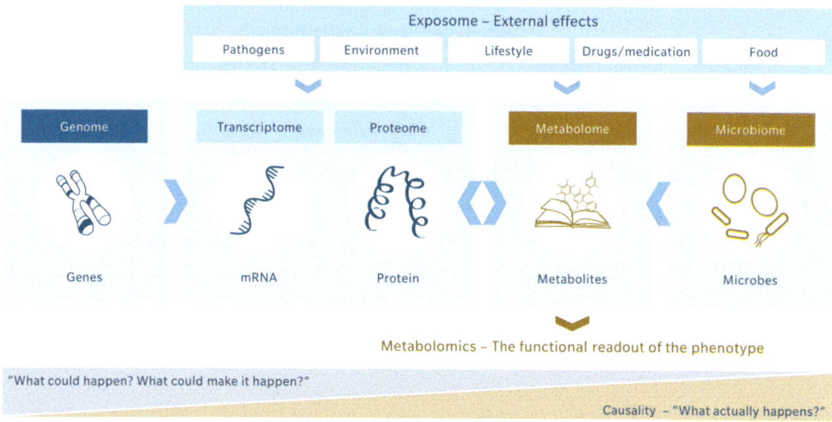

Fig. 7.1 Metabolomics in relationship to other—omics technologies

dynamic changes with factors such as age, lifestyle, and even circadian rhythmicity. This dynamic regulatory landscape provides meaningful functional insights as metabolite alterations are assigned to pathophysiological processes (Fig. 7.1).

Metabolites such as creatinine and glucose have been in diagnostic use for decades. Thanks to the rapid development of analytical technology, a broader range of metabolites can now be analyzed in parallel, enabling researchers to investigate diverse sets of molecules in relation to each other rather than relying on few selected substances.

7.3 How Metabolomics Provides Actionable Insights into Disease Pathophysiology

Here, we explore why metabolomics is so well-suited to the study of the nutrition-microbiome-disease axis, and how it can provide the basis for personalized therapies.

Most tissues have a large degree of metabolic flexibility. In other words, an organism can remain healthy even if the optimal set of nutrients is unavailable or cannot be produced endogenously. For example, glucose oxidation is the preferred route for energy production in many tissues, but if sufficient glucose is unavailable, most cell types can continue to function by utilizing other sources of energy. The switch from glucose consumption to lipid mobilization and oxidation in endurance training is a good example of this principle.

It has been hypothesized that, while homeostasis could be maintained for a certain period imbalances occur long before there's any evidence of manifest and symptomatic disease. These imbalances eventually cause subtle changes in the biochemical profiles of cells, tissues, and biofluids that could be detected with metabolomics (Fig. 7.2).

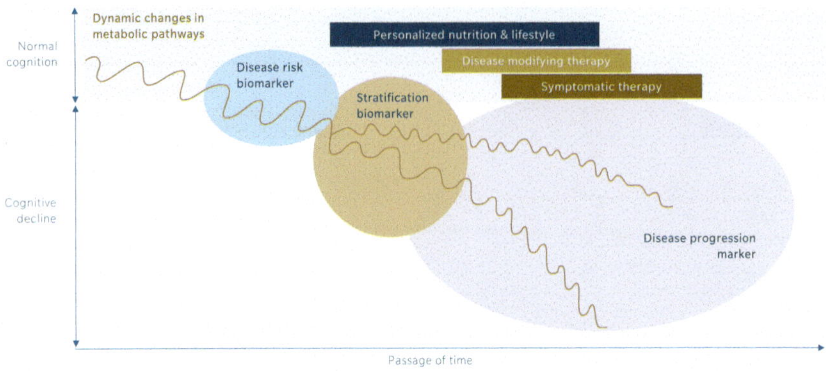

Fig. 7.2 Use of metabolomics in biomarker research; Adapted from D. I. Ellis et al., Metabolic fingerprinting as a diagnostic tool, Pharmacogenomics (2007)

Knowing what drives early disease pathophysiology can also inform pharmaceutical approaches, nutrition/lifestyle interventions, and choices to help prevent or delay the onset of symptomatic disease. Metabolomics has already been instrumental in improving our understanding of both advanced and early stages in the pathophysiology of the most common chronic diseases.

Here, we discuss Alzheimer's disease as a case study, though similar observations could be made in a wide variety of diseases. In neurodegenerative diseases such as AD, pharmaceutical therapy remains challenging, and personalized nutrition and lifestyle recommendations could be instrumental in fighting the rise in incidence and disease-related socioeconomic cost.

7.4 Western-Style Diet, Metabolism, and the Epidemiology of Chronic Disease

In the USA, six in ten adults have a chronic disease, and unhealthy lifestyle choices are a known contributor (CDC 2021). As a consequence, nutritional guidelines now tend to focus on encouraging nutritional choices that help avoid chronic disease, rather than securing the sufficient intake of essential nutrients (Hite 2018).

A comprehensive review on nutritional epidemiology is beyond the scope of this chapter, but it is worth mentioning a few recent notable findings.

Body mass index (BMI), while an imperfect measure of both body composition and lifestyle, provides insights into the interaction between the Western-style diet and chronic disease. Field et al. (2001) showed that the risk for multiple chronic diseases rises with increasing BMI. Unsurprisingly, multiple metabolomics studies have shown vast metabolic consequences of overweight and obesity in both adult and adolescent populations (Rangel-Huerta et al. 2019; Handakas et al. 2021). In a recent study, Frigerio et al. (2021) confirmed an association between increasing BMI and reduced levels of glutamine, a precursor of the excitatory neurotransmitters glutamate and aspartate.

Levels of many other amino acids, including branched-chain amino acids (BCAAs), have also been shown to increase with BMI. High levels of the BCAAs valine, leucine, and isoleucine (which are essential food components) have been identified as a causal factor in diabetes pathogenesis and insulin resistance (Lotta et al. 2016; Magnusson et al. 2013). BCAAs and ketogenic acids such as leucine and isoleucine have been shown to function as insulin analogs, thus inducing insulin secretion or insulin resistance if chronically elevated (Wishart 2019), and potentially contributing to a pro-inflammatory phenotype (Ruge et al. 2009). In addition, elevated saturated chain triglycerides (TAGs) (Rhee et al. 2011) have been implicated as a major driver of diabetes.

Diabetes is an established risk factor for AD (Leibson et al. 1997; Luchsinger et al. 2001; Ott et al. 1999). Obesity has also been implicated as a risk factor for the development of AD, with glutamate thought to be an important factor (Ma et al. 2020; Lloret et al. 2019).

Finally, the results implicating a typical Western lifestyle with AD have sparked significant interest in the role of the microbiome. In a review of recent literature, Jiang et al. (2017) considered how alterations in the microbiome affect the risk for and pathogenesis of AD.

7.5 Western-Style Diet, the Association with Intracellular Malnutrition, and Nutritional Interventions in Alzheimer's Disease

For these reasons, nutrition has become a matter of growing interest for the prevention and management of chronic conditions, including AD and other neurodegenerative diseases. For example, Więckowska-Gacek et al. (2021) hypothesized that the Western-style diet could trigger Alzheimer's disease through both systemic inflammation and neuroinflammation, leading to neurodegeneration. Kim et al. (2015) showed that adhering to Korean national nutritional guidelines can reduce the risk for developing AD.

As outlined above, excessive BCAA levels are associated with obesity and insulin resistance. They also affect brain biochemistry, as circulating levels of BCAAs inhibit the uptake of other amino acids in the brain, altering the production of various neurotransmitters (Pardridge 1977; Fernstrom 2005). High-BCAA diets have also been associated with lower cortical levels of threonine and tryptophan, with a negative impact on cognition (Tournissac et al. 2018).

These findings show how malnutrition on a systemic level can lead to metabolic profile alterations in tissues, and malnutrition in cells that must maintain homeostasis under less-than-ideal circumstances.

Naturally, reports of an association between nutritional factors and AD risk have prompted investigations into how specific modes of nutrition influence risk.

One nutritional approach that has received a great deal of attention is the ketogenic diet. Recent findings on its effects have been reviewed at length in de la Rubia et al. (2021), Broom et al. (2019) and Rusek et al. (2019). As the name suggests, a ketogenic diet induces a state of ketosis. It simulates a state of fasting, potentially inducing opposite effects of the usual Western-style diet. A recent randomized trial showed that following a ketogenic diet for 12 weeks led to improvements in quality of life and daily function for patients with AD (Phillips et al. 2021).

The Mediterranean-style diet has also been investigated as a potential nutritional approach to reducing AD risk. Gu et al. (2010) hypothesized that inflammatory and metabolic pathways could change the association between the Mediterranean diet and reduced AD risk, but the classical inflammation marker high-sensitivity C-reactive protein (hsCRP), adiponectin, and fasting insulin, did not seem to mediate the effect. Berti et al. (2018) found that long-term adherence to a Mediterranean diet slows progression of selected AD biomarkers. More sensitive biomarkers are needed to capture the biochemical processes that mediate the interactions of the nutrition-microbiome-AD axis.

7.6 A Metabolic Model of Alzheimer's Disease, and Its Relationship with Nutritional and Microbiome-Related Factors

This body of evidence shows conclusively that malnutrition contributes to obesity, changes in the microbiome, and metabolic risk factors involved in the pathophysiological processes of AD development, such as insulin resistance, diabetes, and inflammation. We have also hinted at the role of selected metabolites and shown that classical biomarkers are insufficient to fully elucidate the biochemical processes at play.

This is where metabolomics comes in.

The first studies of the interaction between metabolome and nutrition in the context of AD were published in the early 2010s. Grimm et al. (2011) found that certain lipids are strongly reduced in post-mortem brain tissue from AD patients compared to that of non-AD individuals. Interestingly, the structure of those lipids suggests that a diet rich in polyunsaturated fatty acids might be protective against AD. These findings led to clinical trials investigating the effect of dietary supplementation with these nutrients. Recent results indicate that this approach could improve outcomes by slowing disease progression (Soininen et al. 2021). However, these results also show that a prolonged intervention is required for significant effects.

Confirming the role of complex lipids, Mapstone et al. (2014) used metabolomics to identify the first blood-based biomarker signature for prediction of a future AD diagnosis with very high accuracy, based on 10 lipid metabolites.

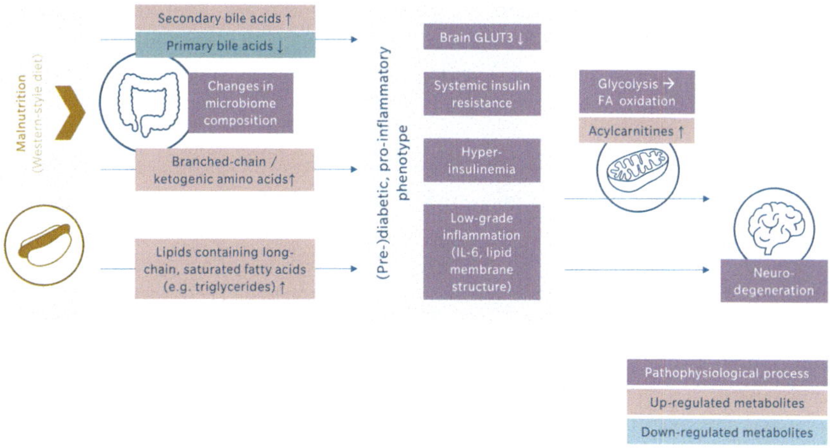

Fig. 7.3 Metabolic model of AD: An event cascade describing AD pathophysiology by the means of metabolomics

This study was ground-breaking in describing a biomarker signature for a neurodegenerative order that could be easily obtained through blood sampling, rather than through the invasive collection of cerebrospinal fluid.

In the following paragraphs, we will discuss how the field has evolved since those early findings and describe the clinical phenotype of Alzheimer's disease. Figure 7.3 shows a metabolomics model of AD, demonstrating how the Western diet drives cellular malnutrition and thus contributes to AD pathophysiology. These processes are driven by a (pre-) diabetic phenotype and changes in the gut microbiome. Changes in circulatory and tissue metabolites can be observed occurring in parallel and in sequence. Lipids, glucose, branched-chain amino acids, ketogenic amines, and bile acids all play a role in AD risk and development.

7.6.1 Lipid Metabolism

As we've established, a Western-style diet, typically rich in fat and carbohydrates and low in polyunsaturated fatty acids (PUFAs), is thought to influence AD pathophysiology (Gustafson et al. 2020). A diet that's too low in PUFAs could lead to a reduction in cell membrane fluidity (Abedi and Sahari 2014). Low-grade inflammation may contribute to the association between Western lifestyle, changes in the gut microbiome, (lipid) metabolism, and development of AD (Sochocka et al. 2019; Gentile and Weir 2018). In a recent study, mice fed a high-fat diet showed increased levels of several important lipid metabolites, but decreased levels of eicosapentaenoic acid (EPA) in serum, brain, and other tissues. EPA is important in maintaining a healthy membrane composition and a precursor to eicosanoids, which are intimately involved in immune regulation (Pakiet et al. 2019). In an analysis of three cohorts within the Alzheimer's disease neuroimaging initiative (ADNI), Toledo et al. (2017) confirmed changes in phospholipid levels as early metabolic events in the AD pathophysiology, citing changes in membrane composition as a probable explanation.

7.6.2 Glucose Homeostasis and Alterations in Energy Supply

As we have described above, diabetes (characterized by high circulatory glucose levels and peripheral insulin resistance) is a risk factor for AD. In fact, Alzheimer's disease has been dubbed "type 3" diabetes. The fact that branched-chain and ketogenic amino acids, such as valine and α-aminoadipic acid, have been found in symptomatic stages of AD (Toledo et al. 2017) adds to the hypothesis that diabetes and pre-diabetic phenotypes are involved in the development of AD. This association exists even though the supply of glucose to the brain, which constitutes the main energy source, mainly depends on an insulin-independent glucose transporter GLUT3.

A study by An et al. (2018) showed that increased brain glucose levels and reduced neuronal expression of insulin-independent glucose transporters (GLUT3) were associated with the severity of AD in post-mortem brains. The authors proposed that lower levels of glucose transporters reduced glycolytic flow, forcing the brain to utilize sources other than glucose for energy production. The hypothesis of altered glucose metabolism as a hallmark of AD pathogenesis was later confirmed in a large-scale proteomics study (Johnson et al. 2020).

To satisfy their energy requirements, the metabolism of cells probably shifts from glycolysis to fatty acid (FA) oxidation. These changes in energy metabolism are probably a late event in AD pathogenesis (Toledo et al. 2017). Metabolically, this is reflected by altered circulatory levels of multiple acylcarnitines in advanced stages, which are involved in mitochondrial beta oxidation of fatty acids. Peripheral insulin resistance and hyperinsulinemia may lead to increased lipolysis in adipose tissue. Triacylglycerol can cross the blood–brain barrier and supply energy demand through fatty acids.

7.6.3 Bile Acid Metabolism

Although multiple microbiota-related metabolites have been associated with AD, here, we will focus on bile acids (BAs).

Bile acids are one endpoint of cholesterol metabolism. Primary BAs are produced in the liver and secreted into the intestine via enterohepatic circulation. In the large intestine, bile acids are irreversibly modified by certain gut bacteria. BAs not only play a role in fat absorption, but also act as important hormone-like signaling molecules. They are involved in insulin signaling (Ahmad and Haeusler 2019) and have toxic effects on mitochondria (Krähenbühl et al. 1994). BAs are involved in multiple signaling pathways, for example, through the TGR5 and FXR receptors. FXR has also been found to be involved in immune regulation (Schote et al. 2007; Vavassori et al. 2009; Fiorucci et al. 2018).

In routine clinical biochemistry, bile acids are typically measured as a sum, but primary and secondary bile acids have very different actions on FXR: some induce strong activation, some have no affinity at physiological levels, and others have antagonistic effects (Lefebvre et al. 2009; Sayin et al. 2013). Mahmoudian-Dehkordi et al. (2019) showed that serum levels of secondary bile acids, produced by specific gut bacteria, increase the risk of AD. Specifically, the authors showed that the ratio between the secondary bile acid deoxycholic acid and the primary bile acid cholic acid was associated with cognitive decline. Bile acid signatures are also correlated with classical AD biomarkers (Amyloid, Tau and Neurodegeneration, referred to as ("A/T/N")) (Nho et al. 2019). As bile acids regulate their own production, it is plausible that both bile acid metabolism and bile acid synthesis are affected in AD (Baloni et al. 2020).

7.6.4 The Metabolic Model of Alzheimer's Disease (AD)

Taking account of the processes described above, the extant literature on metabolomics in Alzheimer's disease indicates a cascade of events. A Western-style diet increases the risk of AD by creating an unfavorable lipid composition, contributing to a diabetic phenotype and sub-clinical inflammation. This phenotype is also promoted by an overload of branched-chain and ketogenic amino acids, through the induction of insulin resistance and hyperinsulinemia.

In addition, altered profiles of gut bacteria lead to an increased production of secondary bile acids with effects via FXR, which further deteriorates glucose homeostasis. Although a direct link has not been proven experimentally, this may contribute to a reduced GLUT3 expression in neurons. These processes create cellular malnutrition and a shift of energy supply from glycolysis to fatty acid oxidation. In the long run, this will cause an overload and partial dysfunction of mitochondria. Consequently, incompletely metabolized acylcarnitines will accumulate.

The proposed model is not only in line with recent metabolomics literature in AD, but also with the hypothesis that we have stated in the beginning of the chapter. While metabolic signatures might be helpful in diagnosis and stratification in an advanced stage of the disease, perturbations in selected pathways probably accumulate over years and even decades. This provides a basis for earlier intervention and preventive strategies, such as with personalized nutrition.

7.7 Conclusion

Metabolomics has great potential in elucidating metabolic risk factors for chronic diseases including Alzheimer's disease (AD). Gaining insights into the metabolic processes that lead to the development and progression of AD will also help build a better understanding of how malnutrition, obesity, insulin resistance and diabetes, and dysbiosis affect other chronic diseases. The literature discussed in this chapter demonstrates how malnutrition contributes to AD on a systemic and cellular level, and the metabolic impact of various approaches for supportive nutrition in AD. Specifically, we can conclude that current knowledge points to metabolic alterations in AD in the context of Western-style diet, as depicted in the graphical model.

References

Abedi E, Sahari MA (2014) Long-chain polyunsaturated fatty acid sources and evaluation of their nutritional and functional properties. Food Sci Nutr 2(5):443–463. https://doi.org/10.1002/fsn3.121

Ahmad TR, Haeusler RA (2019) Bile acids in glucose metabolism and insulin signaling—mechanisms and research needs. Nat Rev Endocrinol 15(12):701–712. https://doi.org/10.1038/s41574-019-0266-7

An Y, Varma VR, Varma S et al (2018) Evidence for brain glucose dysregulation in Alzheimer's disease. Alzheimers Dement 14(3):318–329. https://doi.org/10.1016/j.jalz.2017.09.011

Baloni P, Funk CC, Yan J et al (2020) Metabolic network analysis reveals altered bile acid synthesis and metabolism in Alzheimer's disease. Cell Rep Med 1(8):100138. https://doi.org/10.1016/j.xcrm.2020.100138

Berti V, Walters M, Sterling J et al (2018) Mediterranean diet and 3-year Alzheimer brain biomarker changes in middle-aged adults. Neurology 90(20):e1789–e1798. https://doi.org/10.1212/WNL.0000000000005527

Broom GM, Shaw IC, Rucklidge JJ (2019) The ketogenic diet as a potential treatment and prevention strategy for Alzheimer's disease. Nutrition 60:118–121. https://doi.org/10.1016/j.nut.2018.10.003

CDC (2021) Chronic Diseases in America|CDC; Accessed 02.12.21

de la Rubia Ortí JE, Fernández D, Platero F, García-Pardo MP (2021) Can ketogenic diet improve Alzheimer's disease? Association with anxiety, depression, and glutamate system. Front Nutr 8:744398. https://doi.org/10.3389/fnut.2021.744398

Fernstrom JD (2005) Branched-chain amino acids and brain function. J Nutr 135(6):1539S-S1546. https://doi.org/10.1093/jn/135.6.1539S

Field AE, Coakley EH, Must A et al (2001) Impact of overweight on the risk of developing common chronic diseases during a 10-year period. Arch Intern Med 161(13):1581–1586. https://doi.org/10.1001/archinte.161.13.1581

Fiorucci S, Biagioli M, Zampella A, Distrutti E (2018) Bile acids activated receptors regulate innate immunity. Front Immunol 9:1853. https://doi.org/10.3389/fimmu.2018.01853

Frigerio G, Favero C, Savino D et al (2021) Plasma metabolomic profiling in 1391 subjects with overweight and obesity from the SPHERE study. Metabolites 11(4):194. https://doi.org/10.3390/metabo11040194

Gentile CL, Weir TL (2018) The gut microbiota at the intersection of diet and human health. Science 362(6416):776–780. https://doi.org/10.1126/science.aau5812

Grimm MO, Grösgen S, Riemenschneider M, Tanila H, Grimm HS, Hartmann T (2011) From brain to food: analysis of phosphatidylcholins, lyso-phosphatidylcholins and phosphatidylcholin-plasmalogens derivates in Alzheimer's disease human post mortem brains and mice model via mass spectrometry. J Chromatogr A 1218(42):7713–7722. https://doi.org/10.1016/j.chroma.2011.07.073

Gu Y, Luchsinger JA, Stern Y, Scarmeas N (2010) Mediterranean diet, inflammatory and metabolic biomarkers, and risk of Alzheimer's disease. J Alzheimers Dis 22(2):483–492. https://doi.org/10.3233/JAD-2010-100897

Gustafson DR, Bäckman K, Scarmeas N et al (2020) Dietary fatty acids and risk of Alzheimer's disease and related dementias: observations from the Washington Heights-Hamilton Heights-Inwood Columbia Aging Project (WHICAP). Alzheimers Dement 16(12):1638–1649. https://doi.org/10.1002/alz.12154

Handakas E, Lau CH, Alfano R et al (2021) A systematic review of metabolomic studies of childhood obesity: state of the evidence for metabolic determinants and consequences [published online ahead of print, 2021 Nov 19]. Obes Rev e13384. https://doi.org/10.1111/obr.13384

Hite AH (2018) Nutritional epidemiology of chronic disease and defining "healthy diet." Global Food History 4(2):207–225. https://doi.org/10.1080/20549547.2018.1498256

Jiang C, Li G, Huang P, Liu Z, Zhao B (2017) The gut microbiota and Alzheimer's disease. J Alzheimers Dis 58(1):1–15. https://doi.org/10.3233/JAD-161141

Johnson ECB, Dammer EB, Duong DM et al (2020) Large-scale proteomic analysis of Alzheimer's disease brain and cerebrospinal fluid reveals early changes in energy metabolism associated with microglia and astrocyte activation. Nat Med 26(5):769–780. https://doi.org/10.1038/s41591-020-0815-6

Kim J, Shin S, Lee D, Park J, Hong E, Joung H (2015) Association between compliance with dietary guidelines and Alzheimer's disease in Korean elderly. J Nutr Health. 48:221.https://doi.org/10.4163/jnh.2015.48.3.221

Krähenbühl S, Talos C, Fischer S, Reichen J (1994) Toxicity of bile acids on the electron transport chain of isolated rat liver mitochondria. Hepatology 19(2):471–479. https://doi.org/10.1002/hep.1840190228

Lefebvre P, Cariou B, Lien F, Kuipers F, Staels B (2009) Role of bile acids and bile acid receptors in metabolic regulation. Physiol Rev 89(1):147–191. https://doi.org/10.1152/physrev.00010.2008

Leibson CL, Rocca WA, Hanson VA, Cha R, Kokmen E, O'Brien PC et al (1997) Risk of dementia among persons with diabetes mellitus: a population-based cohort study. Am J Epidemiol 145:301–308

Lloret A, Monllor P, Esteve D, Cervera-Ferri A, Lloret MA (2019) Obesity as a risk factor for Alzheimer's disease: implication of Leptin and glutamate. Front Neurosci 13:508. https://doi.org/10.3389/fnins.2019.00508

Lotta LA, Scott RA, Sharp SJ et al (2016) Genetic predisposition to an impaired metabolism of the branched-chain amino acids and risk of type 2 diabetes: a Mendelian randomisation analysis. PLoS Med 13(11):e1002179. doi:https://doi.org/10.1371/journal.pmed.1002179

Luchsinger JA, Tang MX, Stern Y, Shea S, Mayeux R (2001) Diabetes mellitus and risk of Alzheimer's disease and dementia with stroke in a multiethnic cohort. Am J Epidemiol 154:635–641

Ma Y, Ajnakina O, Steptoe A, Cadar D (2020) Higher risk of dementia in English older individuals who are overweight or obese. Int J Epidemiol 49(4):1353–1365. https://doi.org/10.1093/ije/dyaa099

Magnusson M, Lewis GD, Ericson U et al (2013) A diabetes-predictive amino acid score and future cardiovascular disease. Eur Heart J 34(26):1982–1989. https://doi.org/10.1093/eurheartj/ehs424

MahmoudianDehkordi S, Arnold M, Nho K et al (2019) Altered bile acid profile associates with cognitive impairment in Alzheimer's disease—an emerging role for gut microbiome [published correction appears in Alzheimers Dement. 15(4):604]. Alzheimers Dement 15(1):76–92. https://doi.org/10.1016/j.jalz.2018.07.217

Mapstone M, Cheema AK, Fiandaca MS et al (2014) Plasma phospholipids identify antecedent memory impairment in older adults. Nat Med 20(4):415–418. https://doi.org/10.1038/nm.3466

Nho K, Kueider-Paisley A, Mahmoudian Dehkordi S et al (2019) Altered bile acid profile in mild cognitive impairment and Alzheimer's disease: relationship to neuroimaging and CSF biomarkers. Alzheimers Dement 15(2):232–244.https://doi.org/10.1016/j.jalz.2018.08.012

Ott A, Stolk RP, van Harskamp F, Pols HA, Hofman A, Breteler MM (1999) Diabetes mellitus and the risk of dementia: the Rotterdam study. Neurology 53:1937–1942

Pakiet A, Jakubiak A, Czumaj A, Sledzinski T, Mika A (2019) The effect of western diet on mice brain lipid composition [published correction appears in Nutr Metab (Lond). 17:30]. Nutr Metab (Lond) 16:81. https://doi.org/10.1186/s12986-019-0401-4

Pardridge WM (1977) Kinetics of competitive inhibition of neutral amino acid transport across the blood-brain barrier. J Neurochem 28(1):103–108. https://doi.org/10.1111/j.1471-4159.1977.tb07714.x

Phillips MCL, Deprez LM, Mortimer GMN et al (2021) Randomized crossover trial of a modified ketogenic diet in Alzheimer's disease. Alzheimers Res Ther 13(1):51. https://doi.org/10.1186/s13195-021-00783-x

Rangel-Huerta OD, Pastor-Villaescusa B, Gil A (2019) Are we close to defining a metabolomic signature of human obesity? A systematic review of metabolomics studies. Metabolomics 15(6):93. https://doi.org/10.1007/s11306-019-1553-y

Rhee EP, Cheng S, Larson MG et al (2011) Lipid profiling identifies a triacylglycerol signature of insulin resistance and improves diabetes prediction in humans. J Clin Invest 121(4):1402–1411. https://doi.org/10.1172/JCI44442

Ruge T, Lockton JA, Renstrom F et al (2009) Acute hyperinsulinemia raises plasma interleukin-6 in both nondiabetic and type 2 diabetes mellitus subjects, and this effect is inversely associated with body mass index. Metabolism 58(6):860–866. https://doi.org/10.1016/j.metabol.2009.02.010

Rusek M, Pluta R, Ułamek-Kozioł M, Czuczwar SJ (2019) Ketogenic diet in Alzheimer's disease. Int J Mol Sci 20(16):3892. https://doi.org/10.3390/ijms20163892

Sayin SI, Wahlström A, Felin J et al (2013) Gut microbiota regulates bile acid metabolism by reducing the levels of tauro-beta-muricholic acid, a naturally occurring FXR antagonist. Cell Metab 17(2):225–235. https://doi.org/10.1016/j.cmet.2013.01.003

Schote AB, Turner JD, Schiltz J, Muller CP (2007) Nuclear receptors in human immune cells: expression and correlations. Mol Immunol 44(6):1436–1445. https://doi.org/10.1016/j.molimm.2006.04.021

Sochocka M, Donskow-Łysoniewska K, Diniz BS, Kurpas D, Brzozowska E, Leszek J (2019) The gut microbiome alterations and inflammation-driven pathogenesis of Alzheimer's disease-a critical review. Mol Neurobiol 56(3):1841–1851. https://doi.org/10.1007/s12035-018-1188-4

Soininen H, Solomon A, Visser PJ et al (2021) 36-month LipiDiDiet multinutrient clinical trial in prodromal Alzheimer's disease [published correction appears in Alzheimers Dement. 17(5):909]. Alzheimers Dement 17(1):29–40. https://doi.org/10.1002/alz.12172

Toledo JB, Arnold M, Kastenmüller G et al (2017) Metabolic network failures in Alzheimer's disease: a biochemical road map. Alzheimers Dement 13(9):965–984. https://doi.org/10.1016/j.jalz.2017.01.020

Tournissac M, Vandal M, Tremblay C et al (2018) Dietary intake of branched-chain amino acids in a mouse model of Alzheimer's disease: effects on survival, behavior, and neuropathology. Alzheimers Dement 4:677–687. https://doi.org/10.1016/j.trci.2018.10.005

Vavassori P, Mencarelli A, Renga B, Distrutti E, Fiorucci S (2009) The bile acid receptor FXR is a modulator of intestinal innate immunity. J Immunol 183(10):6251–6261. https://doi.org/10.4049/jimmunol.0803978

Więckowska-Gacek A, Mietelska-Porowska A, Wydrych M, Wojda U (2021) Western diet as a trigger of Alzheimer's disease: from metabolic syndrome and systemic inflammation to neuroinflammation and neurodegeneration. Ageing Res Rev 70:101397. https://doi.org/10.1016/j.arr.2021.101397

Wishart DS (2019) Metabolomics for investigating physiological and pathophysiological processes. Physiol Rev 99(4):1819–1875. https://doi.org/10.1152/physrev.00035.2018

Precision Nutrition and Cognitive Decline

8

Peterlin Borut, Zalar Bojan, and Peterlin Ana

Contents

Abstract

Cognitive decline, both due to ageing or disease, presents one of Western societies' most significant disease burdens. Personalized preventive actions and public health interventions and policies can contribute to the prevention of dementia and healthy ageing. Nutrition is an integral part of the exposome,

P. Borut (✉)
Clinical Institute of Genomic Medicine, University Medical Center Ljubljana, Ljubljana, Slovenia
e-mail: borut.peterlin@guest.arnes.si

Z. Bojan
University Psychiatric Clinic Ljubljana, Ljubljana, Slovenia

P. Ana
Faculty of Medicine, Institute of Histology and Embryology, University of Ljubljana, Ljubljana, Slovenia

A. G. Haslberger (ed.), *Advances in Precision Nutrition, Personalization and Healthy Aging*, https://doi.org/10.1007/978-3-031-10153-3_8

acting either as a risk factor or preventive intervention. Precision nutrition approach for prevention of cognitive decline involves the ability to profile the population for the risk of dementia, knowledge of the impact of nutrients and dietary patterns on biological processes involved in the pathogenesis of cognitive decline and the ability to predict individual heterogeneity response to diet. Genetic factors, either as rare monogenic causes or polygenic risk scores and known risk factors for dementia, can be used to assess the individual risk for dementia. In addition, several nutrients and dietary patterns influencing common mechanisms involved in dementia pathogenesis like oxidative stress, neuroinflammation, and hypoxia have been investigated as potential preventive interventions—the evidence is still accumulating. Finally, genomic variation, epigenetics, and microbiome have been proposed to modulate individual response to diet, and different »omic« surrogate biomarkers could improve the monitoring of individual response to nutrition. Large population data sets covering comprehensive information on mentioned variables are needed for the system medicine approach to dementia prevention. In addition, there is an urgent need for standardization of the methodological inventory and study design on the international level for that purpose.

8.1 Introduction

The cognitive decline that progresses to dementia is currently the fifth most significant contributor to the global burden of disease (Prince et al. 2015)) and substantially affects patients' lives, their families, and society. It affects around 1% of the population aged 30–64 years (Hendriks et al. 2021). Due to the ageing population worldwide, it is projected that the prevalence will triple by 2050 (Collaborators et al. 2016); therefore, dementia is considered a global public health priority. It was estimated that a five-year delay in the onset of dementia would reduce the number of people with a disease by 33–50% by 2050 (Jennings et al. 2020).

Dementia is not a specific disease but rather an overarching term to describe a group of symptoms affecting memory, thinking, and social abilities which interfere with daily life. It is associated with several distinct diseases with different aetiology and pathophysiology. Alzheimer disease is the most common cause, followed by vascular dementia, dementia with Lewy bodies, frontotemporal degeneration, and dementias associated with brain injury, infections, alcohol abuse, and genetic causes. Different entities are associated with specific etiological factors and neuropathologic hallmarks, while etiological factors and neuropathologic signs are often shared among dementias (Raz et al. 2015). Alzheimer's disease is characterized by the extracellular accumulation of senile plaques composed of a peptide and intraneuronal accumulation of neurofibrillary tangles composed of hyperphosphorylated microtubule-binding protein-tau. Vascular dementia is a heterogeneous disorder associated with endothelial dysfunction, atherosclerosis, small vessel disease, ischemia, and haemorrhage. Dementia with Lewy bodies is characterized by abnormal aggregation of the synaptic protein alpha-synuclein, while frontotemporal dementia may be associated with either cellular p-tau inclusions, ubiquitin-positive tau-negative neuronal inclusions, or neurodegeneration without ubiquitin or tau inclusions. There is significant heterogeneity in the comorbidity, and the cognitive impact of age-related neuropathology and mixed neuropathologies is the most common cause of dementia in the population (Boyle et al. 2021). Thus, the contribution of neuropathologies to cognitive decline is person-specific which has an essential impact on prevention strategies (Boyle et al. 2018).

Common mechanisms involved in dementias include neuroinflammation, neurodegeneration, autophagy, hypoxia with cerebrovascular dysfunction, blood–brain barrier dysfunction, and oxidative stress (Raz et al. 2015; Dominguez and Barbagallo 2018). Equally important for designing preventive interventions are functional pathways associated with cognitive resilience, including inflammation, amyloid degradation, memory function, and neurotransmission (Pérez-González et al. 2021).

Understanding pathology and mechanisms associated with cognitive decline provides a basis for prevention. In addition to public health interventions and policies, personalized preventive actions can contribute to dementia prevention. Preliminary evidence has not demonstrated that population screening for dementia had clear benefits or harm in quality of life, mood, or improved diagnostics (Fowler et al. 2020). Therefore, individualized strategies to identify risk factors and planning intervention strategies are expected to impact future prevention significantly.

Precision nutrition interventions could impact several biological processes involved in the pathogenesis of cognitive decline one hand. Also, there is individual heterogeneity in response to diet. To explore the potential individualized nutritional interventions, we will discuss current evidence of dietary effects on cognition, individual differences in response to nutrition in the context of cognition, as well as our understanding of risk factors for dementia which could lead to an assessment of individualized risk for dementia or stratification individuals for precision nutrition interventions (Fig. 8.1).

8.2 Nutrients and Dietary Patterns

Many nutritions are associated with mechanisms and risk factors for cognitive decline and dementia. However, most of the evidence linked to potential interventions is still lacking or is contradictory. Moreover, in several cases, potential favourable effects are limited to specific subpopulations, which requires further scientific confirmation and stipulates the potential of individualized nutritional interventions.

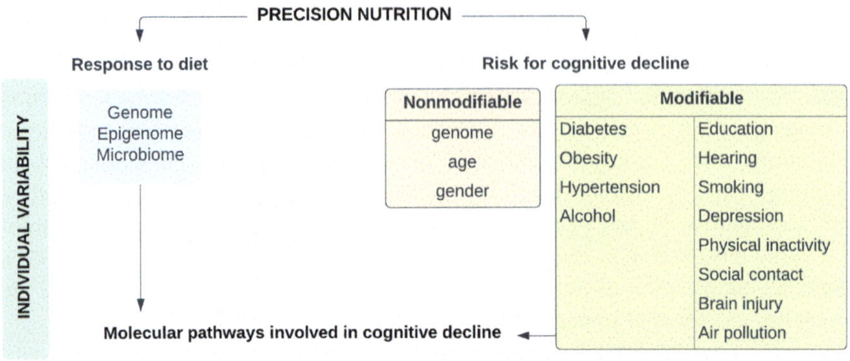

Fig. 8.1 Potential applications of precision nutrition on cognitive decline

8.2.1 Antioxidants

The contribution of oxidative stress to neurodegenerative disorders is not clearly understood, and several mechanisms have been proposed (Cobley et al. 2018). The primary antioxidant defences include vitamin C, vitamin E, carotenoids, flavonoids, and polyphenols.

The meta-analysis demonstrated low-certainty evidence for a positive effect of vitamin C and carotenoids on overall cognitive function (Rutjes et al. 2018). A recent meta-analysis of randomized intervention trials suggested that carotenoids are associated with better cognitive performance (Davinelli et al. 2021). Furthermore, there is partial evidence of synergism between carotenoids and vitamin E regarding basic cognitive performance (Beydoun et al. 2020). The results on vitamin E impact on cognition have been mixed (Lakhan et al. 2021).

Flavonoids were associated with a beneficial effect for maintaining cognitive function (Yeh et al. 2021; Gardener et al. 2021).

Similarly, polyphenols have been reported to improve cognitive function in both healthy middle-aged volunteers and students (Philip et al. 2019; Carrillo et al. 2021).

8.2.2 Vitamins

Both vitamin B12 and vitamin B6 are involved in the metabolism of homocysteine. Most systematic reviews showed no overall evidence that oral B vitamin supplementation prevented cognitive decline (Rutjes et al. 2018; Behrens et al. 2020; Markun et al. 2021). Pooled post hoc analysis of two randomized clinical trials showed that B vitamins had favourable effects on global cognitive functioning and whole-brain atrophy in older people with mild cognitive impairment (Wu et al. 2021). Furthermore, a combination of B6 and B12, folate, and n-3 fatty acids contributed to preserving semantic memory in a subgroup of women and men with previous coronary artery disease or ischaemic stroke aged 45–80 years in a randomized clinical trial (Andreeva et al. 2011).

Observational studies on folic acid supplementation provided mixed results (Scarmeas et al. 2018). A recent randomized clinical trial provided evidence that combining folic acid and docosahexaenoic acid therapy might improve cognitive function and reduce Aβ production in patients with mild cognitive impairment (Bai et al. 2021).

Few evidence on vitamin D supplementation demonstrated no significant effect on cognition; there is some evidence of modest effect in older black adults (Kang et al. 2021).

8.2.3 Omega-3 Fatty Acids

Randomized clinical trials reported mixed findings with supplementation of docosahexaenoic acid (DHA), however, in the LipiDiDiet trial Souvenaid (medicinal food with docosahexaenoic acid as one of the bioactive ingredients) improved clinical dementia rating score (Soininen et al. 2017). Recently, it was suggested that the supplementation of eicosapentaenoic acid but not DHA improved global cognitive function (Patan et al. 2021), while no treatment effect was found after omega-3 supplementation in young adults (Marriott et al. 1854). Combining DHA and folic acid therapy might be more beneficial in improving cognitive function and reducing Aβ-related biomarkers in older adults with mild cognitive impairment (Bai et al. 2021).

8.2.4 Dietary Patterns

Due to the complex interactions among different nutrients on one side and complex evolution of cognitive decline on the other, dietary patterns rather than single nutrients could better address the therapeutic potential of nutrients. Three dietary patterns were most extensively studied with cognitive decline: Mediterranean diet, dietary approaches to stop hypertension (DASH) and Mediterranean—DASH intervention for neurodegenerative delay (MIND).

While several observational studies reported beneficial effects of mostly Mediterranean and MIND diets on cognition, few randomized controlled trials with mixed results do not provide conclusive evidence (Duplantier and Gardner 2021). Furthermore, standardized approaches are needed both for nutrition and cognitive assessment (Scarmeas et al. 2018; Duplantier and Gardner 2021).

8.3 Individualized Response to Diet

There is considerable individual variation in how nutrients and food-derived bioactive molecules are absorbed and metabolized in humans. Understanding and assessing this variation may lead to better dietary recommendations and personalized dietary interventions. Sources for metabolic heterogeneity may include the human genome, epigenome, and microbiome (Zeisel 2020).

8.3.1 Human Genome

Genetic variation in the human genome may be associated with differential responses to nutrients (nutrigenetics). Few studies were addressing nutrigenetic implications in cognitive disfunction. In four studies, a healthy lifestyle including diet, fatty fish consumption, fruits and vegetables, and moderate intake of polyunsaturated fats were associated with reduced dementia risk (Samieri et al. 2021). Additionally, genetic variation in the MTHFR gene was associated with the total plasma homocysteine concentrations following B vitamin supplementation in the SU.FOL.OM3 trial (Fezeu et al. 2018). Polymorphisms in genes involved in vitamin uptake, transport, and metabolism were associated with vitamin status (Niforou et al. 2020). Similarly, data on the genetic variation of antioxidant enzymes relevant to nutritional components accumulate (Birk 2021).

8.3.2 Epigenome

The field of epigenetics is concerned with changes in gene expression that are not related to changes in DNA sequence. Epigenetic control of gene expression involves several mechanisms, including DNA methylation and hydroxymethylation, histone modifications, non-protein-coding RNA molecules, RNA editing, chromatin remodelling, and telomere control. Epigenetic marks are reversible and influenced by several environmental factors, including nutrition, infections, chemicals, stress and age, to name only a few.

Some dietary factors, such as folate, vitamins B6 and B12, choline and methionine, are involved in metabolic pathways directly related to DNA methylation. However, diet effects also other epigenetic mechanisms, such as histone modifications and non-coding RNAs (Dauncey 2014). Consequently, dietary factors such as deficiency in folate, vitamins B6 and B12 effect cognition via alterations in DNA methylation or in the case of vitamins A, E and C via histone acetylation (Polverino et al. 2021).

8.3.3 Microbiome

The microbiome is involved in nutrient metabolism and releases specific diet-microbial metabolites in the gut and the bloodstream. These include neurotransmitters, pro-inflammatory factors, and short-chain fatty acids. There are marked person-specific diet-microbiome interactions in the population, and thus, the gut microbiome contributes to the variation of subject-specific responses to diet (Leshem et al. 2020). In addition, several gut-derived metabolites linked to the brain metabolism, which are associated with specific diet—bacterial strains interactions, were identified (Samieri et al. 2021).

8.4 Risk Factors for Dementia

Risk factors for dementia include both nonmodifiable such as age, sex, and genetic predisposition as well as modifiable risk factors including less education, hypertension, hearing impairment, smoking, obesity, depression, physical inactivity, diabetes, low social contact, excessive alcohol consumption, traumatic brain injury, and air pollution (Livingston et al. 2020).

While diabetes and obesity may present direct targets for nutritional prevention or intervention and will be discussed in the next chapter, genetic predisposition may be used to estimate the risk of developing the disease. Moderately raised concentrations of homocysteine were reported to be associated with an increased risk of dementia in men and women over 65 years (Smith et al. 2018). It has been estimated that 25% of individuals aged 55 years and older have a family history of dementia (Loy et al. 2014) which might be associated with a monogenic or polygenic predisposition and complex interactions with environmental factors.

Monogenic forms of dementia may be due to pathogenic genetic variations in the APP, PSEN1 and PSEN2 genes related to Alzheimer disease, MAPT, GRN, C9orf72, CHMP2B, FUS, VCP, SQSTM1, OPTN, UBQLN2 and TBK1 genes related to frontotemporal dementia, NOTCH3, GLA, TREX1, COL4A1 and HTRA1 genes associated with the vascular type of dementia, and more than 168 genes related to rare syndromic forms of dementia (Huq et al. 2021). In addition to rare monogenic forms of dementia, polygenic risk scores can be used to predict the risk of developing dementia, as has been shown for Alzheimer disease (Rojas et al. 2021; Escott-Price et al. 2017; Leonenko et al. 2021).

Prognostic models may assess different risk factors and provide the likelihood that an individual will develop dementia; however, there is no golden standard to evaluate the dementia risk yet (Geethadevi et al. 2021; Goerdten et al. 2019).

8.5 Challenges and Future Directions

Presented data show evidence that, on the one hand, evidence is accumulating about the effects of nutrition on pathological processes involved in cognitive decline and for the assessment of individual responses to diet as well as risks for developing dementia. On the other hand, current evidence is mainly based on a single or small number of observed interactions. We are just at the beginning of understanding single elements that are part of much more complex and heterogeneous networks, necessary to understand and model in order to design comprehensive dietary interventions for an individualized approach to cognitive decline prevention (Ommen et al. 2017; Ebaid and Crewther 2020).

To advance the field, comprehensive data on multiple combinations of nutrition exposures for different, specific domains of cognitive function are needed (Samieri et al. 2021). High throughput technologies will provide several levels of nutrigenomic biomarkers, including transcriptomic, proteomic, metabolomic, lipidomic, and immunomic, which will provide data for a systems approach to understanding nutritional effects on molecular pathways. At the same time, data on potential sources influencing metabolic heterogeneity in response to nutrition, including individual genetic and microbiome data, should be available to personalize intervention strategies. Furthermore, tools for dementia risk assessment based on a combination of risk factors will provide information on individualized risk for cognitive decline to inform precision nutrition interventions. Finally, due to the heterogeneous and complex aetiology and pathogenesis of cognitive decline, precision nutrition is one of the multidomain interventions, which should simultaneously target several risk factors and mechanisms to achieve optimal preventive effects (Solomon et al. 2021).

Large cohorts from different world populations, standardized study designs, and analytical methods are needed to provide adequate power for evidence-based recommendations.

References

Andreeva VA, Kesse-Guyot E, Barberger-Gateau P, Fezeu L, Hercberg S, Galan P (2011) Cognitive function after supplementation with B vitamins and long-chain omega-3 fatty acids: ancillary findings from the SU.FOL.OM3 randomized trial. Am J Clin Nutr 94:278–86

Bai D, Fan J, Li M, Dong C, Gao Y, Fu M, Huang G, Liu H (2021) Effects of folic acid combined with DHA supplementation on cognitive function and amyloid-β-related biomarkers in older adults with mild cognitive impairment by a randomized, double blind. Placebo-controlled trial. J Alzheimer's Dis 81:155–167

Behrens A, Graessel E, Pendergrass A, Donath C (2020) Vitamin B—can it prevent cognitive decline? A systematic review and meta-analysis. Syst Rev 9:111

Beydoun MA, Canas JA, Fanelli-Kuczmarski MT, Maldonado AI, Shaked D, Kivimaki M, Evans MK, Zonderman AB (2020) Association of antioxidant vitamins A, C, E and carotenoids with cognitive performance over time: a cohort study of middle-aged adults. Nutrients 12:3558

Birk R (2021) Nutrigenetics of antioxidant enzymes and micronutrient needs in the context of viral infections. Nutr Res Rev 34:174–184

Boyle PA, Yu L, Wilson RS, Leurgans SE, Schneider JA, Bennett DA (2018) Person-specific contribution of neuropathologies to cognitive loss in old age. Ann Neurol 83:74–83

Boyle PA, Wang T, Yu L, Wilson RS, Dawe R, Arfanakis K, Schneider JA, Bennett DA (2021) To what degree is late life cognitive decline driven by age-related neuropathologies? Brain 144:awab092

Carrillo JÁ, Arcusa R, Zafrilla MP, Marhuenda J (2021) Effects of fruit and vegetable-based nutraceutical on cognitive function in a healthy population: placebo-controlled, double-blind, and randomized clinical trial. Antioxidants 10:116

Cobley JN, Fiorello ML, Bailey DM (2018) 13 reasons why the brain is susceptible to oxidative stress. Redox Biol 15:490–503

Collaborators GD, Nichols E, Szoeke CEI, Vollset SE, Abbasi N, Abd-Allah F, Abdela J, Aichour MTE, Akinyemi RO, Alahdab F et al (2016) Global, regional, and national burden of Alzheimer's disease and other dementias, 1990–2016: a systematic analysis for the Global Burden of disease study 2016. Lancet Neurol 18:88–106

Dauncey MJ (2014) Nutrition, the brain and cognitive decline: insights from epigenetics. Eur J Clin Nutr 68:1179–1185

Davinelli S, Ali S, Solfrizzi V, Scapagnini G, Corbi G (2021) Carotenoids and cognitive outcomes: a meta-analysis of randomized intervention trials. Antioxidants 10:223

de Rojas I, Moreno-Grau S, Tesi N, Grenier-Boley B, Andrade V, Jansen IE, Pedersen NL, Stringa N, Zettergren A, Hernández I et al (2021) Common variants in Alzheimer's disease and risk stratification by polygenic risk scores. Nat Commun 12:3417

Dominguez LJ, Barbagallo M (2018) Nutritional prevention of cognitive decline and dementia. Acta Bio-Medica Atenei Parmensis 89:276–290

Duplantier SC, Gardner CD (2021) A critical review of the study of neuroprotective diets to reduce cognitive decline. Nutrients 13:2264

Durga J, van Boxtel MP, Schouten EG, Kok FJ, Jolles J, Katan MB, Verhoef P (2007) Effect of 3-year folic acid supplementation on cognitive function in older adults in the FACIT trial: a randomized, double blind, controlled trial. Lancet 369:208–216

Ebaid D, Crewther SG (2020) Time for a systems biological approach to cognitive aging?—a critical review. Front Aging Neurosci 12:114

Escott-Price V, Myers AJ, Huentelman M, Hardy J (2017) Polygenic risk score analysis of pathologically confirmed Alzheimer disease. Ann Neurol 82:311–314

Fezeu LK, Ducros V, Guéant J-L, Guilland J-C, Andreeva VA, Hercberg S, Galan P (2018) MTHFR 677C → T genotype modulates the effect of a 5-year supplementation with B-vitamins on homocysteine concentration: The SU.FOL.OM3 randomized controlled trial. Plos One 13:e0193352

Fowler NR, Perkins AJ, Gao S, Sachs GA, Boustani MA (2020) Risks and benefits of screening for dementia in primary care: The Indiana University Cognitive Health Outcomes Investigation of the Comparative Effectiveness of Dementia Screening (IU CHOICE)Trial. J Am Geriatr Soc 68:535–543

Gardener SL, Rainey-Smith SR, Weinborn M, Bondonno CP, Martins RN (2021) Intake of products containing anthocyanins, flavanols, and flavanones, and cognitive function: a narrative review. Front Aging Neurosci 13:640381

Geethadevi GM, Quinn TJ, George J, Anstey K, Bell JS, Cross AJ (2021) Multidomain prognostic models used in middle aged adults without known cognitive impairment for predicting subsequent dementia. Cochrane Db Syst Rev. https://doi.org/10.1002/14651858.cd014885

Goerdten J, Čukić I, Danso SO, Carrière I, Muniz-Terrera G (2019) Statistical methods for dementia risk prediction and recommendations for future work: a systematic review. Alzheimer's Dementia Transl Res Clin Interventions 5:563–569

Hendriks S, Peetoom K, Bakker C, Flier WM van der, Papma JM, Koopmans R, Verhey FRJ, Vugt M de, Köhler S, Group Y-ODES et al (2021) Global prevalence of young-onset dementia. Jama Neurol 78:1080–1090

Huq AJ, Sexton A, Lacaze P, Masters CL, Storey E, Velakoulis D, James PA, Winship IM (2021) Genetic testing in dementia-a medical genetics perspective. Int J Geriatr Psych 36:1158–1170

Jennings A, Cunnane SC, Minihane AM (2020) Can nutrition support healthy cognitive ageing and reduce dementia risk? BMJ 369:m2269

Kang JH, Vyas CM, Okereke OI, Ogata S, Albert M, Lee I-M, D'Agostino D, Buring JE, Cook NR, Grodstein F et al (2021) Effect of vitamin D on cognitive decline: results from two ancillary studies of the VITAL randomized trial. Sci Rep-Uk 11:23253

Lakhan R, Sharma M, Batra K, Beatty FB (2021) The role of vitamin E in slowing down mild cognitive impairment: a narrative review. Healthc 9:1573

Leonenko G, Baker E, Stevenson-Hoare J, Sierksma A, Fiers M, Williams J, de Strooper B, Escott-Price V (2021) Identifying individuals with high risk of Alzheimer's disease using polygenic risk scores. Nat Commun 12:4506

Leshem A, Segal E, Elinav E (2020) The gut microbiome and individual-specific responses to diet. Msystems 5:e00665-e720

Livingston G, Huntley J, Sommerlad A, Ames D, Ballard C, Banerjee S, Brayne C, Burns A, Cohen-Mansfield J, Cooper C et al (2020) Dementia prevention, intervention, and care: 2020 report of the lancet commission. Lancet 396:413–446

Loy CT, Schofield PR, Turner AM, Kwok JB (2014) Genetics of dementia. Lancet 383:828–840

Markun S, Gravestock I, Jäger L, Rosemann T, Pichierri G, Burgstaller JM (2021) Effects of vitamin B12 supplementation on cognitive function, depressive symptoms, and fatigue: a systematic review, meta-analysis, and meta-regression. Nutrients 13:923

Marriott BP, Turner TH, Hibbeln JR, Newman JC, Pregulman M, Malek AM, Malcolm RJ, Burbelo GA, Wismann JW (1854) Impact of fatty acid supplementation on cognitive performance among United States (US) military officers: the ranger resilience and improved performance on phospholipid-bound omega-3's (RRIPP-3) study. Nutrients 2021:13

Niforou A, Konstantinidou V, Naska A (2020) Genetic variants shaping inter-individual differences in response to dietary intakes—a narrative review of the case of vitamins. Front Nutr 7:558598

Patan MJ, Kennedy DO, Husberg C, Hustvedt SO, Calder PC, Khan J, Forster J, Jackson PA (2021) Supplementation with oil rich in eicosapentaenoic acid, but not in docosahexaenoic acid, improves global cognitive function in healthy, young adults: results from randomized controlled trials. Am J Clin Nutrition 114:914–924

Pérez-González M, Badesso S, Lorenzo E, Guruceaga E, Pérez-Mediavilla A, García-Osta A, Cuadrado-Tejedor M (2021) Identifying the main functional pathways associated with cognitive resilience to Alzheimer's disease. Int J Mol Sci 22:9120

Philip P, Sagaspe P, Taillard J, Mandon C, Constans J, Pourtau L, Pouchieu C, Angelino D, Mena P, Martini D et al (2019) Acute intake of a grape and blueberry polyphenol-rich extract ameliorates cognitive performance in healthy young adults during a sustained cognitive effort. Antioxidants 8:650

Polverino A, Sorrentino P, Pesoli M, Mandolesi L (2021) Italy I of D and THC Naples, Italy D of M and WS University of Naples "Parthenope", Naples, France I de N des S Aix-Marseille University, Marseille, Italy I of AS and IS National Research Council, Pozzuoli, Italy D of HS University of Naples Federico II, Naples,: Nutrition and cognition across the lifetime: an overview on epigenetic mechanisms. Aims Neurosci 8:448–476

Prince M, Wimo A, Guerchet M et al. World Alzheimer Report 2015. Alzheimers Disease International, London

Raz L, Knoefel J, Bhaskar K (2015) The neuropathology and cerebrovascular mechanisms of dementia. J Cereb Blood Flow Metabolism 36:172–186

Rutjes AW, Denton DA, Nisio MD, Chong L, Abraham RP, Al-Assaf AS, Anderson JL, Malik MA, Vernooij RW, Martínez G et al (2018) Vitamin and mineral supplementation for maintaining cognitive function in cognitively healthy people in mid and late life. Cochrane Db Syst Rev 2019:CD011906

Samieri C, Yassine HN, van Lent DM, Lefèvre-Arbogast S, van de Rest O, Bowman GL, Scarmeas N (2021) Personalized nutrition for dementia prevention. Alzheimer's Dementia J Alzheimer's Assoc. https://doi.org/10.1002/alz.12486

Scarmeas N, Anastasiou CA, Yannakoulia M (2018) Nutrition and prevention of cognitive impairment. Lancet Neurol 17:1006–1015

Smith AD, Refsum H, Bottiglieri T, Fenech M, Hooshmand B, McCaddon A, Miller JW, Rosenberg IH, Obeid R (2018) Homocysteine and dementia: an international consensus statement1. J Alzheimer's Dis 62:561–570

Soininen H, Solomon A, Visser PJ, Hendrix SB, Blennow K, Kivipelto M, Hartmann T, group L clinical study, Hallikainen I, Hallikainen M et al (2017) 24-month intervention with a specific multinutrient in people with prodromal Alzheimer's disease (LipiDiDiet): a randomized, double-blind, controlled trial. Lancet Neurol 16:965–975.

Solomon A, Handels R, Wimo A, Antikainen R, Laatikainen T, Levälahti E, Peltonen M, Soininen H, Strandberg T, Tuomilehto J et al (2021) Effect of a multidomain lifestyle intervention on estimated dementia risk. J Alzheimer's Dis 82:1461–1466 van Ommen B, van den Broek T, de Hoogh I, van Erk M, van Someren E, Rouhani-Rankouhi T, Anthony JC, Hogenelst K, Pasman W, Boorsma A et al (2017) Systems biology of personalized nutrition. Nutr Rev 75:579–599

Wu Y, Smith AD, Refsum H, Kwok T (2021) Effectiveness of B vitamins and their interactions with aspirin in improving cognitive functioning in older people with mild cognitive impairment: pooled post-hoc analyses of two randomized trials. J Nutrition Heal Aging 25:1154–1160

Yeh T-S, Yuan C, Ascherio A, Rosner BA, Willett WC, Blacker D (2021) Long-term dietary flavonoid intake and subjective cognitive decline in US men and women. Neurology 97:e1041–e1056

Zeisel SH (2020) Precision (personalized) nutrition: understanding metabolic heterogeneity. Annu Rev Food Sci T 11:1–22

Algorithms for and Challenges in the Analysis of Markers in Personalized Health Care

9

Clemens Heitzinger

Contents

C. Heitzinger (✉)
Center for AI and ML (CAIML) and Department for Mathematics and Geoinformation, TU Wien, Vienna, Austria
e-mail: Clemens.Heitzinger@TUWien.ac.at

A. G. Haslberger (ed.), *Advances in Precision Nutrition, Personalization and Healthy Aging*, https://doi.org/10.1007/978-3-031-10153-3_9

Abstract

Nowadays, the various omics disciplines such as genomics, proteomics, metabolomics, metagenomics, and transcriptomics generate a plethora of data. At the same time, a multitude of omics markers may be accompanied by a multitude of diseases. Hence, finding relationships between omics markers and disease in their early stages is a challenge that is at the very core of predictive or personalized medicine. In this chapter, an overview of algorithms for solving these problems of supervised learning is given, and challenges in this problem domain are discussed. Questions of learnability should be considered, and the quality and precision of the predictions should be assessed critically and quantitatively. Therefore, quality metrics for the assessment of the predictions are discussed as well.

9.1 Introduction

Modern sequencing technologies make it possible to generate genomic data of patients routinely and at reasonable cost. In addition to sequences of the nucleobases, epigenetic modifications and the regulation of genes may be associated with diseases (Tan et al. 2021), which motivates the inclusion of epigenetic states in the collected data. The effects of healthy nutrition (Mainardi et al. 2019) are also expected to be observable in patients' data. More generally, overviews of the role machine learning in health care are shown in Figs. 9.1 and 9.2.

But generating all these data is just a necessary prerequisite to solve problems of biological, medical, or clinical relevance. Having made the data available, the challenge can be stated as follows: Given a certain disease, is it possible to find or calculate a function (the predictor or prediction model) that takes genetic, epigenetic, or other markers as inputs and that predicts the disease as early as possible and as reliably as possible? Part of the challenge is to identify any markers that are expedient for finding such a predictor. Also, the data or markers used should be obtainable in a manner that is as non-intrusive as possible. In the ideal case, only blood samples (Chen et al. 2018; Cohen 2020; Elena Tomeva et al. 2022) or harmless radiation is used.

It goes without saying that answering this question is of great importance in health care and also in view of aging populations in developed countries. As usual in economy, it can be expected that technological progress and commercial interest will drive the price of determining relevant markers down, even if the markers are expensive to obtain at present. Hence, the major challenge in this area is most likely to identify expedient (genetic) markers first. As discussed below, there are good theoretic reasons why the identification of useful markers is a hard learning problem.

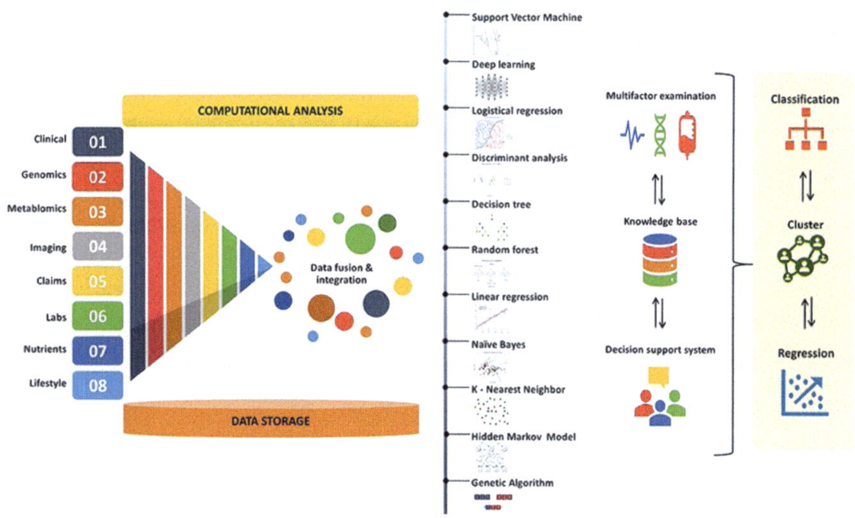

Fig. 9.1 Overview of machine learning in medicine. From https://academic.oup.com/database/article/doi/10.1093/database/baaa010/5809229

Data collection and storage	⟹	Machine learning	⟹	Usage and benefits
∗omics		Supervised learning		Intelligence augmentation:
Clinical data		Unsupervised learning		Decision-support systems
Imaging		Reinforcement learning		Education and training
Lab data				Expert systems
Lifestyle				Knowledge bases
Nutrient data				etc.
etc.				

Fig. 9.2 The role of machine learning in health care. Data from various sources are collected in databases. Methods from the three pillars of machine learning are then used to extract information. Finally, the extracted information and knowledge is made available to patients and healthcare specialists

Clearly, large benefits are obtained whenever the predictor is so reliable that it can be employed at a large scale for multiple diseases and that it can reliably trigger the use of more conventional means to obtain clinically relevant and actionable diagnoses by possibly more intrusive means if the test result is positive for a certain disease.

Early and reliable detection are both stated as desirable goals in the problem statement above. These two goals are, however, certainly conflicting requirements. As an extreme example, a predictor that always returns a positive answer never misses a case of a disease, but is utterly useless, as all healthy individuals become false positives. Therefore, various metrics to assess the quality of a predictor are needed and discussed below.

The data available for determining or finding a predictor are fraught with uncertainties due to measurement errors, recording and transmission errors, the possible lack of significant markers, and the possible rareness of the disease. All these factors contribute to the challenge of finding such predictors.

If there were an infinite data stream and a physiological basis of disease and all relevant aspects were available in the data and sufficient time were available to perform the calculations, it would certainly be possible to learn predictors. However, some of these assumptions are violated in reality. In particular, the available data are always finite. Therefore, questions arise: How well can a predictor be learned (a priori estimate)? How well has it been learned having done the calculations (a posteriori estimate)? These questions motivate the development of a theory of learning underlying and informing implementations of learning algorithms. One of the most fundamental and important theories in this regard is probably approximately correct (PAC) learning and discussed below.

In summary, this chapter provides an overview of the challenges inherent in the analysis of markers in personalized health care and an introduction to fundamental learning theory and algorithms for finding predictors.

The rest of this chapter is organized as follows. In Sect. 9.2, the prediction problem is stated within the context of machine learning and supervised learning in particular. In Sect. 9.4, the inherent challenges in prediction based on (epi-)genetic markers are discussed. In Sect. 9.3, an overview of algorithms for supervised learning is given. In Sect. 9.5, metrics to assess the quality of predictors are presented. In Sect. 9.6, the theory of probably approximately correct learning is shortly discussed, since it provides the intellectual foundation and framework to formulate learning problems, to discuss learnability, and to assess whether and how well a learning problem has been solved. Finally, in Sect. 9.7, conclusions are drawn.

9.2 Supervised Learning

Supervised learning is the field of machine learning that is most relevant for finding predictors of disease.

9.2.1 Basic Definitions and the Learning Task

In general, there is a vector of variables that are called *independent variables* or *inputs* or *features*. These variables are used to predict another vector of variables, the *dependent variables* or *outputs* or *responses*. *Discrete* variables (taking values from a finite set) are also called *categorical* variables or *qualitative* variables or *factors*. *Continuous* variables (taking real values) are also called *quantitative* variables.

We discern two types of *prediction* tasks. A *classification* task is concerned with predicting discrete or qualitative outputs, while a *regression* task is concerned with predicting continuous or quantitative outputs.

We usually denote inputs by X and outputs by Y (when they are referred to as random variables) or by x and y (when we mean concrete values or vectors of values). The capital letters signify random variables which are a useful notion here as we usually make the assumption that the values that are available for learning are drawn from underlying and fixed distributions.

The *measurements, observations, realizations,* or *samples* of the random variables, i.e., the data points used for training/learning, testing, and validation, are denoted by

$$(x_i, y_i) \in S_{\text{in}} \times S_{\text{out}}, \quad i \in \{1, \ldots, N\}.$$

Depending on the purpose for which a set

$$\big\{(x_i, y_i) \in S_{\text{in}} \times S_{\text{out}} \mid i \in \{1, \ldots, N\}\big\}$$

consisting of such realizations is used, it is called the *training data,* the *testing data,* or the *validation data.*

We make the basic assumption that a *statistical model* of the form

$$Y = f(X) + \epsilon, \quad \mathbb{E}[\epsilon] = 0, \tag{9.1}$$

holds, where the function

$$f\colon \quad S_{\text{in}} \to S_{\text{out}}$$

is the true relationship between the inputs (which are elements of the set S_{in}) and the outputs (which are elements of the set S_{out}). The random variable

$$\epsilon \sim E$$

is i.i.d. observation *noise*, whose distribution is E. Otherwise, there would be nothing to learn.

Now the learning task in supervised learning is to find a *prediction model* or *predictor*

$$\hat{f}\colon \quad S_{\text{in}} \to S_{\text{out}}$$

that resembles the unknown true relationship f as closely as possible by using only realizations.

In order to quantify this resemblance, loss functions and errors are commonly used.

9.2.2 Loss Functions

Loss functions and errors are used in order to quantify how well a prediction model resembles the truth. Various different loss functions and errors may be used for regression and classification tasks.

9.2.2.1 Regression Tasks

We denote (vector valued) inputs by $\mathbf{x}$, (vector valued) outputs by $\mathbf{y}$, and the prediction model by $\hat{f}$ as usual. For regression tasks, common choices for the loss function L are the p-norms

$$L(\hat{f}(\mathbf{x}), \mathbf{y}) := \|\hat{f}(\mathbf{x}) - \mathbf{y}\|_p$$

including the special cases of the absolute error ($p = 1$), the squared Euclidean error

$$L(\hat{f}(\mathbf{x}), \mathbf{y}) := \|\hat{f}(\mathbf{x}) - \mathbf{y}\|_2^2$$

($p = 2$ and squared), and the maximum error ($p = \infty$).

9.2.2.2 Classification Tasks

For classification tasks, different loss functions are usually used. We suppose that the qualitative or categorical response $y \in S_{\text{out}}$ takes values k from the set $S_{\text{out}} := \{1, \ldots, K\}$, which represents K classes. Often, but not always, the probabilities

$$\hat{p}_k(\mathbf{x}) := \mathbb{P}[k = y \mid \mathbf{x}], \quad k \in S_{\text{out}},$$

are modeled and the most likely value is chosen, i.e., the prediction model $\hat{g}$ is

$$\hat{g}(\mathbf{x}) := \operatorname*{arg\,max}_{k \in S_{\text{out}}} p_k(\mathbf{x}) = \operatorname*{arg\,max}_{k \in S_{\text{out}}} \mathbb{P}[k = y \mid \mathbf{x}] \in S_{\text{out}}.$$

Typical loss functions are the zero-one loss

$$L(\hat{g}(\mathbf{x}), y) := [\![\hat{g}(\mathbf{x}) = y]\!]$$

and -2 times the log-likelihood

$$L(\hat{g}(\mathbf{x}), y) := -2 \sum_{k=1}^{K} [\![k = y]\!] \log \hat{p}_k(\mathbf{x}) = -2 \log \hat{p}_y(\mathbf{x}).$$

9.2.3 Errors

Having decided on a loss function, we define various types of errors based on the loss function. In the following, the prediction model $\hat{f}_{\mathcal{T}}$ has been estimated using a training set

$$\mathcal{T} := \big\{(x_i, y_i) \in S_{\text{in}} \times S_{\text{out}} \mid i \in \{1, \ldots, N\}\big\}$$

with size $N = |\mathcal{T}|$. We imagine that the samples have been drawn from the underlying data distribution

$$(X, Y) \sim D.$$

To make the dependence of the prediction model on the training data explicit in our notation, we add the index $\mathcal{T}$ to $\hat{f}$ and write $\hat{f}_{\mathcal{T}}$.

We now make this dependence even more explicit by christening the (unfortunately usually unnamed) learning algorithm. A learning algorithm a takes a training set as input and outputs a prediction model, i.e., it is a function

$$a\colon \operatorname{sets}(N, S_{\text{in}} \times S_{\text{out}}) \to F(S_{\text{in}}, S_{\text{out}}), \tag{9.2}$$

where $\operatorname{sets}(N, S)$ denote the set of all sets with N elements in S, and $F(A, B)$ denotes the set of all functions from the set A to the set B. Using the learning algorithm a, we have

$$\hat{f}_{\mathcal{T}} := a(\mathcal{T}).$$

This allows us to make the dependencies of the errors defined below on the learning algorithm and the training set explicit.

The *training error* TE is defined as the (empirical) expected value

$$\operatorname{TE}(a, \mathcal{T}) = \operatorname{TE}(\hat{f}_{\mathcal{T}}, \mathcal{T}) := \frac{1}{N} \sum_{i=1}^{N} L(\hat{f}_{\mathcal{T}}(x_i), y_i)$$

of the loss calculated on the training set $\mathcal{T}$. The training error is available immediately during and after training/learning, but it is important to note that minimizing the training error is not our ultimate goal (see Sect. 9.2.5). Since the size of the training set $\mathcal{T}$ is finite, the training error can always be reduced to zero (for a fixed set $\mathcal{T}$) by increasing the model complexity, i.e., by enlarging the class of functions from which the prediction model is calculated.

The *generalization error* or *test error* GE is defined as

$$\operatorname{GE}(a, \mathcal{T}) = \operatorname{GE}(\hat{f}_{\mathcal{T}}, \mathcal{T}) := \mathbb{E}_{(X,Y)\sim D}[L(\hat{f}_{\mathcal{T}}(X), Y)],$$

where the random variables X and Y are drawn from their joint distribution D over $S_{\text{in}} \times S_{\text{out}}$. The generalization error depends on the training set $\mathcal{T}$ via the prediction model $\hat{f}_{\mathcal{T}}$ and hence is the error for this particular prediction model and training set. The index $(X, Y) \sim D$ of the expectation indicates that the samples to be used for calculating the expectation and hence the generalization error are to be (newly) drawn from the underlying distribution D of all data available to us and hence that these samples are *independent* of the training data $\mathcal{T}$ used for finding $\hat{f}_{\mathcal{T}}$.

To solve learning problems, the generalization error is much more important than the training error. The generalization error tells us how small we can expect the loss to be when we make predictions using our prediction model $\hat{f}_{\mathcal{T}}$. Still, the generalization error depends on the training data $\mathcal{T}$, and we may have been lucky in our choice of training data resulting in a particularly well working prediction model and hence in an exaggerated confidence in our ability to solve the prediction problem.

So far, the training set $\mathcal{T}$ has been chosen once and remained fixed. In order to assess our ability to solve a learning problem in a *reproducible* manner with respect to new choices of the training set, we must consider what happens when we generate new training sets (by drawing samples $(X, Y) \sim D$ from the underlying data distribution D). We hence extend the expectation in the generalization error also over the sampling of the training set, i.e., also over $\mathcal{T} \sim D^{|\mathcal{T}|}$, where the notation indicates that each of the $N = |\mathcal{T}|$ elements of the set $\mathcal{T}$ is sampled as $(X, Y) \sim D$.

The *expected prediction error* or *expected test error* is defined as

$$\mathrm{EPE}(a) := \mathbb{E}_{\mathcal{T}\sim D^{|\mathcal{T}|}}[\mathrm{GE}(a, \mathcal{T})] = \mathbb{E}_{\mathcal{T}\sim D^{|\mathcal{T}|},\ (X,Y)\sim D}[L(\hat{f}_{\mathcal{T}}(X), Y)], \tag{9.3}$$

where the expectation now also averages over all training sets. Therefore, the expected prediction error assesses how well a learning algorithm can solve a prediction problem.

Finally, we can consider the expected value $\mathbb{E}[\mathrm{EPE}]$ of the expected prediction error over all algorithms. Such considerations result in no-free-lunch theorems (Wolpert and Macready 1997) which state that while an algorithm may outperform others on a particular class of learning problems, it cannot outperform other algorithms consistently on all learning problems. No-free-lunch theorems are of no immediate interest here.

9.2.4 Model Selection, Model Assessment, and Datasets

In the previous section, we have already mentioned that learning should be reproducible with respect to the training set, and we have seen that the expected prediction error is a function of the learning algorithm only, allowing us to shed any dependence (of the generalization error) on the training set and also allowing us to assess the usefulness of a learning algorithm.

In order to solve a learning problem, there are two separate goals that need to be achieved. The first is *model selection.* Since, there is a huge choice of learning algorithms of the form (9.2) and of representations of the prediction model (see Sect. 9.3), we must estimate the performance of different models and algorithms in order to choose the best one for our learning task.

The second goal is *model assessment.* Having decided on the prediction model and learning algorithm for our task, we would like to estimate the generalization error (which is always calculated using new data) of the final prediction model that we consider the solution to our learning problem so that we know how well we have solved the given learning task.

Whenever enough data are available, the whole dataset is randomly divided into three parts, and the typical procedure comprises the following three steps. In each step, one of the errors defined in Sect. 9.2.3 is used.

1. Model fitting: The *training set* is used to fit the prediction models. The training error is useful in this regard.
2. Model selection: The *validation set*, which has not been used yet, is used to estimate the expected prediction error and to select the model.

3. Model assessment: Finally, the *test set,* which has not been used yet, is employed to assess the generalization error of the finally chosen model. If the performance is satisfactory, the learning task has been solved.

A typical split may be to use 50% of the data for the training set, 25% for the validation set, and 25% for the test set, but this split is far from being a general rule.

9.2.5 Learnability and Data

Loss functions and errors make it possible to formulate problems in supervised learning as optimization problems. The training error is minimized such that a suitable prediction model $\hat{f}$ is found, typically by calculating the parameter vector θ that corresponds to a parameterized prediction model $\hat{f}_\theta$ (the minimizer) whenever we search for prediction models in a class of functions parameterized by θ.

In this manner, part of learning becomes an optimization problem, for which many computational approaches are available. For computational purposes, it is expedient that the loss functions and the optimization algorithms harmonize. However, this implies that the minimizer $\hat{f}$ is *not* a unique function, as our particular choices of the class of functions from which it is sought, of loss function, and of optimization algorithm may yield different solutions or solution candidates, i.e., prediction models. Importantly, as already discussed in the previous sections, learning is more than an optimization problem, as it is not sufficient to just minimize the training error.

In this regard, model selection and model assessment (see Sect. 9.2.4) are essential, and sufficient data are necessary. In other words, the ability to learn from data, i.e., the learnability, depends on the availability of data. Therefore, we discuss the amount of data and some aspects of learnability in the following. (Learnability can also be investigated in the view of computational restraints, but this is not of immediate interest here.)

The basic assumption is that a stream of data points

$$z_1, z_2, z_3, \ldots$$

is available, where the z_i are realizations of a random variable

$$Z \sim D, \tag{9.4}$$

where D is the data distribution. The data distribution is often assumed to be constant. But this is not always the case; in certain type learning, e.g., in reinforcement learning, it may change over time, and then solutions that adopt to the changing conditions are sought.

In the case of supervised learning, the random variable Z has the (vector) structure $Z = (X, Y)$, where X is an input and Y is an output. In other types of learning, the random variable Z may have a different structure.

Depending on when the data or realizations becomes available and when the learning algorithm is executed, we can discern two types of learning, namely online and offline learning.

- In offline learning, all realizations are available before learning. In other words, historic data are collected, and then the learning algorithm is executed.
- In online learning, data collection and running the learning algorithm are interleaved. In other words, the learning algorithm is run again or continues whenever a new data point or a batch of data points becomes available. Note that learning using simulations is online learning.

Depending on the length of the stream of data points, we discern two kinds of learning results.

- In the first case, we assume that an infinite number of data points, i.e., a sequence $\langle z_i \rangle_{i\in\mathbb{N}}$, is available for learning. Then learning results state that the solution calculated by a learning algorithm converges (in a specific sense) to the true solution or an optimal solution. We call these learning results qualitative ones, as they provide no indication how fast convergence occurs.
- In the second case, we assume that a finite number of data points, i.e., a vector $(z_1, \ldots, z_N)$, is available for learning. Then learning results state that the solution calculated by a learning algorithm solves the problem with a given maximal error (in other words, to a prescribed accuracy) with a certain probability; the learned statement is probably approximately correct.
  Both limitations mentioned in the last sentence, namely that we solve the problem to a given accuracy (only approximately) and with a certain probability (only probably), are inherent in learning [at least in the sense of the assumption leading to (9.4)].
  The fundamental limitation that we can only learn with a certain probability stems from the problem that we may not have observed all relevant cases. For example, if our task is to learn the colors of swans, we have never observed a sample in Australia, we are led to believe that all swans are white, because we have never observed a *Cygnus atratus,* a black swan.
  The fundamental limitation that we can only learn to a certain accuracy stems from the problem that we may have observed an unluckily large number of outliers. For example, if we want to estimate the expectation of a real-valued random variable and observe many outliers, the accuracy of the sample mean as the estimator is reduced.
  Of course, as the number of observations increases, the probability that a statement is correct increases for a given maximal error or, vice versa, the maximal error decreases for a given probability.
  These considerations lead to probably approximately correct learning (see Sect. 9.6). If such a learning result is available, the learning problem is solved for all intents and purposes. Still, the practical question may remain whether the error

bounds provided by the theory for a certain probability are sharp or not. If they are not sharp, the error or the number of required samples is overestimated.

In the second case, we can furthermore discern the two cases whether the data fit into memory or not. If they do not fit, the fact that a part of the data needs to be fetched from a slower source influences the run time and the design of the learning algorithm. The case that not all available data fit into memory also provides a good definition of the field of big data.

Theoretic results for the first case are already very beneficial, as they show that a learning algorithm calculates what it is supposed to calculate, although we do not know quantitatively how well it performs. Results about probably approximately correct learning in the second case are even better, as they also yield the convergence speed as a function of the number of available data points.

In Sect. 9.2.4, learning under the assumption that enough data are available was discussed. However, restrictions on data availability are quite common, as collecting, storing, cleaning, and imputing data are associated with cost (cf. Sect. 9.4.2). The most notable exception may be generating the data $Z \sim D$ from simulations.

In order to discuss artificial intelligence and machine learning in a historical context, we recall that Leibniz believed that most of human reasoning can and should be reduced to calculations:

> The only way to rectify our reasonings is to make them as tangible as those of the mathematicians so that we can find our error at a glance, and when there are disputes among persons, we can simply say: let us calculate [calculemus] without further ado to see who is right. Leibniz (2022)

Hence the imperative "calculemus!" set out a whole research program in a single word.

By understanding learning as reliably extracting information from observations as in (9.4), we note, however, that uncertainties must additionally be dealt with and included in the learning process. Examples are the statistical model in (9.1) and the discussions of probably approximately correct learning above and in Sect. 9.6. In this sense, artificial intelligence and machine learning must go beyond deterministic models, which have been prevalent during most of the history of science.

To emphasize this notion and to paraphrase Leibniz, the imperative

> "datis discamus!" ("let us learn from the data!")

summarizes artificial intelligence and machine learning. Our goal is to learn from observations in the face of uncertainties, building on Leibniz' program, and accepting new challenges.

9.2.6 Bias and Variance

After these general considerations, we elucidate the structure of the expected prediction error (9.3) for a particular loss function, namely the squared loss

$$L(a, b) := (a - b)^2,$$

which is useful for regression problems.

While expectations over the error $\epsilon \sim E$ were implicit in Sect. 9.2.3, we now mention them explicitly to make the calculations clearer.

For the squared loss, the expected prediction error (9.3) becomes

$$\text{EPE} = \mathbb{E}_{\epsilon \sim \mathbb{E},\ \mathcal{T} \sim D^{|\mathcal{T}|},\ (X,Y) \sim D}[(\hat{f}_{\mathcal{T}}(X) - Y)^2]. \tag{9.5}$$

The important result below means that this expected prediction error can be decomposed into three parts. For the decomposition, the expectation over $(X, Y) \sim D$ is not necessary, and we hence perform the calculations using the *expected squared error*

$$\text{ESE} := \mathbb{E}_{\epsilon \sim E,\ \mathcal{T} \sim D^{|\mathcal{T}|}}[(\hat{f}_{\mathcal{T}}(X) - Y)^2].$$

With this definition, the equality

$$\text{EPE} = \mathbb{E}_{(X,Y) \sim D}[\text{ESE}]$$

holds, and hence the decomposition below translates to the expected prediction error.

Theorem 2.1 (bias–variance decomposition for squared loss function) *Suppose that the statistical model* (9.1) *holds with* $\mathbb{E}[\epsilon] = 0$ *and that the noise* ϵ *is independent of the random variables* X *and* Y. *Then the expected squared error* (9.5) *is equal to*

$$\begin{aligned} \text{ESE} &= \mathbb{E}_{\epsilon \sim E,\ \mathcal{T} \sim D^{|\mathcal{T}|}}[(\hat{f}_{\mathcal{T}}(X) - Y)^2] \\ &= \underbrace{\mathbb{V}_{\mathcal{T} \sim D^{|\mathcal{T}|}}[\hat{f}_{\mathcal{T}}(X) - \bar{f}(X)]}_{\text{variance}} + \underbrace{(\bar{f}(X) - f(X))^2}_{\text{bias}^2} + \underbrace{\mathbb{V}[\epsilon]}_{\text{noise}}, \end{aligned}$$

where

$$\bar{f}(X) := \mathbb{E}_{\mathcal{T} \sim D^{|\mathcal{T}|}}[\hat{f}_{\mathcal{T}}(X)] \tag{9.6}$$

is the mean (over all sets $\mathcal{T}$*) prediction.*

A proof is given in Appendix. The assumption that $\mathbb{E}[\epsilon] = 0$ is benign, because otherwise a nonzero constant can be absorbed into the true model f.

The three terms in the decomposition have specific meanings.

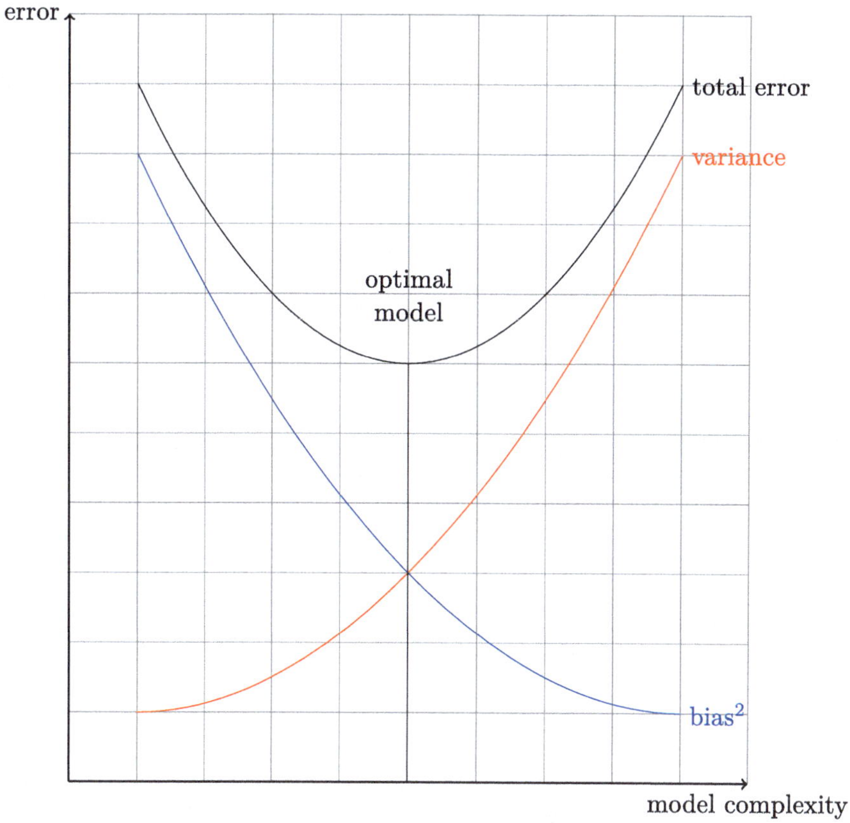

Fig. 9.3 Schematic diagram of the behavior of the bias squared and the variance as functions of model complexity or degrees of freedom

1. The first term is the *variance* of the learning algorithm. Models with low complexity (i.e., a low number of degrees of freedom) have low variance, and models with high complexity have high variance.
2. The second term is called the *bias squared,* it does not depend on the training set $\mathcal{T}$, and it is due to the model class. Models with low complexity have high bias, and models with high complexity have low bias.
3. The third term is the observation *noise.* It is inherent in the statistical model (9.1) and hence irreducible, i.e., we cannot reduce it (e.g., by our choice of model class or learning algorithm).

Generally speaking, bias–variance decompositions decompose the expected loss into the three terms above, i.e., the variance, the bias, and the noise. The theorem above states the standard decomposition for the squared loss. The proof works only for the squared loss, and a generalization to other loss functions is not obvious. For example, a number of different decompositions have been proposed for the zero-one loss (Domingos 2000).

The typical behavior of the bias squared and the variance as functions of model complexity or degrees of freedom is shown in Fig. 9.3. The observation noise is not shown as it is constant and does not influence the optimal model which minimizes the total error.

9.3 Algorithms

Various types of algorithms have been developed for calculating prediction models. In this section, some of the most commonly used ones are introduced.

9.3.1 KNN

The k-nearest neighbors algorithm (KNN) first finds the k neighbors of the given point that are closest in the training set and then returns the prediction based on these k neighbors. When used for classification, a plurality vote of the neighbors is returned. When used for regression, the expectation of the values of the k nearest neighbors or a weighted mean of these values is returned.

The only parameter in the KNN algorithm is the number of neighbors to be used. If k is small, the local structure of the data is predominant in the prediction; if k is large, the prediction model is smoother.

In implementations of the KNN algorithm, it is often expedient to use k-dimensional trees, feature selection, and/or dimensionality reduction.

9.3.2 Linear Regression

In linear regression, linear combinations of regressors are fit to the known data. Regressors are functions of the independent variables, such as the commonly used constants, that are chosen by the user with the intent of facilitating learning by incorporating prior knowledge into the prediction models. Many variants of this general approach have been developed.

9.3.3 Logistic Regression

Logistic models (or logit models) model probabilities. If we denote the probability of a class in a classification problem by p, the logit or log-odds (the logarithm with base b of the odds $p/(1-p)$) is defined as

$$l(p) := \log_b \frac{p}{1-p}, \quad p \in (0, 1).$$

The logit is then modeled, for example, using a linear regression to find

$$l(p) = \sum_i \alpha_i r_i,$$

where the α_i are the model parameters and the r_i are the regressors (see Sect. 9.3.2). After having found the model parameters using the data, the equation is solved for p to obtain

$$p = \frac{1}{1 + b^{-\sum \alpha_i r_i}},$$

which can be written as

$$p = \sigma_b \left(\sum \alpha_i r_i \right),$$

where σ_b is the sigmoid function

$$\sigma_b(x) := \frac{1}{1 + b^{-x}}$$

with base b.

9.3.4 Artificial Neural Networks

The study of artificial neural networks (ANN), which are computational models for neural networks, began in the middle of the past century. In their simplest form, in feed-forward ANN, the input is propagated through layers of neurons. Going from one layer to the next, an affine transformation is applied to the output of the previous layer and a nonlinear function is applied element-wise at each neuron. It goes without saying that many variations exist. ANN have a huge number of parameters, i.e., the model complexity is high, and therefore, overfitting (see Sect. 9.4.1) should be prevented.

Training algorithms are generally based on stochastic gradient descent. The gradient of the ANN required in the training algorithms is calculated by backpropagation, which is another name for automatic backward differentiation. Backpropagation makes it possible to calculate the gradient of the ANN at a given point at about twice the computational cost of evaluating the ANN, which is a great improvement compared to the naive finite-difference approximation of the gradient, whose computational cost increases linearly with the number of parameters.

9.3.5 Naive Bayes Classifiers

Naive Bayes classifiers are called naive because of the strong assumption that all independent variables or features are independent given the output variable. Denoting the independent variables by $\mathbf{x} = (x_1, \ldots, x_n)$ and the event that the sample is in class $k \in \{1, \ldots, K\}$ by C_k, naive Bayes classifiers estimate the conditional probability $p(C_k|\mathbf{x})$. By Bayes' theorem, it is equal to

$$p(C_k|\mathbf{x}) = \frac{p(C_k)p(\mathbf{x}|C_k)}{p(\mathbf{x})}.$$

In Bayesian terminology, this equation means

$$\text{posterior} = \frac{\text{prior} \times \text{likelihood}}{\text{evidence}}.$$

The denominator is only important as a constant of proportionality.

The nominator is equal to the joint probability $p(C_k, \mathbf{x})$ and can be rewritten using the definition of the conditional probability as

$$\begin{aligned} p(C_k, \mathbf{x}) &= p(x_1, \ldots, x_n, C_k) \\ &= p(x_1 \mid x_2, \ldots, x_n, C_k) p(x_2, \ldots, x_n, C_k) \\ &= \cdots \\ &= p(x_1 \mid x_2, \ldots, x_n, C_k) p(x_2 \mid x_3, \ldots, x_n, C_k) \cdots \\ &\quad p(x_{n-1} \mid x_n, C_k) p(x_n|C_k) p(C_k). \end{aligned}$$

By the (naive) assumption that all independent variables x_i are mutually independent conditional on the class C_k, we have

$$p(x_i \mid x_{i+1}, \ldots, x_n, C_k) = p(x_i|C_k) \quad \forall i \in \{1, \ldots, n-1\}.$$

This yields

$$p(C_k, \mathbf{x}) = p(C_k) \prod_{i=1}^{n} p(x_i|C_k)$$

and therefore the equation

$$p(C_k|\mathbf{x}) = \frac{p(C_k) \prod_{i=1}^{n} p(x_i|C_k)}{p(\mathbf{x})}.$$

for the posterior probability.

To find the predicted class, the maximum-a-posteriori rule is often used. It assigns the class k with the largest probability to a given sample $\mathbf{x}$. Since the denominator is constant for this purpose, the predicted class is

$$\operatorname*{arg\,max}_{k \in \{1,\ldots,K\}} p(C_k|\mathbf{x}) = \operatorname*{arg\,max}_{k \in \{1,\ldots,K\}} p(C_k) \prod_{i=1}^{n} p(x_i|C_k),$$

where the likelihoods $p(x_i|C_k)$ and $p(C_k)$ have been learned from the training data.

9.3.6 Decision Trees and Random Forests

A decision tree is a certain kind of tree. Each internal node has two children and is labeled with an independent variable and a rule or decision such as "follow the left child of the node if the independent variable is less equal than a specific value and the right child if not." Each leaf of the tree is labeled with a specific class. By following the children starting at the root according to the rules, a classification of each sample is achieved.

A decision tree is constructed by partitioning the training data into two subsets, which then constitute the two children and the corresponding decision. This process is repeated recursively for each subset or child and is stopped when partitioning no longer adds value the classification. This algorithm is called top-down induction of decision trees and is the most common one for learning decision trees.

The construction of a decision tree is not unique. Techniques that construct more than one decision tree are called ensemble methods. They include boosted trees (e.g., AdaBoost), bootstrap aggregated decision trees (e.g., random forests), and rotation forests.

9.4 The Challenges

Learning predictive models for healthcare applications is challenging as already mentioned in the introduction in Sect. 9.1. The reasons are discussed in this section. While is also possible to consider problems in *unsupervised learning* by asking questions such as which genes are similar or which samples are similar, we focus on supervised learning for the purposes of predicting diseases here.

9.4.1 Overfitting

Having summarized basic notions of supervised learning in Sect. 9.2, we start this section with a discussion of a general point always to be kept in mind in supervised learning. Recalling the goal of minimizing the expected prediction error defined in Sect. 9.2.3 and the bias–variance decomposition for the squared loss in Sect. 9.2.6, just minimizing the bias is not sufficient to solve a learning problem (see Fig. 9.3). Overfitting means that the model complexity has become too large, and hence the increasing variance increases the total error despite the decrease of the bias. The model complexity must be chosen prudently such that the total error is minimized and an optimal model is found.

One way to reduce overfitting is regularization. Regularization means that a term that penalizes complex models is added to the loss function, which effectively reduces the size of the model class. For example, when the models are artificial neural networks, the regularization term may penalize large parameters.

9.4.2 (Epi-)genetic Problems

A fundamental problem that hinders the search for genetic and/or epigenetic features (or biomarkers) is that the number of samples (or patients) is usually much, much smaller than the number of possible (epi-)genetic features such as single-nucleotide polymorphisms and other mutations. Therefore, correlations between the disease and one of the many possible features may occur purely by chance and without any predictive value. As the number of samples (or patients) increases, these spurious correlations vanish and only features with predictive value remain.

Another point that should be kept in mind when screening features is that the features may only be screened using the training dataset. The wrong way is to use the whole dataset for screening features, which would violate the separation between training, validation, and test datasets (see Sect. 9.2.4) (Hastie et al. 2009, Sect. 7.10.2).

9.4.3 A Numerical Example

To illustrate the effect of the huge number of possibly relevant features in (epi-)genetic problems and to estimate the number of samples (patients) that suffices to identify relevant features (biomarkers) and to achieve reliable classification, a numerical experiment was conducted. The main assumption is that there is a (small) number of relevant features hidden among the (large) total number of features.

In the numerical experiment, a certain number of features is generated as standard normally distributed random numbers, and it is assumed that a patient has a tumor if and only if the values of a certain (small) number of features are positive. In other words, the rest of the features is ignored when determining the health status of the simulated patients. This results in about 1/2 to the power of the number of relevant features of the total number of patients having a tumor in the numerical experiments. Then the Pearson correlation coefficient of each feature with the health status is calculated, the correlations are sorted by their absolute values, and a certain number of features with largest correlations (by absolute value) is employed to predict the health status using k nearest neighbors.

Numerical results are given in Table 9.1. A limitation of this example that we use the number of relevant features in the learning process, which is known here; in the real world, the number of features to be used must be learned. It is seen in the table that about 300 training samples are sufficient in this example to correctly identify the three relevant features. The remaining incorrect predictions are due to the KNN prediction.

It is also observed that the informedness lives up to its name and can discern much clearer than the accuracy when the features are identified to inform the prediction.

Table 9.1 Numerical results for various numbers of features, relevant features, and best correlations used

Total features	Relevant features	Used features	Training samples	Found all relevant features (%)	Accuracy	Informedness
100	3	3	50	4	0.834	0.138
100	3	3	100	36	0.881	0.424
100	3	3	150	72	0.922	0.625
100	3	3	200	90	0.941	0.720
100	3	3	250	99	0.953	0.783
100	3	3	300	100	0.956	0.800
100	3	3	400	100	0.960	0.816
100	3	3	500	100	0.963	0.830
100	3	3	1 000	100	0.970	0.864
200	3	3	50	0	0.826	0.103
200	3	3	100	18	0.862	0.338
200	3	3	150	63	0.910	0.574
200	3	3	200	86	0.937	0.703
200	3	3	250	94	0.948	0.756
200	3	3	300	99	0.955	0.794
200	3	3	400	100	0.960	0.813
200	3	3	500	100	0.962	0.821
200	3	3	1 000	100	0.970	0.864
400	3	3	50	1	0.822	0.060
400	3	3	100	15	0.856	0.277
400	3	3	150	57	0.903	0.538
400	3	3	200	80	0.932	0.680
400	3	3	250	95	0.949	0.761
400	3	3	300	99	0.956	0.800
400	3	3	400	100	0.960	0.813
400	3	3	500	100	0.963	0.832
400	3	3	1 000	100	0.970	0.864

The numbers of training samples is also given. The number of samples for testing is always equal to 100 000, and 100 runs were used to average the results the last three columns

In summary, such simulations make it possible to roughly estimate the relationship between the numbers of training samples and of the relevant features.

9.5 Quality Metrics

Errors, loss functions, optimization algorithms, etc., are only intermediate goals. The ultimate goal is the correctness and usefulness of the prediction model. Therefore, commonly used quality metrics are summarized in this section. It is prudent to not only consider a single quality metric, as it is hard to summarize all information about the quality of a prediction in a single number.

9.5.1 True and False Positives and Negatives

We denote the number of true positive samples by P and the number of true negative samples by N. Therefore, the total number of samples is $P + N$.

The number TP of true positives is the number of predictions that correctly classify a positive sample as positive. The number TN of true negatives is the number of predictions that correctly classify a negative sample as negative.

The number FP of false positives is the number of predictions that wrongly classify a negative sample as positive. The number FN of false negatives is the number of predictions that wrongly classify a positive sample as negative.

Clearly, the equalities

$$\begin{aligned} P &= \mathrm{TP} + \mathrm{FN}, \\ N &= \mathrm{TN} + \mathrm{FP}, \\ P + N &= \mathrm{TP} + \mathrm{TN} + \mathrm{FP} + \mathrm{FN} \end{aligned}$$

hold.

9.5.2 Positive and Negative Rates

The *sensitivity* or *hit rate* or *true positive rate* is defined as

$$\mathrm{TPR} := \frac{\mathrm{TP}}{P} = \frac{\mathrm{TP}}{\mathrm{TP} + \mathrm{FN}} = 1 - \mathrm{FNR}.$$

The *specificity* or *selectivity* or *true negative rate* is defined as

$$\mathrm{TNR} := \frac{\mathrm{TN}}{N} = \frac{\mathrm{TN}}{\mathrm{TN} + \mathrm{FP}} = 1 - \mathrm{FPR}.$$

The *false positive rate* is defined as

$$\mathrm{FPR} := \frac{\mathrm{FP}}{N} = \frac{\mathrm{FP}}{\mathrm{FP} + \mathrm{TN}} = 1 - \mathrm{TNR}.$$

The *miss rate* or *false negative rate* is defined as

$$\mathrm{FNR} := \frac{\mathrm{FN}}{P} = \frac{\mathrm{FN}}{\mathrm{FN} + \mathrm{TP}} = 1 - \mathrm{TPR}.$$

9.5.3 Predictive Values Etc.

The *precision* or *positive predictive value* is defined as

$$\mathrm{PPV} := \frac{\mathrm{TP}}{\mathrm{TP} + \mathrm{FP}} = 1 - \mathrm{FDR}.$$

The *negative predictive value* is defined as

$$\mathrm{NPV} := \frac{\mathrm{TN}}{\mathrm{TN} + \mathrm{FN}} = 1 - \mathrm{FOR}.$$

The *false discovery rate* is defined as

$$\mathrm{FDR} := \frac{\mathrm{FP}}{\mathrm{FP} + \mathrm{TP}} = 1 - \mathrm{PPV}.$$

The *false omission rate* is defined as

$$\mathrm{FOR} := \frac{\mathrm{FN}}{\mathrm{FN} + \mathrm{TN}} = 1 - \mathrm{NPV}.$$

9.5.4 Prevalence, Accuracy, and Informedness

The *prevalence* is defined as

$$\frac{P}{P + N},$$

and the *accuracy* is defined as

$$\mathrm{ACC} := \frac{\mathrm{TP} + \mathrm{TN}}{P + N} = \frac{\mathrm{TP} + \mathrm{TN}}{\mathrm{TP} + \mathrm{TN} + \mathrm{FP} + \mathrm{FN}}.$$

The *(bookmaker) informedness* or *Youden's index* (Youden 1950) is defined as

$$\begin{aligned}
\mathrm{INF} &:= \frac{1}{2}\left(\frac{\mathrm{TP} - \mathrm{FN}}{\mathrm{TP} + \mathrm{FN}} + \frac{\mathrm{TN} - \mathrm{FP}}{\mathrm{TN} + \mathrm{FP}}\right) \\
&= \frac{\mathrm{TP}}{\mathrm{TP} + \mathrm{FN}} + \frac{\mathrm{TN}}{\mathrm{TN} + \mathrm{FP}} - 1 \\
&= \mathrm{TPR} + \mathrm{TNR} - 1 \\
&= \frac{\mathrm{TP}}{P} + \frac{\mathrm{TN}}{N} - 1.
\end{aligned}$$

It has some desirable features (Youden 1950). Its value is always in the interval $[-1, 1]$, i.e.,

$$-1 \leq \text{INF} \leq 1,$$

as TP/P and TN/N are always in the interval $[0, 1]$.

The value $\text{INF} = 1$ implies that $\text{FN} = 0$ and $\text{FP} = 0$, i.e., the prediction works perfectly.

The value $\text{INF} = 0$ is obtained, e.g., when $\text{TP} = \text{FN}$ and $\text{TN} = \text{FP}$, which means that the prediction is useless.

The value $\text{INF} = -1$ implies that $\text{TP} = 0$ and $\text{TN} = 0$, meaning that all samples are misclassified. In this case or, more generally, whenever $\text{INF} \in [-1, 0)$, the two classes in the prediction should be swapped to obtain a better prediction and an informedness in the interval $(0, 1]$. Swapping the two classes means that TP is swapped with FN and TN is swapped with FP. Therefore, the informedness after swapping is

$$\overline{\text{INF}} := \frac{\text{FN}}{\text{FN} + \text{TP}} + \frac{\text{FP}}{\text{FP} + \text{TN}} - 1.$$

Hence, we have $\text{INF} + \overline{\text{INF}} = 0$ and thus

$$\overline{\text{INF}} = -\text{INF} \in [-1, 1].$$

Since we have assumed $\text{INF} \in [-1, 0)$, we find $\overline{\text{INF}} \in (0, 1]$, and swapping the predicted classes indeed improves the prediction.

In summary, larger (absolute) informedness values are better. False positives and false negatives are equally undesirable in the informedness. The informedness can be interpreted as the probability of an informed decision as opposed to a random guess taking into account all predictions (Powers 2011). Finally, it can be used in conjunction with receiver-operating-characteristic analysis.

9.5.5 Weighted Informedness or Weighted Youden's Index

In the (bookmaker) informedness or Youden's index defined in the previous section, false positives and false negative are equally undesirable. In certain applications (such as tumor prediction), false negatives are to be avoided more than false positives. It is, therefore, desirable to generalize the informedness such that false positives and false negatives can be assigned different weights.

We, therefore, propose and define the *weighted informedness* or the *weighted Youden's index* as

$$\text{WINF} := \alpha \frac{\text{TP} - \text{FN}}{\text{TP} + \text{FN}} + (1 - \alpha) \frac{\text{TN} - \text{FP}}{\text{TN} + \text{FP}},$$

where $\alpha \in [0, 1]$ is the weight of the positives and $1 - \alpha$ the weight of the negatives. The first fraction $(\text{TP} - \text{FN})/(\text{TP} + \text{FN})$ is the proportion $\text{TP}/(\text{TP} + \text{FN})$ of positive samples correctly classified as positive minus the proportion $\text{FN}/(\text{TP} + \text{FN})$

Table 9.2 Confusion matrix

Samples	Predicted positive	Predicted negative	Total
True positive	TP	FN	$\mathrm{TP} + \mathrm{FN} = P$
True negative	FP	TN	$\mathrm{FP} + \mathrm{TN} = N$

The rows contain the true classification and the columns contain the predictions

of incorrectly classified positive samples. The analogous interpretation for negative samples motivates the second fraction.

The equation

$$\frac{a-b}{a+b} = 2\frac{a}{a+b} - 1 \quad \forall a \in \mathbb{R}, \quad \forall b \in \mathbb{R}$$

applied to both fractions in the definition of WINF yields

$$\begin{aligned}
\mathrm{WINF} &= 2\alpha\frac{\mathrm{TP}}{\mathrm{TP}+\mathrm{FN}} + 2(1-\alpha)\frac{\mathrm{TN}}{\mathrm{TN}+\mathrm{FP}} - 1 \\
&= 2\alpha\mathrm{TPR} + 2(1-\alpha)\mathrm{TNR} - 1 \\
&= 2\alpha\frac{\mathrm{TP}}{P} + 2(1-\alpha)\frac{\mathrm{TN}}{N} - 1.
\end{aligned}$$

9.5.6 Confusion Matrix

By convention, the true and false positives and negatives are arranged in a matrix as given in Table 9.2.

9.5.7 A Numerical Example

Table 9.3 gives the quality metrics defined in this section for a single classification with the indicated parameters as in Sect. 9.4.3.

9.6 Probably Approximately Correct

In view of these challenges, it is not clear whether a prediction model for a certain disease can be learned reliably at all in situations with many features and few samples. Therefore, questions of learnability arise. The term learnability often also refers to questions of computability, but these are not of primary interest in the present context.

In addition to learning optimal prediction models, one is, therefore, also interested in answering the question how reliable the learned prediction model is given the (limited) number of available training data. A fundamental theory that addresses

Table 9.3 Quality metrics for a single classification as in Sect. 9.4.3

Name	Value	
Inputs		
Number of training samples	500	
Number of test samples	100,000	
Number of features	100	
Number of relevant features	3	
Number of used features	3	
Quality metrics		
Total number of samples	100,000	
Number of positives	12,307	
Number of negatives	87,693	
Number of true positives	10,944	
Number of true negatives	85,542	
Number of false positives	2151	
Number of false negatives	1363	
True positive rate	0.889	
True negative rate	0.975	
False positive rate	0.111	
False negative rate	0.111	
Positive predictive value	0.836	
Negative predictive value	0.984	
False discovery rate	0.164	
False omission rate	0.016	
Prevalence	0.123	
Accuracy	0.965	
Informedness	0.865	
Weighted informedness for $\alpha = 0$	0.951	
Weighted informedness for $\alpha = 0.1$	0.934	
Weighted informedness for $\alpha = 0.2$	0.916	
Weighted informedness for $\alpha = 0.3$	0.899	
Weighted informedness for $\alpha = 0.4$	0.882	
Weighted informedness for $\alpha = 0.5$	0.865	
Weighted informedness for $\alpha = 0.6$	0.847	
Weighted informedness for $\alpha = 0.7$	0.830	
Weighted informedness for $\alpha = 0.8$	0.813	
Weighted informedness for $\alpha = 0.9$	0.796	
Weighted informedness for $\alpha = 1$	0.779	
Confusion matrix	10,944	1363
	2151	85,542

this question is probably approximately correct (PAC) learning, which is shortly discussed in the following.

Probably, approximately correct statements or estimates is among the most important in learning theory. Since learning problems are usually stochastic in nature due to the fact that the samples obtained for learning are stochastic, any statements about the accuracy and validity of learning results must also be stochastic.

To discuss a leading example, we consider the problem of calculating the expected value $\mathbb{E}[X]$ of a random variable X (i.e., $\mathbb{E}[X]$ is the true value) using samples of the random variable X (the values that we can observe). We would hence like to minimize the absolute value $|X - \mathbb{E}[X]|$ of the error. Then a PAC statement as the solution of a learning problem has the form

$$\forall \delta \in [0, 1]: \quad \mathbb{P}[|X - \mathbb{E}[X]| < \epsilon(\delta)] > 1 - \delta,$$

where the challenge is to find the function ϵ of the probability δ. Such a statement is interpreted as now being able to give confidence intervals: with probability at least $1 - \delta$ (the P in PAC), the absolute value $|X - \mathbb{E}[X]$ of the error is at most $\epsilon(\delta)$ (the A in PAC). Thus $\epsilon(\delta)$ gives the size of the (here two-sided) confidence interval around X in which the true value $\mathbb{E}[X]$ lies.

Concentration inequalities are a source of PAC estimates. Good bounds are available if the random variable is bounded or if it is bounded and has finite variance. Finding better bounds is still an active area of research, as simpler inequalities only yield loose bounds.

9.7 Conclusions

This chapter has given an overview of supervised learning with genetic, epigenetic, and proteomic applications in mind. Caveats have been pointed out along the way.

Although immense amounts of genetic, epigenetic, and proteomic data are available and more will become available, prediction of diseases, and especially of rare diseases remains a challenge. While sequencing easily produces many features for each patient at relatively low cost, the number of patients is limited, and increasing their number is costly. This combination of many features and few samples is a major challenge. On the other hand, since genetic, epigenetic, and proteomic information should suffice to detect diseases, it is expected that careful application of machine learning methods will enhance our capabilities for early detection of disease significantly in the coming years, improving quality of life.

Proof of Theorem 2.1

Proof First, we use the statistical model (9.1) and expand the square to find

$$\begin{aligned}
\text{ESE} &= \mathbb{E}_{\mathcal{T}\sim D^{|\mathcal{T}|},\, \epsilon\sim E}[(\hat{f}_{\mathcal{T}}(X) - Y)^2] \\
&= \mathbb{E}_{\mathcal{T}\sim D^{|\mathcal{T}|},\, \epsilon\sim E}[(\hat{f}_{\mathcal{T}}(X) - f(X) - \epsilon)^2] \\
&= \mathbb{E}_{\mathcal{T}\sim D^{|\mathcal{T}|},\, \epsilon\sim E}[(\hat{f}_{\mathcal{T}}(X) - f(X))^2] \\
&\quad - 2\mathbb{E}_{\mathcal{T}\sim D^{|\mathcal{T}|},\, \epsilon\sim E}[\epsilon(\hat{f}_{\mathcal{T}}(X) - f(X))] \\
&\quad + \mathbb{E}_{\mathcal{T}\sim D^{|\mathcal{T}|},\, \epsilon\sim E}[\epsilon^2] \\
&= \mathbb{E}_{\mathcal{T}\sim D^{|\mathcal{T}|}}[(\hat{f}_{\mathcal{T}}(X) - f(X))^2] + \mathbb{V}_{\epsilon\sim E}[\epsilon].
\end{aligned}$$

The middle term vanishes, since the independence of ϵ from X and Y implies

$$\mathbb{E}_{\mathcal{T}\sim D^{|\mathcal{T}|},\, \epsilon\sim E}[\epsilon(\hat{f}_{\mathcal{T}}(X) - f(X))] = \mathbb{E}_{\epsilon\sim E}[\epsilon]\mathbb{E}_{\mathcal{T}\sim D^{|\mathcal{T}|}}[\hat{f}_{\mathcal{T}}(X) - f(X)] = 0$$

because of $\mathbb{E}_{\epsilon\sim E}[\epsilon] = 0$ by assumption.

Next, we split the first remaining term into two parts. In order to do so, we define the intermediate (9.6), which is the mean (over all sets $\mathcal{T}$) prediction. Using $\bar{f}(X)$, we can write

$$\begin{aligned}
\text{ESE} &= \mathbb{E}_{\mathcal{T}\sim D^{|\mathcal{T}|}}[(\hat{f}_{\mathcal{T}}(X) - \bar{f}(X) + \bar{f}(X) - f(X))^2] + \mathbb{V}_{\epsilon\sim E}[\epsilon] \\
&= \mathbb{E}_{\mathcal{T}\sim D^{|\mathcal{T}|}}[(\hat{f}_{\mathcal{T}}(X) - \bar{f}(X))^2] \\
&\quad + 2\mathbb{E}_{\mathcal{T}\sim D^{|\mathcal{T}|}}[(\hat{f}_{\mathcal{T}}(X) - \bar{f}(X))(\bar{f}(X) - f(X))] \\
&\quad + \mathbb{E}_{\mathcal{T}\sim D^{|\mathcal{T}|}}[(\bar{f}(X) - f(X))^2] + \mathbb{V}_{\epsilon\sim E}[\epsilon].
\end{aligned}$$

The first term is equal to the variance $\mathbb{V}_{\mathcal{T}\sim D^{|\mathcal{T}|}}[\hat{f}_{\mathcal{T}}(X) - \bar{f}(X)]$, since

$$\mathbb{E}_{\mathcal{T}\sim D^{|\mathcal{T}|}}[\hat{f}_{\mathcal{T}}(X) - \bar{f}(X)] = \bar{f}(X) - \bar{f}(X) = 0.$$

Since the second factor in the middle term does not depend on $\mathcal{T}$, the middle term becomes

$$\begin{aligned}
&\mathbb{E}_{\mathcal{T}\sim D^{|\mathcal{T}|}}[(\hat{f}_{\mathcal{T}}(X) - \bar{f}(X))(\bar{f}(X) - f(X))] \\
&= (\bar{f}(X) - f(X))\mathbb{E}_{\mathcal{T}\sim D^{|\mathcal{T}|}}[\hat{f}_{\mathcal{T}}(X) - \bar{f}(X)] \\
&= (\bar{f}(X) - f(X))(\bar{f}(X) - \bar{f}(X)) \\
&= 0.
\end{aligned}$$

In summary, we have

$$\begin{aligned}
\text{ESE} &= \mathbb{E}_{\mathcal{T}\sim D^{|\mathcal{T}|}}[(\hat{f}_{\mathcal{T}}(X) - \bar{f}(X))^2] + (\bar{f}(X) - f(X))^2 + \mathbb{V}[\epsilon] \\
&= \mathbb{V}_{\mathcal{T}\sim D^{|\mathcal{T}|}}[\hat{f}_{\mathcal{T}}(X) - \bar{f}(X)] + (\bar{f}(X) - f(X))^2 + \mathbb{V}[\epsilon],
\end{aligned}$$

which concludes the proof. □

References

Chen X et al (2020) Non-invasive early detection of cancer four years before conventional diagnosis using a blood test. Nat Commun 11:3475/1–3475/110

Cohen Joshua D et al (2018) Detection and localization of surgically resectable cancers with a multi-analyte blood test. Science 359:926–930

Domingos P (2000) A unified bias-variance decomposition and its applications. In: Langley P (ed) Proceedings of 17th international conference on machine learning (ICML 2000). Morgan Kaufmann, pp 231–238

Elena Tomeva OJ, Switzeny CH, Hippe B, Haslberger AG (2022) Comprehensive approach to distinguish patients with solid tumors from healthy controls by combining androgen receptor mutation p.H875Y with cell-free DNA methylation and circulating miRNAs. Cancers 14(TBD):462/1–462/114 (Impact factor of *Cancers:* 6.639)

Hastie T, Tibshirani R, Friedman J (2009) The elements of statistical learning: data mining, inference, and prediction, 2nd edn. Springer, Berlin

Leibniz GW (1685) The art of discovery

Mainardi F, Drewnowski A, Green H (2019) Personalized nutrient profiling of food patterns: Nestlé's nutrition algorithm applied to dietary intakes from NHANES. Nutrients 11(2):379/1–379/122

Powers DMW (2011) Evaluation: from precision, recall and F-score to ROC, informedness, markedness and correlation. J Mach Learn Technol. 2(1):37–63

Tan K-T, Ding L-W, Wu C-S, Tenen DG, Yang H (2021) Repurposing RNA sequencing for discovery of RNA modifications in clinical cohorts. Sci Adv 7(32):eabd2605/1–eabd2605/126

Wolpert DH, Macready WG (1997) No free lunch theorems for optimization. IEEE Trans Evolution Comput 1(1)

Youden WJ (1950) Index for rating diagnostic tests. Cancer 3(1):32–35

Precise Nutrition and Functional Foods

10

Ivanovic Dj Nevena, Berit Hippe, Stephanie Lilja, and Alexander G. Haslberger

Contents

Abstract

The different views and tensions between foods for growth of the body and foods for health and disease prevention can be seen since philosophers of the antique and middle ages. The efforts to develop foods for the improvement of health for persons at risk, e.g., elderly have resulted in exciting scientific and

I. Dj Nevena
Department of Bromatology, Faculty of Pharmacy, University of Belgrade, Vojvode Stepe 450, Belgrade, Serbia

B. Hippe · S. Lilja · A. G. Haslberger (✉)
Department of Nutritional Science, University of Vienna, Vienna, Austria
e-mail: alexander.haslberger@univie.ac.at

A. G. Haslberger (ed.), *Advances in Precision Nutrition, Personalization and Healthy Aging*, https://doi.org/10.1007/978-3-031-10153-3_10

medical innovations. But developments also resulted in legal- and trading problems between different regions of the world. Also quite different expectations of different groups of consumers in their food developed, with relevance to their regional and social background. The scientific progress to understand the connections between genetics, epigenetics, microbiota, and metabolomics as well as considerably individual metabolic differences resulted in the increasing acceptance of personal different needs of individuals in their foods. Consequently, foods and diets with different functions have been developed and need to be approved for effectivity, bioavailability, stability, and safety.

10.1 Novel Food, Food Supplements, Nutraceuticals, Phytoceuticals, Medicinal Foods

The role of food in maintaining good health has been known since ancient times. Confirmed by the phrase left by Hippocrates, the Greek physician father of modern medicine: "Let food be your medicine and medicine your food". It is well known that nutrition plays an essential role in maintaining health which is why it is integrated into public health strategies established to promote optimal health conditions throughout life. Modern consumers have placed new demands on food, which provides them with additional health benefits beyond the adequate nutritional effects. This has led to the development of the "functional food" concept, which, although existing since the 1980s, is still not precisely defined, and in many countries, this food category is not even recognized by national regulations. The primary function (nutritional value) of food is to provide us with food energy and the necessary macro and micronutrients. Foods also contribute to our well-being and can influence health status and prevent illnesses (health value) [2] (Cencic and Chingwaru 2010). Food that fulfills this function to an extent or in a special way is referred to as "functional food". Differentiation between food additives and functional foods is delicate and regional different, Fig. 10.1, https://www.burdockgroup.com/dietary-supplements-vs-functional-foods-safety-and-labeling

In the development of precision nutrition concepts, specific functions of diverse diets or foods need to be reflected, despite the considerable problems of definition and legal aspects in this area. For the use of foods as "precise nutrition", either the chemical composition of food should be taken into account, or the effect it has on the human metabolism [1]. To give individual tailored advice, a cumulative diagnosis of each individual must be done including microbiota composition, metabolomics, bioavailable tests for phytoceuticals, genetic testing, metabolomics, and many more tests. Or people get assigned to the individual symptoms for which there is an increased need for different nutrients like increased infection risk, diet-gut microbiome interplay, increased food cravings or accelerated aging, and many more, Fig. 10.2.

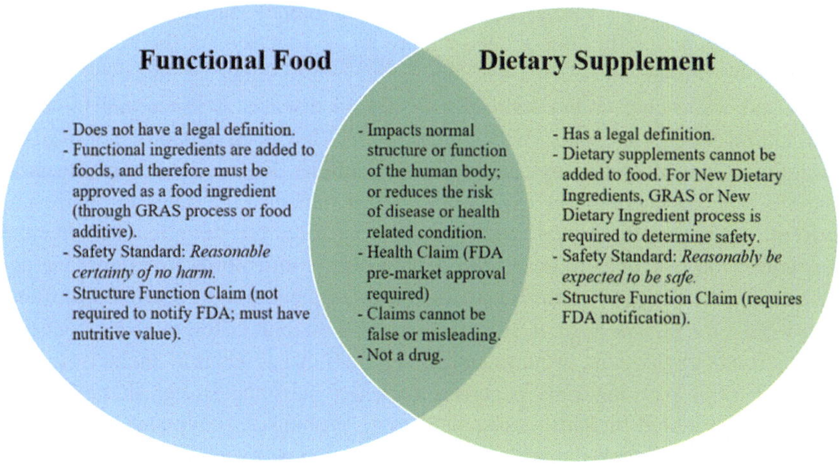

Fig. 10.1 Functional foods and dietary supplements, https://www.burdockgroup.com/dietary-supplements-vs-functional-foods-safety-and-labeling

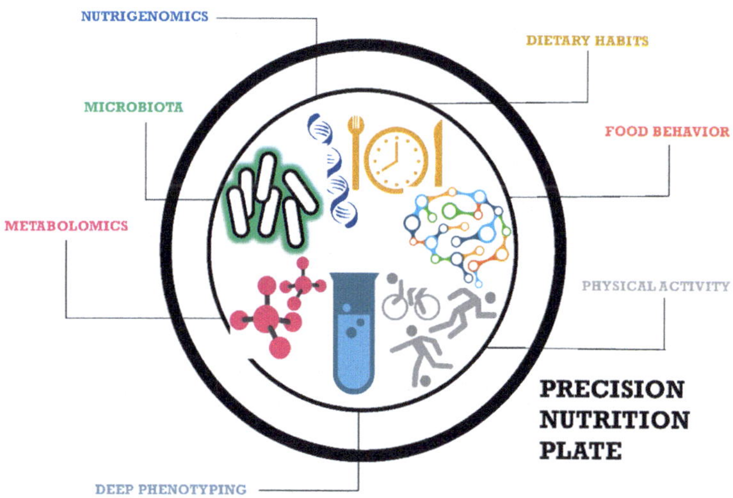

Fig. 10.2 Analyzing most relevant aspects affecting an individual response to lifestyle/nutritional interventions [1] (de Toro-Martín et al. 2017)

The concept of functional food was first introduced in 1984 in Japan to improve the consumer's health through the diet fortified with functional ingredients. The so-called FOSHU concept (FOSHU = foods for specified health use) was established, and placed 1991 in Japanese law. Food manufacturers who want to label their food with a special FOSHU seal must undergo a corresponding approval process. In 2002, 293 foods were already approved in this area in Japan (Shimizu 2002).

The European Food Safety Authority (EFSA) defines functional foods as: "A food, which beneficially affects one or more target functions in the body, beyond adequate nutritional effects, in a way that is relevant to either an improved state of health and well-being and/or reduction of risk of disease. A functional food can be a natural food or a food to which a component has been added or removed by technological or biotechnological means, and it must demonstrate their effects in amounts that can normally be expected to be consumed in the diet". In contrast to conventional foods, functional foods have demonstrated physiological benefits and can reduce the risk of chronic disease beyond basic nutritional functions, including maintenance of gut health. It provides health effects that go beyond traditional nutritional effects. As such it is closely related to but different from concepts such as food supplements or nutraceuticals. Functional foods may include (i) conventional foods with naturally occurring bioactive substances (e.g., dietary fiber); (ii) food with added bioactive substances (e.g., probiotics, polyphenols); and (iii) obtained from food ingredients (prebiotics, bioactive peptides).

In the USA, the functional foods concept was initiated and coined by Stephen DeFelice and Steve McNamara in 1989. They suggested the term "nutraceuticals" based on the 1983 Orphan drug. "Nutraceuticals are natural, bioactive chemical compounds that have health-promoting, disease-preventing, or general medicinal properties" (del Castillo and Iriondo-DeHond 2018). The definition of nutraceuticals includes "functional foods"—i.e., foods that provide a specific health benefit based on their ingredients.

Experts belonging to the Functional Food Center, USA (FFC) currently define functional foods as "natural or processed foods that contain known or unknown biologically active compounds, which, in defined, effective, and non-toxic amounts, provide a clinically proven and documented health benefit utilizing specific biomarkers for the prevention, management, or treatment of chronic disease or its symptoms". In this context, bioactive compounds, which are considered as a backbone of the functional foods, are understood as "primary and secondary metabolites of nutritive and non-nutritive natural components generating health benefits by preventing or are functional foods essential for "sustainable health".

The general confusion on definitions and legal uncertainties is reflected in the scientific literature, e.g.; "unfortunately, there are no internationally agreed definitions of "nutraceuticals" and "functional foods", or of similar terms, such as "health foods", or of terms related to herbal products, which are sometimes referred to as "nutraceuticals", compounding the confusion. "Nutraceuticals" and "functional foods" are vague, nondiscriminatory, unhelpful terms; the evidence suggests that they should be abandoned in favor of more precise terms. The term "dietary supplement" is widely used to designate formulations that are also called "nutraceuticals", but it would be better restricted to individual compounds used to treat or prevent deficiencies. "Fortified foods", sometimes called "designer foods", are foods to which compounds of proven therapeutic or preventive efficacy (e.g., folic acid) have been added" (Aronson 2017).

Botanical dietary supplements are generally available as whole plants, plant parts, powdered plant material, or plant extracts. These supplements are marketed

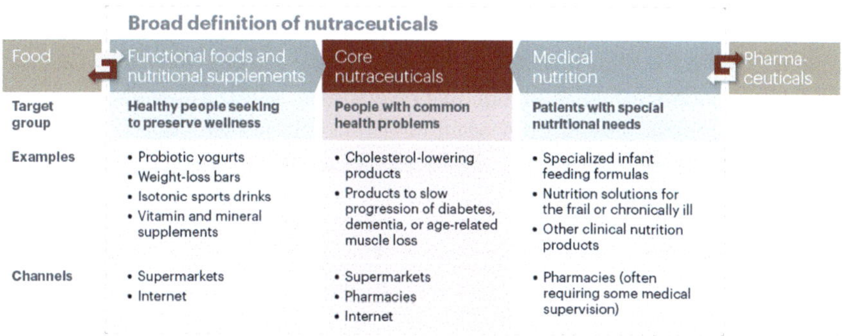

Source: A.T. Kearney analysis

Fig. 10.3 Continuum between nutraceuticals and pharmaceuticals, https://www.es.kearney.com/consumer-retail/article/?/a/nutraceuticals-the-front-line-of-the-battle-for-consumer-health

in various forms, including as powders, tablets, capsules, gummies, teas, tinctures, and essential oils. A variety of botanical dietary supplements are used in complementary and integrative health practices (https://nccih.nih.gov/health/integrative-health). Although there is overlap in the botanical species used in dietary supplements and other forms of complementary medicine, such as Ayurveda and Traditional Chinese Medicine, the applications can vary widely.

Another cause for concern is the overlap between nutraceuticals and pharmaceuticals. At one end of the spectrum is functional foods and beverages, as well as dietary supplements, aimed primarily at maintaining health. At the other, more medical end of the spectrum is products aimed at people with special nutritional needs. In the middle, it is an emerging gray area of products that have a physiological effect to reduce known risk factors, such as high cholesterol, or appear to slow or prevent the progression of common diseases “https://www.es.kearney.com/consumer-retail/article/?/a/nutraceuticals-the-front-line-of-the-battle-for-consumer-health”, Fig. 10.3.

A validated food composition database, also used by European EFSA, including fortifications and claim identification is https://www.eurofir.org. This database will be further broadened by the outcomes of the combined EU action ‘Creation of Open Access EU Food Composition Database (EU FCDB)’.

10.2 Fasting, Caloric Restriction (CR)

One of the most effective diets for health is certainly caloric restriction (CR) and fasting. Consequently, many kinds of foods are under evaluation aiming to mimic beneficial functions or molecular pathways of CR and fasting. To date, caloric restriction (i.e., a reduction in caloric intake without malnutrition) and fasting (abstinence from food which may be complete or partial, lengthy, of short duration, or intermittent) are the only non-genetic intervention that has consistently been found to extend both mean and maximal life span across a variety of species. Early studies in rodents revealed that mice fed 55–65% caloric restricted diets through their life exhibited a 35–65% greater mean and maximal lifespan than mice eating a non-purified ad libitum diet (Weindruch 1996). Although attenuated, these effects remain present even when moderate caloric restriction (20–40%) is implemented in middle-aged mice (Weindruch et al. 2001).

Importantly, prolonged caloric restriction has also been found to delay the onset of age-associated disease conditions such as cancer and diabetes in rodents (Weindruch et al. 2001) and nonhuman primates (Colman et al. 2009). Thus, findings from animal studies, including recent primate studies, suggest prolonged caloric restriction has the potential to extend health span and thereby increase the quality of life. In recent studies conducted in overweight humans, caloric restriction has been shown to improve a number of health outcomes including reducing several cardiac risk factors (Fontana et al. 2004, 2007; Lefevre et al. 2009), improving insulin-sensitivity (Meyer et al. 2006), and enhancing mitochondrial function (Civitarese et al. 2007). Additionally, prolonged caloric restriction has also been found to reduce oxidative damage to both DNA (Heilbronn and Ravussin 2003; Heilbronn et al. 2006; Hofer et al. 2008) and RNA, as assessed through white blood cells (Hofer et al. 2008).

Results from a controlled trial in healthy humans confirmed the health benefits of moderately limiting calorie intake over a period of two years. In humans CR decreased macrophage protein PLA2G7 such inhibiting the Nlrp3 inflammation and slowed down shrinking in of the thymus mice (Spadaro et al. 2022). However, in a number of popular CR concepts, like intermittent fasting or base fasting, almost no ketogenic switch and benefits from increased production of short-chain fatty acids (SCFAs) are likely.

Recently, our group reported beneficial effects of fasting on GI microbiota functions interacting with epigenetic regulation (Lilja et al. 2021).

Additional biological changes associated with CR and fasting that may contribute to the observed increases in health span and longevity include enhanced cellular quality control through autophagy. The metabolic switch from the use of glucose to the use of fatty acids and ketone bodies results in a beneficial modulation of the epigenetic regulation such as activation of DNA methyltransferases (DNMTs) to hypermethylate the promoter regions of aging-promoting genes such as p16 and RAS which induces cellular senescence. Effects of CR and fasting on the epigenetic methylation can be seen using the epigenetic clocks indicating healthy or accelerated aging (Zhang et al. 2020). At the histone level, CR and fasting are widely reported to be able to activate sirtuins in mammals (SIRT1–SIRT7) (Satoh et al. 2011) (Fig. 10.4).

Meanwhile fasting and other ways to increase SCFAs especially butyrate or ketone bodies especially beta-hydroxybutyrate (BHB) are under investigation to improve metabolic diseases as well as specifically neurological disease linked to neuro-inflammation such as depression, Alzheimer disease, and many others (Gough et al. 2021), Fig. 10.5. Butyrate is known to be essential for the development and maintenance of intestinal immunity and has a known role in supporting epithelial integrity (Ortiz et al. 2022).

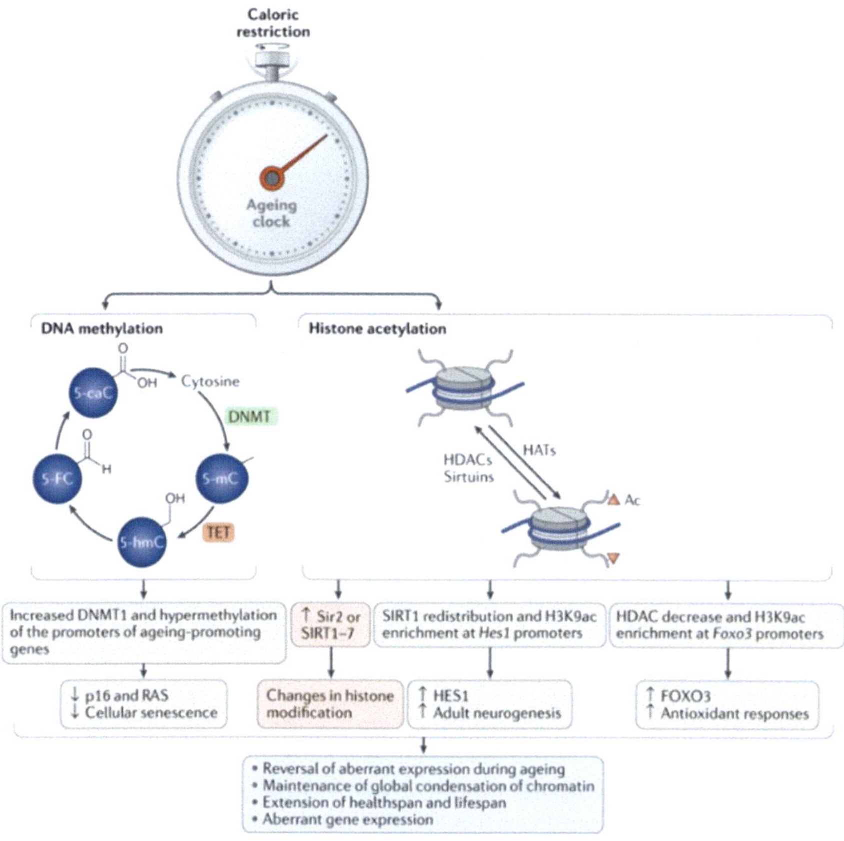

Fig. 10.4 Ac, histone acetylation; 5-caC, 5-carboxycytosine; 5-FC, 5-formylcytosine; HATs, histone acetyltransferases; 5-hmC, 5-hydroxymethylcytosine; 5-mC, 5-methylcytosine (Zhang et al. 2020)

A meta-analysis revealed that fasting interventions improve stress, anxiety, and depressive symptoms (Berthelot et al. 2021). Sodium butyrate (NaB) was shown to attenuate memory deficits and Aβ Plaques in a mouse model (Jiang et al. 2021). In many aspects therefore, functional foods try to mimic health supporting effects of CR and fasting. The concept of fasting mimetics developed.

10.3 Modulating the Diet-Gut Microbiome Interplay

Gut microbiota profiling is a top priority in nutritional interventions, and the impact of specific dietary factors on the ecological diversity of the gut can be supported by pro- and prebiotics. The development of nutritional interventions based on individual profiles is focused on optimizing gut microbial composition, both richness and diversity, and gut microbiota profiling is included as a key feature

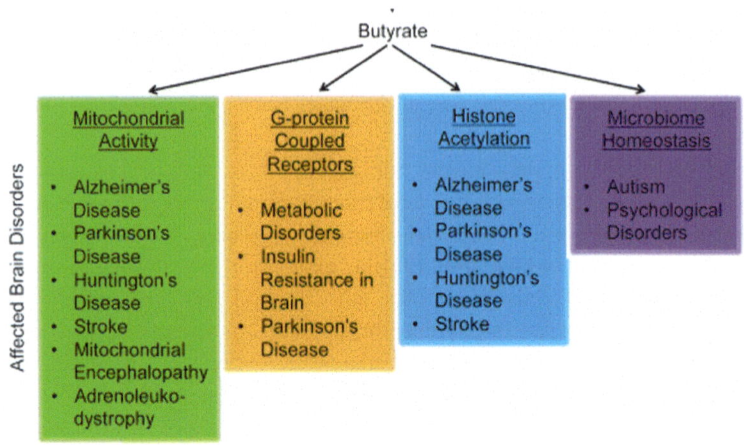

Fig. 10.5 Proposed mechanisms for the neuroprotective effects of ketogenesis, e.g., butyrate and the diseases which may benefit (Gough et al. 2021)

of precision nutrition (Zec 2022). For example, the effectiveness of probiotics are determined (1) by the interactions with prevailing gastrointestinal microbiota, (2) interaction with the host, in achieving a desirable probiotic effect, (3) interaction with diet, to survive, proliferate, and colonize GI, albeit temporary, and in the production of beneficial bioactive metabolites, such as short-chain fatty acids (e.g., butyric acid), bile acid derivatives and trimethylamines. This creates the necessity to design personalized pro- and prebiotics that focuses on treatment of specific disease considering the individual specific gut microbiome (Spacova et al. 2020).

10.4 Probiotics

Probiotics are one of the most common ingredients of functional products, and probiotic-containing foods account for 60–70% of the total functional products market (Kareb and Aider 2019). According to the definition proposed in 2002 by Food and Agriculture Organization of the United Nations (FAO)/World Health Organization (WHO) panel of experts, probiotics are "live microorganisms which when administered in adequate amounts confer a health benefit to the host" (Food Agriculture Organization of the United Nations 2002). After almost two decades, this concept remained in force and was accepted by several organizations, such as the World Gastroenterology Organization (WGO) and the European Food Safety Authority (EFSA) (Hill et al. 2014).

Functional products may contain one or more probiotic microorganisms. The most common genera to which probiotic microorganisms belong are Lactobacillus and Bifidobacterium, while others, less common species, may also belong to genera such as Streptococcus, Lactococcus, Enterococcus, Bacillus. Probiotic microorganisms include one type of yeast, *Saccharomyces cerevisiae* var. boulardii

Table 10.1 Some of the probiotic species used in the food (EFSA BIOHAZ Panel 2021)

Lactobacillus spp.	Other lactic acid bacteria	*Bifidobacterium* spp.	Other microorganisms
L. acidophilus *L. amylolyticus* *L. amylovorus* *L. casei* *L. crispatus* *L. delbrueckii* *L. fermentum* *L. gallinarum* *L. gasseri* *L. helveticus* *L. johnsonii* *L. kefiri* *L. paracasei* *L. plantarum* *L. reuteri* *L. rhamnosus* *L. salivarius*	*Streptococcus. salivarius* subsp. *thermophilus* *Lactococcus lactis* *Leuconostoc mesenteroides*	*B. adolescentis* *B. animalis* *B. bifidum* *B. breve* *B. longum*	*Enterococcus faecium* *Bacillus coagulans* *B. subtilis* *E. coli* strain Nissle *S. cerevisiae var. boulardii*

(EFSA BIOHAZ). Lactic acid bacteria (LAB), including Lactobacillus species, which have been used for thousands of years to extend the shelf life of food, can play a dual role in functional foods, acting as a food fermentation and in addition, impart health benefits. However, the term "probiotic" in these functional fermented products should be strictly reserved for live microorganisms that have shown health benefits in controlled human studies (Guarner et al. 2017, Ivanovic and Đorđević 2018). Table 10.1 shows some of the most commonly used probiotic microorganisms in functional foods and the pharmaceutical industry.

Traditionally, strains belonging to the mentioned bacterial genera have been primarily isolated from fermented products and the fecal microbiome. Fermented products are the most common natural sources of potential probiotic strains of LAB and are considered their most important source in the human gut microbiome. Their consumption is also associated with health benefits such as reducing the risk of developing type 2 diabetes and cardiovascular disease, which is why they represent a significant potential for future production of probiotics (Cunningham et al. 2021).

In order to be characterized as probiotic, bacterial species/strain have to posses specific characteristics: (i) safe for human use; (ii) ability to survive in the digestive tract; (iii) adhesion to the intestinal epithelium; (iv) colonization of the gastrointestinal tract; (v) generation of antimicrobial substances; (vi) stability and viability during technological processes and the storage period (Gutiérrez et al. 2020; Leo et al. 2019; Ivanovic and Đorđević 2018). An adequate selection process for probiotics is essential to ensure health benefits. Probiotics must meet specific safety, functional, and technological criteria, while health effects must be confirmed in clinical studies (Kareb and Aider 2019; Leo et al. 2019). Some of the most important criteria are given in Fig. 10.6.

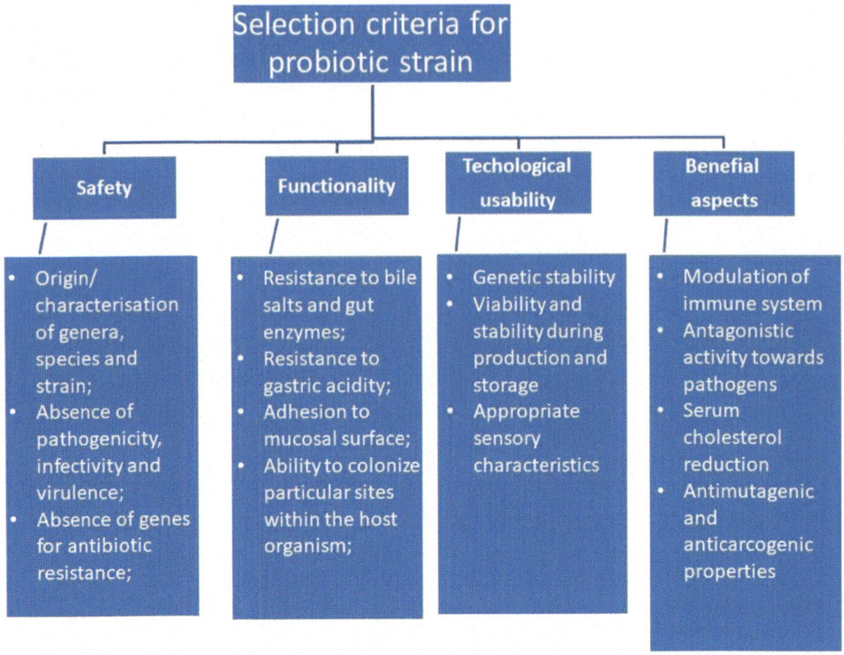

Fig. 10.6 Primary criteria for the selection of probiotics microorganisms in food production. Adapted from Kareb and Aider (2019) and Kumari et al. (2020)

The selection of probiotics begins with an evaluation of their safety. The safety of probiotic microorganisms depends on their origin, the presence of plasmids carrying the antibiotic resistance gene, and their association with other pathogenic cultures. In the United States, the most commonly used and currently available probiotic strains, with a long history of use, have the status of "GRAS" (generally recognized as safe). In the European Union, all microorganisms intended for food and feed production must be granted with "QPS status", which confirms that a particular microorganism does not pose a safety risk to humans and animals (Kareb and Aider 2019). When EFSA grants a microorganism with QPS status, it is included in the QPS list. In order to obtain QPS status, a microorganism must meet the following criteria: (i) Its taxonomic identity must be well defined (on the genus, species, and strain level); (ii) the available body of knowledge must be sufficient to establish its safety; (iii) the lack of pathogenic properties must be established and substantiated; (iv) its intended use must be clearly described (https://www.efsa.europa.eu/en/topics/topic/qualified-presumption-safety-qps). It is important to note that in addition to taxonomic identification, the strain must be deposited in a recognized culture collection (national and international) before use (FAO 2002).

Furthermore, for the effectiveness of probiotics, the microorganisms must be viable and present in the product in an adequate number. However, more recently, a striking beneficial activity was seen with inactivated bacterial cells, possibly via interaction of cell wall constituents with immune cell receptors (Piqué et al. 2019). The number of viable microorganisms should be maintained during the production and storage of the product, until the end of the shelf life, and during the passage through the digestive tract (Hill et al. 2014, Korčok et al. 2018). Although most commercial products contain probiotic strains in an amount between 1 and 10 billion colony-forming units (CFU), a lower or higher dose may be required to exhibit specific effects, making it impossible to establish a general dose; the dose should be based on human studies in which health benefits have been demonstrated (Guarner et al. 2017). However, the Public Health Agency of Canada and the Italian Ministry of Health recommend an intake of a minimum of billion CFU to ensure a beneficial effect (Santos et al. 2020). The ability to adhere to the intestinal epithelium allows probiotic microbes to persist in the digestive tract and exert beneficial effects (Kareb and Aider 2019).

In addition, the FAO/WHO guide from 2002 made recommendations about the labeling of the probiotic-containing products, which should include the following information: (i) genus, species, and strain for each probiotic microorganism present; (ii) the lowest viable number of bacteria at the end of the shelf life; (iii) the recommended dose which must ensure the efficacy of the probiotic concerning the purpose; (iv) recommended storage conditions (Food Agriculture Organization of the United Nations 2002).

A wide range of potentially beneficial effects is associated with the use of probiotics. Some of the beneficial effects of probiotics have been supported by randomized clinical studies and high-quality systematic reviews and meta-analyzes that have resulted in specific recommendations for prophylactic and therapeutic use. Based on the WGO guideline, there are specific indications for probiotic use based on different levels of evidence (following the Oxford Center for Evidence-Based Medicine criteria). For example, in adults, some of these indications are the prevention and treatment of acute diarrhea (*L. paracasei* B21060, *L. rhamnosus* GG or *S. boulardi* CNCM I-745), antibiotic-induced diarrhea (*L. rhamnosus* GG or *S. boulardi* CNCM I-745), as adjunctive therapy in the eradication of *Helicobacter pylori* (*L. rhamnosus* GG alone or in combination with *B. animalis* subspecies lactis), or as treatment of inflammatory bowel disease (*L. plantarum* 299v or *E. coli* DSM17252) (Guarner et al. 2017). According to the EFSA opinion based on a high level of evidence, yogurt with live cultures of *L. delbrueckii* subsp. bulgaricus and *S. thermophilus* effectively reduces symptoms associated with lactose intolerance (EFSA Panel on Dietetic Products and Allergies, 2010). There are also indications of different levels of evidence for the use of certain strains to prevent and treat specific conditions in children (Guarner et al. 2017; Szajewska et al. 2014; WGO 2017; Szajewska 2014). For each of these indications, the probiotic recommended dose is given. Studies indicating the efficacy of probiotics in the prevention and treatment of other diseases can be found in the literature, and some of the proposed mechanisms by which probiotics exhibit beneficial effects are given in Fig. 10.7.

The mechanisms underlying these probiotic effects have not yet been fully elucidated. The detection of mechanisms of probiotic activity is difficult because they are often species- and strain-specific and in most cases, represent a combination

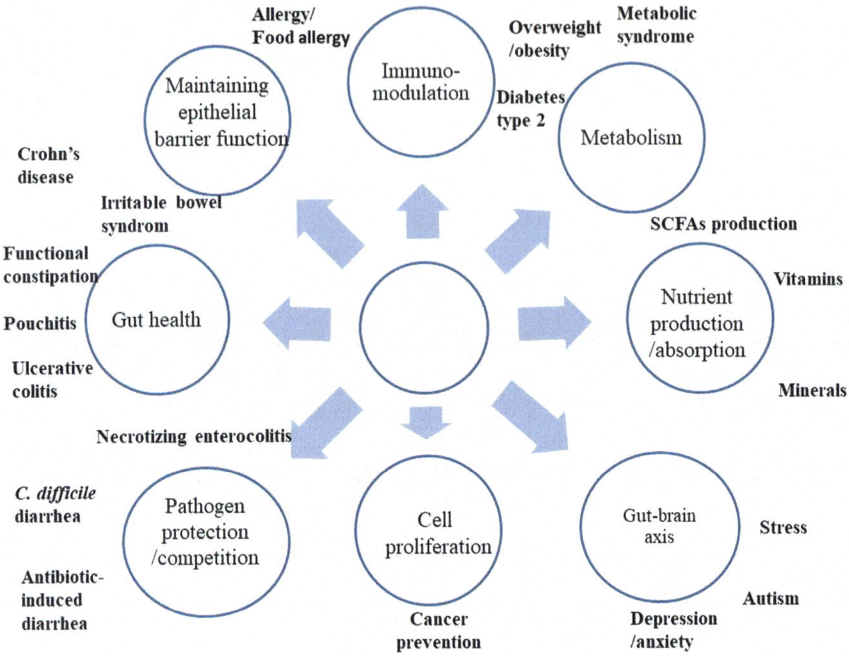

Fig. 10.7 Mechanisms of probiotic actions and potential clinical target. Adopted from Cremon et al. (2018)

of different activities (Ivanovic et al. 2015). For example, the beneficial effects of probiotics in preventing and reducing the duration of infectious diarrhea may be based on different mechanisms of probiotics such as modulation of gut microbiota, production of antimicrobial substances, competition with pathogens for the same adhesive sites, and modulation of immune response (Harzallah and Belhadj 2013). However, four levels of activity can generally be distinguished through which probiotics exhibit health effects: (i) interaction with microorganisms at the site of action or antagonistic activity on pathogenic bacteria by reducing luminal pH, inhibiting pathogen binding to adhesive sites, producing antimicrobial substances, and competing for nutrients; (ii) strengthening the intestinal mucosal barrier by promoting mucus secretion and strengthening intercellular connections; (iii) effect on the host immune system (immunostimulation and immunosuppression), (iv) interaction with the gut-brain axis (Mazhar et al. 2020; Sánchez et al. 2017; Ivanovic and Đorđević 2018).

As most of the probiotic effects of a strain are specific, recommendations for a particular application of probiotics in the clinical setting must be based on probiotic strains that have demonstrated these effects in human studies (Guarner et al. 2017). Namely, different strains within the same species are usually unique and can differ in the specific site of binding in the intestine or how they interact with the host immune system, which is crucial when determining the indication for the selected probiotic (Dimitrijevic et al. 2014; Ivanovic et al. 2015). Some strains will have unique properties that explain certain immunological, neurological, endocrinological, or antimicrobial activities (Guarner et al. 2017; Hill et al. 2014). Specific effects may be species-specific, such as vitamin synthesis, bile salt metabolism, antagonism against pathogenic enzyme activity, or strengthening the intestinal barrier (Ivanovic and Đorđević 2018).

With the expansion of knowledge about the human microbiome and its functions, the development of techniques for complete microbiome sequencing and cultivation methods created conditions for isolation and characterization of new microorganisms from the human microbiota with potential health benefits. Therefore, the conditions for the development of next-generation probiotics are established. Some bacteria isolated from the gut microbiome with probiotic potential are Akkermansia muciniphila, Roseburia intestinalis, Faecalibacterium prausnitzii, Eubacterium spp., Bacteroides spp. (Cunningham et al. 2021; O'Toole et al. 2017). These candidates provide physiological functions such as the production of butyrate, propionate, and other biologically active metabolites that are not always directly conferred by traditional probiotics such as lactobacilli and bifidobacteria (Cunningham et al. 2021). These species belonging to different genera were mainly identified based on comparative analyzes of the gut microbiota composition between healthy and diseased individuals (Martín and Langella 2019). Some authors are currently proposing using next-generation probiotics to avoid some of the side effects that may be accompanied by the intake of exogenous bacteria (Langella and Chatel 2019). The production of commercial products with these species is a challenge as they require certain specific growth media and anaerobic conditions. These difficulties have been partially overcome with A. muciniphila, representing one of the promising candidates for commercialization (Cunningham et al. 2021; Depommier et al. 2019).

Probiotics as a current strategy for treating dysbiosis, repairing the perturbed gut microbiota, and restoring microbial diversity, i.e., directing the microbiome toward health, have become very popular (Diaz et al. 2019). With the development of techniques that have linked taxonomic profiles and specific genera and species to disease and health, there has been significant interest in the precise application of probiotics (Cunningham et al. 2021). Moreover, numerous metabolites produced by intestinal bacteria, such as short-chain fatty acids, amino acids, vitamins, and secondary bile acids, play an essential role in the health and disease of the host. Therefore, the strategy for the precise application of probiotics could be based on the introduction of keystone species that are compromised in the microbiome and closely related to the host's health. Moreover, they could be used to stimulate the production of favorable bacterial metabolites or inhibit the production of metabolites harmful to health (Veiga et al. 2020).

Furthermore, evidence of probiotic efficacy is often very heterogeneous and conflicting between the scientific, medical, and industrial communities (Veiga et al. 2020). Although the term "probiotic" is generally associated with health benefits, the efficacy of probiotics is high strain-, indication- and dose-dependent (Sniffen et al. 2018). However, personal-specific factors, such as diet, age, genetics, and microbiome, also contribute to heterogeneity in probiotic outcomes, often very unpredictable (Shepherd et al. 2018; Veiga et al. 2020). Indeed, the degree to which probiotics can colonize the intestine during supplementation and their effective persistence, which is essential for its health-promoting characteristics, varies significantly between individuals. In addition to host factors, colonization will be primarily influenced by the background microbiota (Ma et al. 2020; Zmora et al. 2018), but the availability of the preferred substrate may also control the persistence of the introduced microbe (Kearney et al. 2018; Shepherd et al. 2018). Although stable colonization is not necessary for the manifestation of beneficial effects for the host, the population must be established at least in transit to exert the effect on the host and the resident gut microbiota (Ojima et al. 2022).

To date, few studies have examined the factors influencing probiotic colonization. Zmora et al. demonstrated that personalized intestinal mucosal resistance to colonization by commercial probiotics is associated with unique host and microbiota characteristics. These host and microbiome factors can lead to different colonization susceptibility to probiotics through the competitive exclusion of probiotic species or site-specific immune responses. Authors have reported the existence of caecal host immune activity against gram-positive bacteria in colonization-resistance subjects. The results of this study indicate the possibility of an individual predisposition to create an unfavorable environment for exogenous probiotics (Zmora et al. 2018).

Further, according to Ojima et al., response to probiotic administration could be predicted according to both availability of ecological niches within the gut microbiota and the relative fitness of the probiotic strain; the success of probiotic colonization is greater with a more significant niche difference between administrated probiotics and the resident bacteria (Ojima et al. 2022). In a human study in which B. longum AH1206 was orally administered, long-term colonization with this strain was found in only 30% of subjects. Microbiome analysis revealed that responders to colonization had a low presence of indigenous B. longum in the basal microbiota and underrepresentation in specific carbohydrate utilization genes (Gomez et al. 2016). Colonization by the exogenous strain in the responders was probably successful due to the absence of the autochthonous microbiota members, occupying the same niche, which are otherwise superior competitors for the same resources (Gomez et al. 2016; Ojima et al. 2022). A similar study on B. lactis V9 found that the high prevalence of the genus Bifidobacterium is the primary factor inhibiting the persistence of consumed B. lactis V9, which the authors attributed to competition for similar substrates and ecological niches as they belong to the same genus (Ma et al. 2020).

Also, the probability of colonization increases with the presence of a substrate for which the probiotic is highly selective, i.e., prebiotics, a substrate that can increase its fitness concerning members of the resident microbiota, provided it has genes for their digestion and utilization (Ma et al. 2020; Shepherd et al. 2018). These studies have highlighted the importance of a precise combination of probiotics and prebiotics in symbiotic products.

Despite increasing evidence of individual differences in susceptibility to probiotic colonization, its impact on clinical outcomes is still unknown. Nevertheless, better characterization of biological responses to probiotics and prebiotics in clinical studies would allow for a deeper understanding of these interventions and increase their potential for precise application.

Several approaches like administration of probiotic, prebiotics, synbiotics, fecal microbiota transplantation have been tried to mitigate the dysbiosis originated ill effects. But the effects of these approaches are highly generic and nonspecific. This creates the necessity to consider the individual specific gut microbiome. The health-promoting commensals could be the new promising prophylactic and therapeutic next-generation probiotics (NGPs). However, their unusual characteristics, unknown identity, and special growth requirements have presented difficulties for researcher, industrial exploitation, and regulatory agencies (Singh and Natraj 2021).

Personalization of probiotics has already been discussed for a long time. Currently, probiotic products are marketed worldwide, with the assumption that probiotics with demonstrated health effects work on all people, irrespective of the genetic (ethnicity), environment (geographical location), dietary habit and lifestyle. However, the benefits acquired from a probiotic are personal, depending on the health status, dietary habit, and prevailing GI microbiota. Personalized probiotics should be established to achieve precision administration of specific probiotic effects for targeted population. Globalization and urbanization of human activities have led to merging of dietary habit, thus effective probiotics should evolve in tandem. Ultimately, the probiotics of choice should be directed at specific physiological stage, health condition, and targeted diseases (Liu et al. 2018).

This concept was tested before when commensal bacterial strains were isolated from the feces of healthy mice and then administered back to the host as a personalized treatment in dextran sodium sulfate (DSS)-induced colitis. The group that received the personalized probiotic showed reduced susceptibility to DSS-colitis as compared to a commercial probiotic. Moreover, the personalized probiotic was more effective in modulating the host immune response, leading to decreased *Il-1β* and *Il-6* and increased *TGF-β* and *Il-10* expression. (Celiberto et al. 2018) The authors summarized that personalized probiotics may possess an advantage over commercial probiotics in treating dysbiotic-related conditions, possibly because they are derived directly from the host's own microbiota.

10.5 Prebiotics

Gibson and Roberfroid originally defined the concept of prebiotics in 1995 as "non-digestible food ingredients that beneficially influence the host's health by stimulating the activity of one or more commensal colon bacteria" (Gibson and Roberfroid 1995). Based on the definition, prebiotic administration is intended to impact the gut ecosystem, dominated by trillions of commensal microorganisms, to benefit the host. With the adoption of new scientific data on their mechanisms, site of action, and selectivity, prebiotics' definition has been revised several times. The last update of the definition was published in 2017 by ISAPP as an expert consensus stating that prebiotics is non-digestible components of food that serve as "substrate that is selectively utilized by host microorganisms conferring a health benefit" (Gibson et al. 2017). This definition has expanded the concept of prebiotics to include potential prebiotic substances other than non-carbohydrates, such as polyphenols and polyunsaturated fatty acids if their documented positive health effects are mediated through the gut microbiota. Moreover, the benefits of prebiotics are not limited to the gut, but the effect can also be manifested in extraintestinal sites, directly or indirectly (Gibson et al. 2017; Sanders et al. 2019).

Currently, there are a limited number of confirmed prebiotic substances, with non-digestible oligosaccharides most widely used in functional food production (Kareb and Aider 2019). The chemical structure of these components makes them resistant to enzymatic digestion in the digestive tract but can serve as nutrient substrates for health-promoting colon microorganisms, increasing their population (Brosseau et al. 2019; Kareb and Aider 2019). These microbiota members possess a wide range of cell-associated and extracellular glycosidases and transport systems that allow them to utilize fermentation products as carbon and energy sources. These are also critical requirements for defining prebiotics because the mentioned mechanisms enable their selectivity in target sites, especially in competitive environments such as the intestinal ecosystem (Gibson et al. 2017; Khangwal and Shukla 2019b; Sanders et al. 2019). Fructans such as fructooligosaccharides (FOS) and inulin are the most studied prebiotics, generally classified as "well established" prebiotics because of their number of commercial applications and regulatory status (Fig. 10.8). They are also recognized as safe food ingredients in the European Union (EU), being in use for more than two decades (Cardoso et al. 2020). These classes of prebiotics are most often naturally present in food, primarily in plant foods such as some vegetables (onions, garlic, chicory, asparagus), fruits (bananas), and cereals (Brosseau et al. 2019; Ivanovic and Đorđević 2018). However, their content in the modern westernized diet is insufficient to achieve the benefits of their intake. Most studies indicate that for achieving positive physiological effects from prebiotics, an intake of 5.5–20 g per day is required, while the average daily dietary intake of prebiotics such as inulin and FOS in the US is estimated at 1–4 g (Green et al. 2020). Due to their low content in food, prebiotics are produced on large industrial scales by various extraction and synthesis processes (Khangwal and Shukla 2019a).

Other carbohydrate-based prebiotics can be used in functional foods, such as isomaltooligosaccharides, xylooligosaccharides, isomaltooligosaccharides, and lactosucrose, which are classified as "emerging" prebiotic, while raffinose, for example, is classified as "under development". which is in detail reviewed in Cardoso et al. (2020). The current prebiotic classification is given in Fig. 10.8.

Gibson et al. proposed the criteria that a particular component should meet to be classified as a prebiotic: (i) resistance to gastric acidity, host enzymes hydrolysis, and gastrointestinal absorption; (ii) being fermentable by a specific

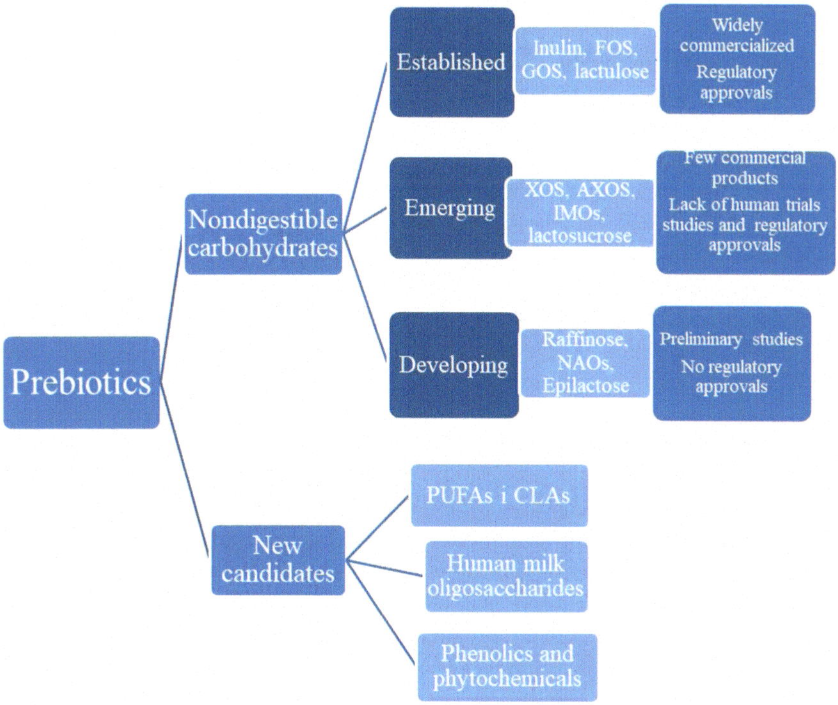

Fig. 10.8 Prebiotics current classification. FOS, fructooligosaccharides; GOS, galactooligosaccharides; XOS, xylooligosaccharides; AXOS, arabino-xylooligosaccharides; IMOs, isomaltooligosaccharides; NAOs, neoagaro-oligosaccharides; PUFAs, polyunsaturated fatty acids; CLAs, conjugated linoleic acid. Adapted from Cardoso et al. (2020)

member of intestinal microbiota; (iii) being able to selectively stimulate the growth and/or activity of intestinal bacteria considered to be beneficial to health (Gibson et al. 2004; Ivanovic and Đorđević 2018). Another essential potential criteria for selecting prebiotics in functional products production are their stability during technological processes so that undigested, chemically unchanged prebiotic reach the site of action (Markowiak and Slizewska 2017).

Selectivity is the most challenging criterion to demonstrate, and at the same time, the main criterion determines the difference between carbohydrate-derived prebiotics and dietary fiber. Namely, the terms prebiotic and dietary fiber are often interchanged when describing components that will not be digested in the upper intestine but are fermented by the action of colon bacteria. However, while fermentable dietary fiber is fermented nonselectively by most colon bacteria, prebiotic fermentation is performed by selected specific microorganisms potentially associated with human well-being and health (Kumari et al. 2020). Moreover, fiber fermentability is not crucial for the action of dietary fibers (Dorna Davani-Davari). Thus, some prebiotics can be considered dietary fiber, but not all dietary fibers are prebiotics (Slavin 2018). It is generally accepted that prebiotics have a highly selective effect on the human gut microbiota, increasing mainly the population of bifidobacteria and lactobacilli while reducing the prevalence of genera with detrimental effects on health (Bindels et al. 2015). However, in recent years, the development of advanced techniques for microbiota analysis has enabled a more profound determination of the range of prebiotic activity, which has led to the expansion of target sites of their selective action, in addition to lactic acid bacteria (LAB), to other candidates of genera with health promotion, such as *Roseburia* spp. Akkermansia and Propionibacterium (Cunningham et al. 2021). Therefore, new candidates for prebiotics have been proposed (Fig. 10.8).

Prebiotics can directly exert a beneficial effect by stimulating the growth and activity of specific groups of microorganisms that use them as substrates, resulting in favorable modulation of the composition and function of the microbiota. The beneficial effect can be manifested indirectly through cross-feeding interaction—their fermentation by specific microorganisms releases one or more metabolic products that serve as growth substrates for other bacteria with health benefits for the host (Cunningham et al. 2021). Changes in microbiota composition and metabolite concentration can affect the epithelial, immune, endocrine, and nerve signaling resulting in health benefits for the host, such as improvement in intestinal tract function, immune response, bone health, lipid, and glucose metabolism, appetite, and satiety regulation (Cunningham et al. 2021; Gibson et al. 2017). The primary metabolites of prebiotic fermentation are short-chain fatty acids (SCFAs), such as acetates, propionates, and butyrates, which interact with these host systems and through which numerous beneficial effects of prebiotics take place (Ashaolu and Ashaolu 2021). To date, the exact mechanisms by which prebiotics benefit remain unclear. The proposed mechanisms supported in vitro and animal studies are given in Table 10.2.

The beneficial effects of prebiotics in defense against pathogens can be achieved in several ways: through the production of SCFAs, which lowers the luminal pH and inhibits the growth of some pathogenic bacteria, and through the establishment of a stable population of commensal microorganisms (Chen and Gänzle 2017; Holscher 2017). Lowering the luminal pH as a consequence of SCFA synthesis can also lead to increased solubility of minerals, such as calcium and magnesium, and thus to their enhanced absorption, which is one of the most studied effects of fructans in both animal and human clinical studies (Porwal et al. 2020). It is important to emphasize that the change in pH in the intestines is one of the mechanisms for changing the composition and population of the intestinal microbiota. A change in pH can alter the population of acid-sensitive species such as Bacteroides and promote butyrate synthesis by Firmicutes, called the butyrogenic effect (Davari et al. 2019). Stimulation of the immune system can be direct or indirect through an increase in the population of beneficial microbes, although the exact mechanism is not understood enough. One of the essential mechanisms by which prebiotics exhibit immunomodulatory activity is a change in cytokine expression (Shokryazdan et al. 2022). There is evidence that many innate and acquired immune response components may be affected by metabolic fermentation

Table 10.2 Potential mechanisms of prebiotics action

Prebiotic effect	Mechanisms/expected results
Changing in gut microbiota composition	Selectively stimulating beneficial members of the gut microbiota with positive effects on the host health
Direct simulation of the immune system	Increasing anti-inflammatory cytokines; decreasing pro-inflammatory cytokines; beneficial effects on the mucosal immune system; increasing the secretory IgA; reducing Th2 responses
Simulating intestinal barrier function	Increasing production of mucus; increasing expression of the tight junction proteins in intestinal epithelial cell lines
Defense against pathogens	Lowering the colonic pH below threshold levels via SCFAs production; antagonism via different inhibitory peptides commonly produced by lactic acid bacteria; limitation in several colonization sites; reducing nutrient availability by establishment of a stable population of commensal microorganisms
Improvements in bowel function	Fecal bulking; regulating the secretion of gut hormones
Improving nutrients absorption	Improving small intestine development Increasing villi height, crypt depth, and number of goblet cells per villus
Increased mineral absorption	Reducing the colonic pH and increasing mineral solubility
Metabolic effects	Improving intestinal barrier function and prevention of inflammatory mediators translocation; regulation of glucose homeostasis and lipid metabolism; regulation of appetite via increased production of anorexigenic hormones such as PYY and GLP-1

Adapted from reports of Sander et al. (2019), Shokryazdan et al. (2022), and Kumari et al. (2020)

products and prebiotics themselves (Kumari et al. 2020). Additionally, intervention with prebiotics can reduce the Th2 immune response and thus affect allergies (Ashaolu 2020). The improvement in bowel function may be due to increased fecal content due to consumption of dietary fiber, although evidence from animal studies indicates that SCFA, fermentation products, can modulate intestinal motility via intestinal hormones (Sanders et al. 2019).

In recent years, the metabolic effects of prebiotics have been the subject of numerous studies. Despite inconsistent results, the generally accepted consensus is that intervention with prebiotics has a positive effect on glucose homeostasis, lipid profile, and inflammation (Sanders et al. 2019). It has been suggested that prebiotic-induced proliferation of commensal bacteria and increased SCFA synthesis positively affects mucosal intestinal barrier function, leading to reduced translocation of bacterial endotoxins from the gut into the circulation, a condition known as metabolic endotoxemia (Green et al. 2020). Metabolic endotoxemia has been suggested as a causative factor in diabetes and obesity. An additional effect on strengthening the mucosal barrier can be exerted by stimulating the tight junction proteins expression and stimulating mucus synthesis (Shokryazdan et al. 2022). The effect of prebiotics on satiety is achieved by synthesizing SCFAs whose interaction with colon L-cells results in the production of anorexigenic hormones such as peptide YY (PYY) and glucagon-like peptide-1 (GLP-1), both involved in the regulation of body weight, appetite, and glucose metabolism (Van Hul and Cani 2019). The effect of probiotics on lipid metabolism can also be manifested by inhibiting the absorption of cholesterol and bile acids, probably through the mechanism of binding prebiotics to these lipid components and promoting their excretion or inhibiting the lipogenic enzymes (Sierra et al. 2019; Pushpass et al. 2021). Increased bile acid excretion may be associated with increased intestinal commensal bacterial proliferation and increased BSH enzyme activity produced by these strains (Pushpass et al. 2021).

Interest in prebiotics as biologically active compounds that can prevent and treat a wide range of conditions and diseases associated with disturbances of the gut microbial ecosystem, named dysbiosis, can be seen in many registered clinical studies (ClinicalTrials.gov). Since October 2021, there have been as many as 268 registered clinical studies that have completed the evaluation of prebiotics (alone or in combination with probiotics) in obesity, type 2 diabetes, irritable bowel syndrome, ulcerative colitis, diarrhea, constipation, colon cancer, infant growth, atopic dermatitis, chronic kidney disease, aging, autism, depression, and other conditions.

However, intervention studies are increasingly incoherent and without statistical power due to the considerable variation in individual responses between participants, primarily when conducted in natural living conditions. A growing body of evidence suggests that the response to food intervention and dietary compounds, including prebiotics, is highly personalized and conditioned by the initial microbiota characteristics. Indeed, individual basal microbiota and health status at the beginning of the intervention affect the magnitude of potential changes in the microbiota and consequently the host (Mills et al. 2019). Theoretically, precise nutrition with existing prebiotics is possible by selecting susceptible individuals or subgroups with appropriate basal commensal microbes which can utilize this substrate. However, such selection is still challenging given the lack of targeted studies, including microbiome data (Vandeputte 2020). It is currently known that the *Prevotella*-dominant (P-enterotype) microbiome can ferment non-digestible carbohydrates more efficiently compared to the *Bacteroides*-dominant (B-enterotype) microbiome, resulting in P-enterotype subjects having more benefits than a fiber-rich diet in terms of losing bodyweight (Christensen et al. 2018). Additionally, in vitro fermentation studies have indicated functional differences between the enterotypes regarding prebiotics or fibers fermentation. Incubation with FOS, arabinoxylans, or fiber from pulse cell wall has led to a higher fermentation rate in P-enterotype inoculum and higher production of SCFAs, especially butyrate and propionate, compared to B-enterotype inoculum (Chen and Gänzle 2017). A similar effect of the initial microbiota composition on the fermentation of prebiotics was found in another in vitro study in which isomaltooligosaccharide was used as a carbon source (Wu et al. 2017). Contrary, B-enterotype microbiota was more efficient in the fermentation of alginate and its derivatives, producing more total SCFAs and butyrates than P-enterotype and Escherichia-enterotype (Fu et al. 2021). These studies have highlighted that the fermentation of prebiotics by the human microbiota is enterospecific, thus the importance of a personalized approach based on enterotype.

In a precise application of prebiotics, an important aspect should be an approach based on the prediction of responders and non-responders (Cunningham et al. 2021). A growing number of studies indicate that the basic microbiota consists of responders and non-responders to dietary intervention as well as the effectors of the host response (Holscher 2017; Holscher et al. 2015; Rodriguez et al. 2020). Thus, individuals with a non-detectible level of bifidobacteria in their gut microbiota failed to respond to agave inulin supplementation (Holscher et al. 2015). Furthermore, the presence or absence of certain species or consortia of microorganisms may also be related to the microbiota responsiveness on the prebiotics. Individuals in whom the presence of *Ruminococcus bromii* was not detected in the microbiota had a significantly reduced capacity for fermentation of resistant starch during ten weeks of intervention, resulting in 20–30% fermentability compared to 100% in individuals in whom R. bromii was detected (Frame et al. 2020; Walker et al. 2011). Also, improvement in metabolic disorders in obese individuals after inulin supplementation was associated with the pre-intervention presence of a specific consortium of bacteria such as Anaerostipes, Akkermansia, and Butyricicoccus but not with basal microbiota diversity (Rodriguez et al. 2020). Contrary, patients who were found to have pre-interventional elevated levels of *Coprococcus* spp. were more likely to benefit from inulin supplementation in terms of mood (Leyrolle et al. 2021). The initial microbiota composition was also related to the degree of fermentation and production of butyrate in response to intervention with resistant starch and hydrolyzed guar gum (Baxter et al. 2019; Reider et al. 2020). According to Ojima et al., response to prebiotics can be predicted by the abundance of enzymes for prebiotics digestion and utilization by the host microorganisms (Ojima et al. 2022).

Long-term dietary habits, especially dietary fiber intake, can also play a significant role in the response of the intestinal microbiota to intervention with prebiotics (Mills et al. 2019; Nogal et al. 2021). Three-week supplementation with inulin-type fructan resulted in a significant increase in the relative presence of *Bifidobacterium* and *Faecalibacterium* in participants classified as high-fiber consumers with a significant reduction in *Coprococcus*, *Dorea,* and *Ruminococcus* compared to low-fiber consumers showing only an increase in *Bifidobacterium*. Moreover, inulin intervention was more effective in increasing bifidobacteria in individuals with a lower basal concentration of bifidobacteria in the gut microbiota (Healey et al. 2018).

What further complicates predictions of response to prebiotic intervention is that their effects on the intestinal microbial community will depend on their structure, including the degree of polymerization, branching, monosaccharide profile, glycosidic linkage, specific structure (Lam and Cheung 2019). For example, in vitro fermentation studies have shown that only slight changes in the structure of arabinoxylans, extracted from different dietary sources, can lead to different changes in the composition and function of the microbiota, suggesting that physiological fiber function is highly dependent on structure (Tuncil et al. 2020). In addition, clinical intervention studies using different sources of either type 2 or type 4 resistant starch, which differs in structural characteristics, reported different effects of these components in modulating microbiota composition and SCFA production (Baxter et al. 2019; Cunningham et al. 2021; Deehan et al. 2020). Also, three weeks' consumption of resistant starch type 3 or type 4 differently affected healthy volunteers' initial fecal microbial population (Martínez et al. 2019).

The health effects of polyphenols as potential prebiotics depend on the ability of individuals to metabolize polyphenols into bioavailable metabolites, which is influenced by the characteristics of the microbiota as the central intermediate between polyphenols and health (Martín et al. 2020; Pushpass et al. 2021). The results of numerous studies have indicated the need to cluster human subjects into metabotypes that have different potentials for polyphenol metabolism and thus manifest their health effects. Therefore, personalized intervention with polyphenols in future will not be conceivable without considering metabotypes. As markers of the ability of individuals to benefit from a given polyphenol, it could serve metabolites produced by polyphenols in the presence of specific microorganisms (Martín et al. 2020).

Postbiotics and personalization

Postbiotics are functional bioactive compounds, generated in a matrix during bacterial fermentation, which may be used to promote health (Wegh et al. 2019). Postbiotics can include all products from microbial fermentation process, such as SCFAs, microbial cell fractions, functional proteins, extracellular polysaccharides (EPS), cell lysates, teichoic acid, peptidoglycan-derived muropeptides, and pili-type structures, Fig. 10.9. Definitions for postbiotics are yet under discussion.

The postbiotic concept would certainly allow various forms of personalization. A decrease of microbial metabolites such as SCFAs or neurological active metabolites have been reported with aging but also in various diseases such as depression or neuroinflammatory processes (Gruendler et al. 2020; Zalar et al. 2018). The origin of inter-individual variability in the action of bioactive small molecules from the diet has been analyzed, e.g., for urinary coffee phenolic acid metabolites. High inter- and intra-individual variation for metabolites produced by the colonic microbiome have been observed (Kerimi et al. 2020). Highly different metabolization of fibers or nutraceuticals due to GI microbiota composition effecting bioavailability as well as activity of food ingredients has been known for a long time. Therefore, analysis of microbiota and relevant metabolites could guide personalized production of postbiotics with improved activity in aging and many complex diseases.

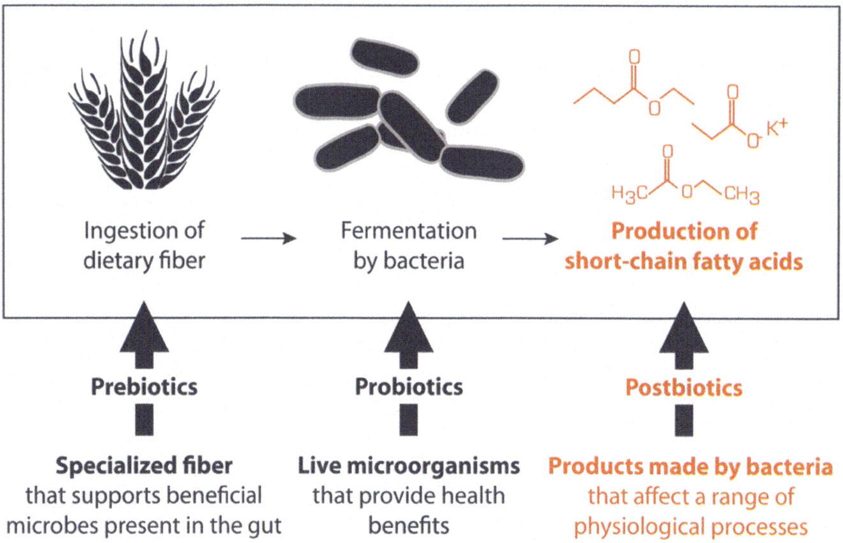

Fig. 10.9 Postbiotic, https://www.thinkbiome.com/postbiotics

10.6 Nutraceuticals

DeFelice proceeded to define nutraceutical as, "a food (or part of a food) that provides medical or health benefits, including the prevention and/or treatment of a disease". When functional food aids in the prevention and/or treatment of disease(s) and/or disorder(s) other than anemia, it is called a nutraceutical. Thus, nutraceuticals differ from dietary supplements in the following aspects: (1) nutraceuticals must not only supplement the diet but should also aid in the prevention and/or treatment of disease and/or disorder; and (2) nutraceuticals are used as conventional foods or as sole items of a meal or diet. Dietary components play beneficial roles beyond basic nutrition, leading to the development of the functional food concept and nutraceuticals. A functional food for one consumer can act as a nutraceutical for another consumer. Examples of nutraceuticals include fortified dairy products (e.g., milk) and citrus fruits (e.g., orange juice). Parameters for the evaluation of a nutraceutical are shown in Fig. 10.10.

Several naturally derived food substances have been studied in complex diseases, e.g., cancer therapies. Vitamin E, selenium, vitamin D, green tea, soy, and lycopene are examples of nutraceuticals widely studied in human health. While many of these compounds have been found to have high-therapeutic potential, future studies should include well-designed clinical trials assessing combinations of these compounds to realize possible synergies for human health.

Polyunsaturated fatty acids (PUFAs) (which include the omega-3 and omega-6 fatty acids) and phytochemicals also play an important role as healthy dietary bioactive compounds (Brower 1998; Zeisel 1999; Kalra 2003). Unfortunately,

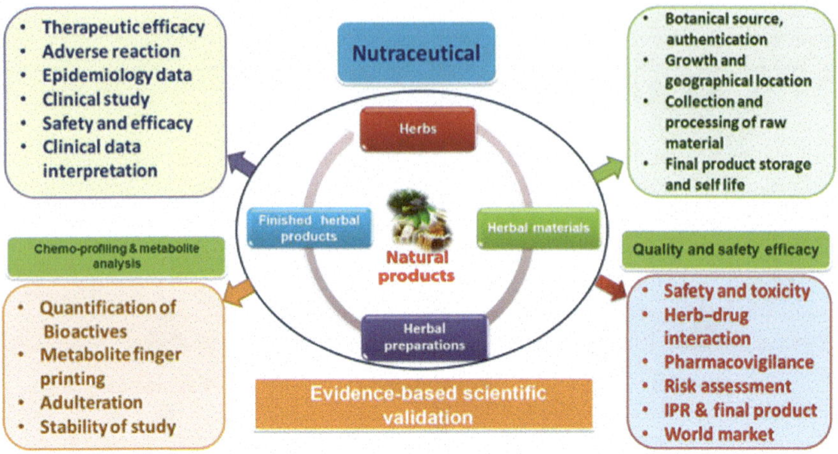

Fig. 10.10 Parameters for the evaluation of a nutraceutical (Mukherjee 2019)

different broad meta-analytical or coherent studies showed strikingly different conclusions on the effectivity on cardio vascular- or aging-related health problems (Abdelhamid AS). In this respect, a problematic stability of products and dose response characteristics have been discussed. Also "consistent and inconsistent responses" of different individuals have been shown, using multi-platform lipidomic approach (Nording et al. 2013). Recently, more studies tend to see positive effects of fish oil in a variety of health problem (Khan et al. 2021; Liao et al. 2022; Zhang et al. 2022).

10.7 Medical Foods

Medical foods are considered to be administered to a "patient receiving active and ongoing medical supervision (e.g., in a health care facility or as an outpatient) by a physician who has determined that the medical food is necessary to the patient's overall medical care". Furthermore, medical foods cannot be used for a condition that can be managed with a simple adjustment of the normal diet, such as diabetes or vitamin and mineral deficiencies. Medical foods include nutritionally complete formulas; nutritionally incomplete formulas containing proteins, carbohydrates, or fats; formulas for metabolic disorders in patients over 12 months of age; and oral rehydration formulas. These foods differ from dietary supplements and FDA-approved drugs in a number of ways. The main difference between medical foods and dietary supplements is that medical foods are used to manage a chronic disease or condition under medical or physician supervision, whereas supplements are intended for healthy individuals and can be obtained over-the-counter (OTC). (Ciampa et al. 2017). The use of medical foods for specific diseases must also be supported by recognized science.

These foods are intended for the exclusive or partial feeding of **people whose nutritional requirements cannot be met by normal foods**. The Directive 1999/21/EC lays down essential requirements on their composition and gives guidance for the minimum and maximum levels of vitamins and minerals.

Nutritional substances that may be used in the manufacture of foods for special medical purposes are laid down in Commission Regulation (EC) No. 953/2009. European EFSA gives guidance for the preparation of dossiers for foods for special medicinal purposes: https://efsa.onlinelibrary.wiley.com/doi/epdf/10.2903/j.efsa.2015.4300

10.8 Mechanistic Aspects of Special Foods

10.8.1 Epigenetic Active Foods

Over the past decade, remarkable breakthroughs in our understanding of epigenetic biology have coincided with an increased public interest in the impact of diet and lifestyle choices on health. It is well established that a balanced diet enhances life expectancy and helps to prevent or treat certain diseases, such as obesity, diabetes, cancer, and mental disorders. However, the biological mechanisms underlying these effects are not yet well understood. One possibility is through directly affecting catalytic activities of the enzymes responsible for "writing" or "erasing" the epigenetic modifications. Wang et al. identified two phytochemicals, dihydrocaffeic acid (DHCA) and malvidin-3′-O-glucoside (Mal-gluc), and metabolic intermediates derived from Concord grape juice, grape seed extract, and trans-resveratrol—that attenuate depression-like behaviors in mice· (Wang et al. 2018).

Nutrients affecting one of the two metabolites of the 1-carbon metabolism, S-Adenosylmethionin, a ubiquitous methyl donor, or S-adenosylhomocysteine, an inhibitor of methyltransferases, potentially alter the methylation of DNA and histones. Methylated promoter and other regulatory regions of a gene are usually associated with gene repression, whereas DNA demethylation within these regions leads to gene activation. Polyphenols, including curcumin, genistein, epigallocatechin gallate (EGCG), resveratrol, and equol, are well known for their beneficial effects via modulation of nuclear factor kappa B (NFkB) expression, chromatin remodeling through regulation of histone deacetylases (HDACs), and DNA methyltransferases activities. Gut microbiota-derived butyrate, sulforaphane, and curcumin affect histone acetyl-transferase (HATs) and/or HDACs activities leading to changes in chromatin structure. Vitamins, like biotin, niacin, and panthothenic acid, influence histone modifications. Resveratrol, butyrate, sulforaphane, and diallyl sulfide inhibit HDACs, whereas curcumin inhibits HATs. Although the action of many bioactive substances is specific to enzymes and proteins involved in the regulation of different components of the epigenome, interaction with other nutrients and lifestyle factors in physiological and pathological conditions must be taken into account. In addition, epigenetic components exert effect over each other. It adds an additional layer of complexity to the action of epigenetically active nutrients. Studies demonstrate that DNA methylation and histone modifications that act together to establish chromatin structure are involved in miRNA regulation and vice versa. Thus, deeper knowledge of bioactive nutrients/diets for characterization of their effects on the epigenome modifying enzymatic activities (acetylation, methylation, phosphorylation, ribosylation, oxidation, ubiquitination, and sumoylation) influencing drug absorption, distribution, metabolism, and excretion is needed (Remely et al. 2015).

Bioactive food components may get increasing importance in the prevention of various diseases, but researchers still face challenges which inhibit the implementation of such compounds into clinical practice. A major factor is the lack of knowledge underlying the mechanisms and effectiveness of these metabolites and

Natural compound	Natural sources	Pharmacological effects	Epigenetic mechanisms of action
Folate, cobalamin, riboflavin, pyridoxine, methionine	Folate and riboflavin: spinach, asparagus, beans, peas, lentils, sunflower seeds, almonds Cobalamin: fish, shellfish, poultry, milk, eggs Pyridoxine and methionine: grains, nuts, dragon fruit, sesame seeds	Anti-cancer, anti-proliferative, chemoprevention of malignant transformation	Regulation of one-carbon metabolism, SAM/SAH ratio, DNMT and MBD expression; regulation of miRNAs (tumour suppressor miR-122, miR-34a, miR-127, and oncogenic miR-21, miR-222)
Retinoic acid	Vietnamese gac, crude palm oil, yellow and orange fruits (mango, papaya), orange root vegetables (carrots), spinach, sweet potatoes	Anti-cancer, anti-proliferative, differentiating, pro-apoptotic	Regulation of DNMTs expression and enzyme activity by affecting p21, AP-1, PTEN and ERs; regulation of miRNAs targeting DNMTs; regulation of tumour suppressor miRNAs (miR-15, miR-16, let-7a, let-7c, miR-34a, miR-342) and oncogenic miRNAs (miR-10a); GNMT regulation; histone acetylation
Vitamin D3	Sun exposure, fish, fish liver oils	Anti-cancer, anti-proliferative, differentiating, pro-apoptotic	Regulation of DNMTs expression and enzyme activity by affecting p21, AP-1, PTEN and ERs; regulation of histone acetylation; regulation of oncogenic miRNAs (miR-181a, miR-181b)
Resveratrol	Roots of hellebore, grapes, mulberries, apricots, pineapples, peanuts	Anti-cancer, antioxidant, anti-proliferative, anti-angiogenesis, anti-inflammatory, pro-apoptotic, cardioprotective	Regulation of DNMTs expression and enzyme activity by affecting p21, AP-1 and PTEN; activation of deacetylase SIRT1 and p300 HAT; down-regulation of UHRF1; regulation of miRNAs
Genistein and daidzein	Soybeans, lupin, kudzu, psoralea, fava beans, coffee	Anti-cancer, antioxidant, antihelminthic, anti-metastatic, cancer protective	Regulation of DNMTs expression and enzyme activity by affecting p21, AP-1 and PTEN; increase in HAT activity; regulation of miRNAs (tumour suppressor miR-1296, miR-16, and oncogenic miR-27a)
EGCG	Green tea	Anti-cancer, antioxidant, anti-proliferative, anti-angiogenesis, anti-inflammatory, pro-apoptotic, cancer protective	Regulation of SAM/SAH ratio by COMT-mediated reactions; direct inhibition of DNMTs by binding to catalytic domain of the enzyme; regulation of tumour suppressor miRNAs (miR-16)
Curcumin	Spice turmeric	Anti-cancer, antioxidant, protects against heart failure	Direct inhibition of DNMTs by binding to catalytic domain of the enzyme; inhibition of HDACs and p300 HAT; regulation of miRNAs (tumour suppressor miR-22, miR-15a, miR-16, and oncogenic miR-21, miR-199a)

Fig. 10.11 Epigenetic active compounds and activities (Stefanska et al. 2012)

whether in vitro results are replicable in vivo in humans. Another concern is the poor **bioavailability** of the nutrients, which may be increased by combining certain bioactive compounds with other nutrients or the use of nanoencapsulation (Di Salle et al. 2016).

Examples of Bioactive Food Components and Epigenome Interactions are given in Fig. 10.11.

10.8.2 Sirtuin Activation by "Sirtfoods"

Sirtuins, commonly also referred to as silent mating type information regulation 2 homologous (SIRT), are a class of proteins which can be found in all living organisms, from bacteria and archaebacteria to mammals, and were first discovered in the 1990s in an effort to find yeast mutants with longer life durations (SINCLAIR 2006). SIRT1 is mainly located in the cell nucleus, but it can also be found in the cytosol. SIRT2 is also located in the cytosol, where it has its main site. SIRT3, SIRT4, and SIRT5 are mitochondrial proteins, but SIRT3 can also be located in the cell nucleus and cytosol under different cellular conditions. SIRT6 and SIRT7 are located in the cell nucleus and nucleolus, respectively (Alhazzazi 2011, p. 80–88) (Fig. 10.12).

Humans possess a total of seven sirtuins (SIRT1–SIRT7). Just like in yeast, they act as energy sensors in our cells and are activated when there is a lack of energy. Thanks to their properties, sirtuins are therefore multifunctional and regulate many metabolic processes as well as the aging process (RAUH 2013). In fact, increased activity of a yeast's sirtuin, silent information regulator two (Sir2), can extend its life. It ensures the silencing of certain chromatin regions by deacetylating histones. This attenuation of chromatin activities, such as during replication, recombination, and transcription, seems to be essential for prolonging the life of Sirt2. Interestingly, an increase in sirtuin activity also prolongs the lifespan of more complex organisms, such as the worm Caenorhabditis elegans or the fruit fly Drosophila melanogaster. Recent studies suggest that sirtuins also play an important role in the regulation of the life span of mammals. In more complex species, sirtuins can deacetylate a number of cellular regulatory proteins, in addition to histones, and can influence their activity in a positive or negative manner. However, the life-enhancing effect of increased sirtuin on mice has not yet been clearly documented.

Strong evidence supports a role for SIRT1 mediating an oxidative stress response by directly deacetylating several transcription factors that regulate antioxidant genes. Notably, SIRT1 activates several members of the FOXO family of transcription factors which promote the expression of stress response genes including SOD2. SIRT1 also promotes mitochondrial biogenesis by activating peroxisome proliferator-activated receptor co-activator 1-α (PGC-1α). PGC-1α increases mitochondrial mass and upregulates the expression of oxidative stress genes including glutathione peroxidase (GPx1), catalase, and manganese SOD (MnSOD). Finally, SIRT1 inactivates the p65 subunit of NF-κB through direct

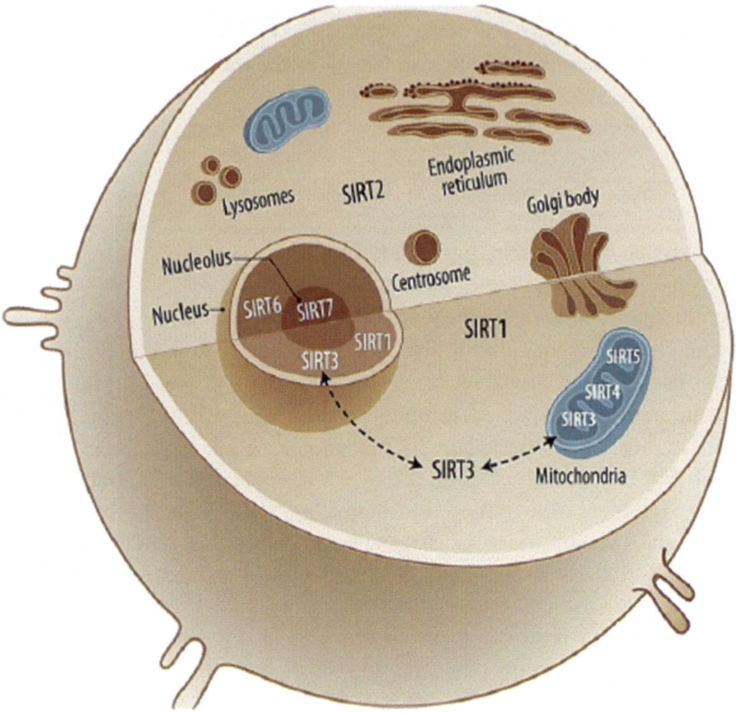

Fig. 10.12 Localization sirtuins (Alhazzazi 2011)

deacetylation. NF-κB inhibition suppresses the inducible nitric oxide synthase (iNOS) and nitrous oxide production and thus may lower the cellular reactive oxigen species (ROS) load. Given its role in the antioxidant response, whether SIRT1 activation contributes to CR mediated lifespan extension has been extensively studied. CR fails to increase the lifespan of SIRT1 knock-out mice, and these mice do not increase their physical activity, a phenotype typically associated with calorically restricted mice. Similarly, SIRT1 overexpression mimics a caloric restriction phenotype. Precisely, how SIRT1 functions during CR remains an open question, but emerging evidence suggests that p53 plays an important role in modulating SIRT1 during CR. Mitochondria account for the majority of cellular ROS production. Mitochondrial SIRT3 deacetylates and activates several enzymes that are critical in maintaining cellular ROS levels. SIRT3 deacetylates SOD2 at two important lysine residues to boost its catalytic activity, and the catalytic activity of SOD2 is diminished when SIRT3 is deleted. SIRT3 knock-out mice fail to reduce their levels of lipid peroxidation and protein carbonylation that are typically observed during caloric restriction indicating that SIRT3 is necessary for caloric restriction to mitigate oxidative stress. Additionally, SIRT3 stimulates the activity of mitochondrial isocitrate dehydrogenase, IDH2, during caloric restriction through direct deacetylation. IDH2 promotes the conversation of NADP+

to NADPH which in turn provides the reducing equivalents for conversion of oxidized to reduced glutathione. In support of this biochemical data, SIRT3 is required to protect calorically restricted mice from age-associated hearing loss. Another link between SIRT3 and oxidative stress comes from the field of oncology. Since ROS can severely damage nucleic acids, it is not surprising that oxidative stress can promote tumorigenesis. SIRT3 knock-out mouse embryonic fibroblasts (MEFs) exhibit higher ROS levels, greater genomic instability, and increased sensitivity to oncogenic transformation compared to wild-type fibroblasts (Kim et al. 2010). Intriguingly, overexpression of SOD2 suppresses oncogenic transformation in SIRT3 knock-out MEFs suggesting that SIRT3 may protect against tumorigenesis through an oxidative stress mechanism. In support of the above in vitro data, mice deficient for SIRT3 is more susceptible to cancer, and many human tumors display reduced SIRT3 levels compared to healthy tissues. In addition to suppressing the formation of cancer, SIRT3 can also combat established tumors (Guarente 2011; Merksamer et al. 2013), Fig. 10.13.

The close link between the sirtuin function and cellular metabolism plays a central role in regulating the lifespan. For example, sirtuins are necessary to activate the life-prolonging effect which occurs during the restriction of calories. Studies have shown that in animals, including mammals, a reduced calorie

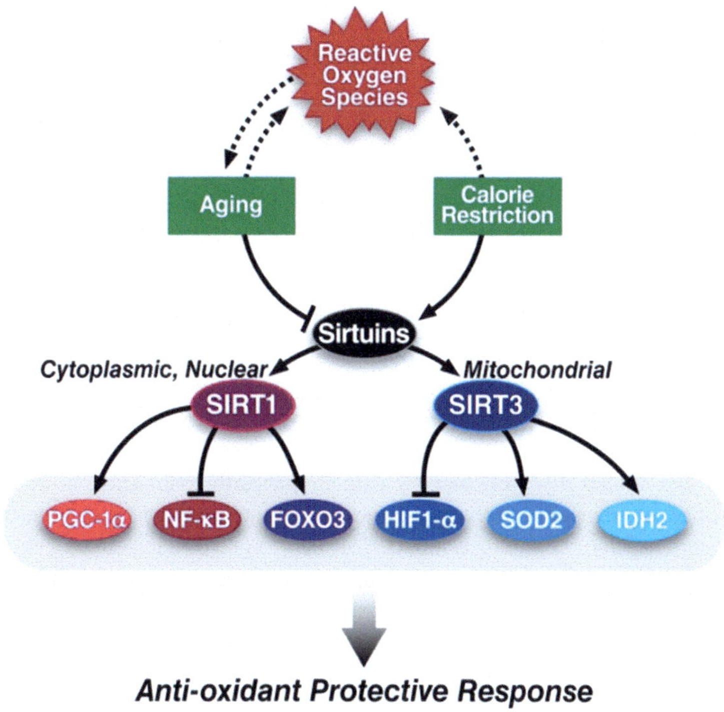

Fig. 10.13 Role of SIRTS, Merksamer et al. (2013)

intake leads to a general increase in fitness and prolonged life span. For example, in mice, Sirt1 activity is increased when calorie restrictions are in place (Bober 2007). Recently, research has focused on ways how to activate these sirtuins without fasting and foods which stimulate sirt enzymes came into broader interest. Sirtfoods are plant-based foods composed of polyphenols—secondary plant substances. In other words, Sirtfoods can activate sirtuins without fasting taking place. In countries such as Japan and India, varying sirtuins are part of the daily diet. Also known as "blue zones", these countries have the lowest incidences of lifestyle-related diseases, including hypertension, obesity, diabetes, fatty liver, and cancer, due to higher sirtuin activating diets. In addition, the polyphenols contained in Sirtfoods have been closely linked to positive effects on epigenetic mechanisms and many other genes known for promoting a long and above all, healthy life. Certain natural plant compounds such as red wine, strawberries, onions, soy, parsley, extra virgin olive oil, dark chocolate (85% cocoa), green tea, buckwheat, turmeric, or walnuts presumably increase the level of these proteins in the body, and foods containing them have been dubbed "Sirtfoods". Sirtfoods have attracted a broad interest in public media: https://www.healthline.com/nutrition/sirtfood-diet#section1; https://www.bbcgoodfood.com/howto/guide/what-sirtfood-diet; https://doi.org/10.1016/B978-0-08-100596-5.22721-2; https://doi.org/10.1016/j.freeradbiomed.2012.10.525

10.9 Senolytic Foods

Accumulating evidence suggests that targeting some of the aging hallmarks, for example, cellular senescence, can significantly improve human health and extend health span. **Cellular senescence** is a phenomenon where normal cells stop dividing. Senescent cells (SCs) accumulate in various tissues during the aging process. On one hand, cellular senescence blocks the propagation of damaged cells in order to maintain tissue homeostasis. On the other hand, it plays a causative role in irreparable, deleterious cellular damage, and loss of tissue homeostasis, which relates to aging and aging-associated diseases. Accumulating evidence demonstrates that elimination of SCs can reduce age-dependent deterioration in tissues and organs, which is useful in improving the treatment of age-associated diseases and alleviating the side effects of therapy-induced. Small molecules that can selectively kill SCs, called **senolytics**, have the potential to both prevent and treat age-related diseases, thereby extending health span. Until now, several classes of senolytic agents, including natural compounds such as quercetin, fisetin, piperlongumine, and curcumin analog EF24, and targeted therapeutics, which are mainly senolytic target inhibitors, have been identified. Compared to the targeted senolytics, natural senolytic compounds are less potent, but have low toxicity. They may also have a better chance of being translated into the clinical setting to treat age-related diseases or used as a lead for the development of more specific and potent senolytic agents (Li et al. 2019; Bielak-Zmijewska et al. 2019) (Fig. 10.14).

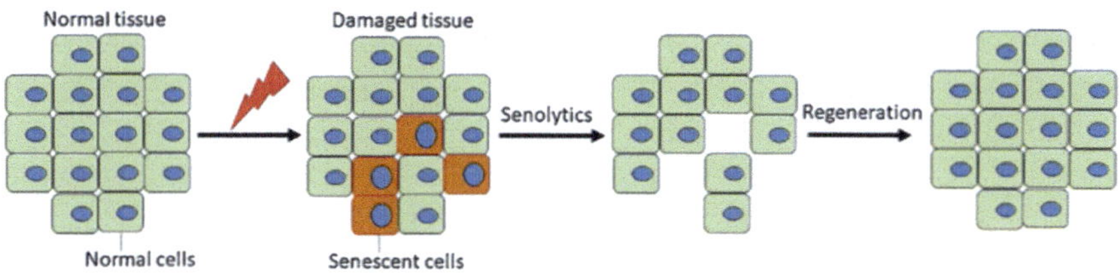

Fig. 10.14 Cellular senescence (Li et al. 2019)

10.9.1 Fasting Mimetics

As the world population ages, chronic diseases such as diabetes, cardiovascular disease, cancer, and neurodegeneration become ever more prevalent. Interventions that favor healthy aging would constitute powerful strategies with which to limit human diseases that have a broad socioeconomic impact. Fasting regimens such as intermittent fasting or dietary adaptations such as caloric restriction are among the few regimens that extend life and beneficially affect health in all tested model organisms, including rodents and nonhuman primates. However, few people seem capable of changing their dietary routines for extended periods. Thus, supplementation with pharmacological or plant derived caloric restriction mimetics (CRMs), mimic the beneficial effects of caloric restriction or fasting, has gained attention as an attractive and potentially feasible strategy (Madeo et al. n.d.).

Various polyphenols such as resveratrol in cells are able to activate biochemical pathways involved in senolysis in mammalian cells. These molecules include resveratrol, catechins, quercetin, and genistein. There is considerable overlap between these pathways and those activated by caloric restriction and mechanisms seem in large part to be mediated by their activation of sirtuin enzymes, Fig. 10.15.

Figure 10.16 shows mechanisms of plant ingredients and pharmaceutics claimed as caoric restriction mimetics and possible senolytic activity. Also combinations of pharmaceuticals and plant ingredients are under development as senolytica. In a clinical trial, the decrease of senescent cells by the tyrosine kinase inhibitor Dasatinib plus quercetin was shown in individuals with diabetic kidney disease (Hickson et al. 2019). The special importance of senolytics for health and economy can be shown by development of antibody-drug conjugates (Poblocka et al. 2021) as well as senolytic CAR T cells (Amor et al. 2020).

10.10 Personalization, Discussion

The ancient philosopher Lucretius recognized that one man's meat is another man's poison. In more modern times, Roger Williams (1956) wrote that "nutrition applied with due concern for individual variations […] offers the solution to many baffling health problems".

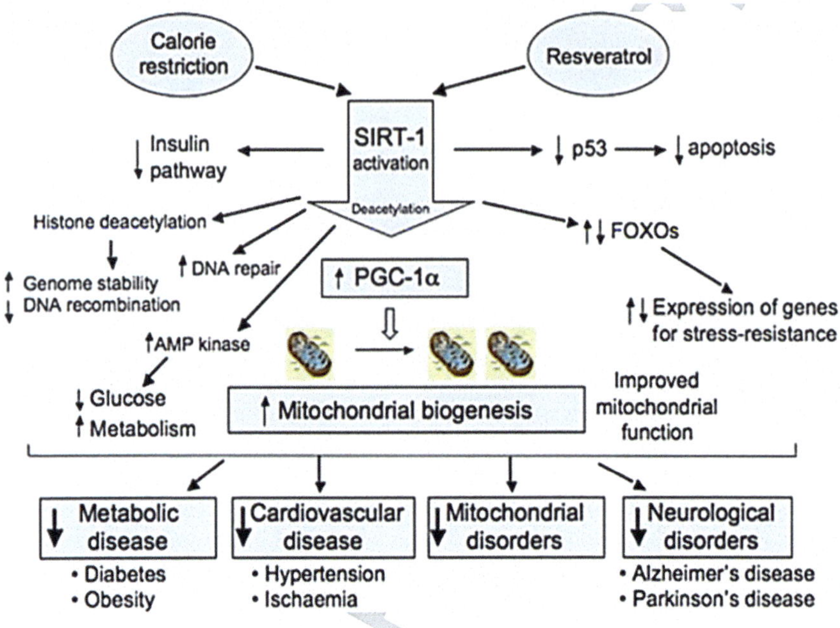

Fig. 10.15 General pathways and processes activated by calorie restriction and resveratrol (Morris 2010)

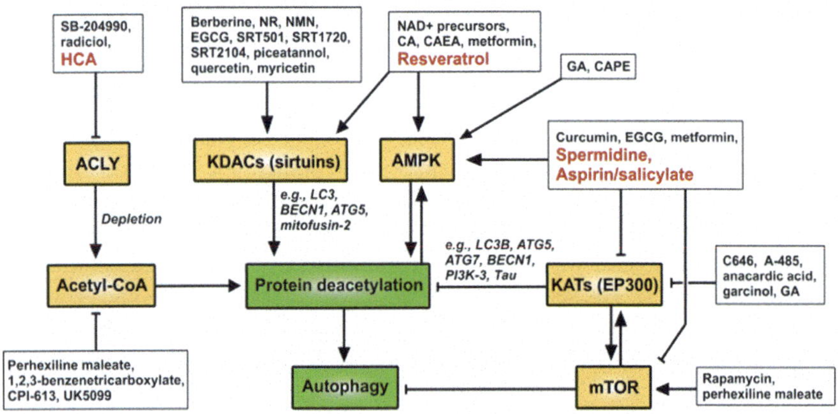

Fig. 10.16 Mechanisms of c caloric restriction mimetics (Madeo et al. 2019)

Clinicians have long recognized that diet, supplement, and lifestyle solutions that are effective for some individuals have no or even deleterious effects on others. Clearly, the need for personalized nutrition exists. The use of nutrigenomics has been discussed as a tool for disease risk intervention (Ferguson 2015). To propose a healthy diet to populations worldwide that must suit high inter-individual variability driven by complex gene–nutrient–environment interactions has been seen as one of the biggest challenges of modern nutrition before. Although a number of functional foods are now proposed in support of a healthy diet, a one-size-fits-all approach to nutrition is inappropriate, and new personalized functional foods are necessary. Metabolic profiling technologies can assist at various levels of the development of functional foods, from screening for food composition to identification of new biomarkers of food intake to support diet intervention and epidemiological studies (Claus 2014).

The wide array of new health products—nutraceuticals, nutritional supplements, functional foods, dietary supplements, foods for special medical purposes, and foods for special dietary uses—as well as different national regulations makes a harmonized understanding of safety evaluation difficult and is often confusing to consumers and industry. Facing often difficult legal categorization of products or unsubstantiated health claims from industry, consumer advice organizations are calling for stricter control mechanisms. On the other hand, stricter legislation and burdens for authorization result in monopolization in food and medical industry often hindering creative new developments for health care.

References

Amor C, Feucht J, Leibold J, Ho YJ, Zhu C, Alonso-Curbelo D, Mansilla-Soto J, Boyer JA, Li X, Giavridis T, Kulick A, Houlihan S, Peerschke E, Friedman SL, Ponomarev V, Piersigilli A, Sadelain M, Lowe SW (2020) Senolytic CAR T cells reverse senescence-associated pathologies. Nature 583(7814):127–132. https://doi.org/10.1038/s41586-020-2403-9

Aronson JK (2017) Defining "nutraceuticals": neither nutritious nor pharmaceutical. Br J Clin Pharmacol 83(1):8–19. https://doi.org/10.1111/BCP.12935

Baxter et al (2019). https://doi.org/10.1128/mBio.02566-18

Berthelot E, Etchecopar-Etchart D, Thellier D, Lancon C, Boyer L, Fond G (2021) Fasting interventions for stress, anxiety and depressive symptoms: a systematic review and meta-analysis. Nutrients 13(11). https://doi.org/10.3390/NU13113947

Bielak-Zmijewska A, Grabowska W, Ciolko A, Bojko A, Mosieniak G, Bijoch Ł, Sikora E (2019) The role of curcumin in the modulation of ageing. Int J Mol Sci 20(5). https://doi.org/10.3390/ijms20051239

Bindels et al (2015). https://doi.org/10.1038/nrgastro.2015.47

Bober (2007). https://doi.org/10.5483/BMBRep.2008.41.11.751. https://doi.org/10.1101/gad.227439.113

Brosseau et al (2019). https://doi.org/10.3390/nu11081841

Brower V (1998) No Title. Nat Biotechnol 16:728–731

Cardoso et al (2021). https://doi.org/10.1016/bs.afnr.2020.08.001

Celiberto LS, Pinto RA, Rossi EA, Vallance BA, Cavallini DCU (2018) Isolation and characterization of potentially probiotic bacterial strains from mice: proof of concept for personalized probiotics. Nutrients 10(11). https://doi.org/10.3390/NU10111684

Cencic A, Chingwaru W (2010) The role of functional foods, nutraceuticals, and food supplements in intestinal health. Nutrients 2(6):611. https://doi.org/10.3390/NU2060611
Christensen et al (2018). https://doi.org/10.1093/ajcn/nqy175
Ciampa BP, Ramos ER, Borum M, Doman DB (2017) The emerging therapeutic role of medical foods for gastrointestinal disorders. Gastroenterol Hepatol 13(2):104–115
Claus (2014). https://doi.org/10.1097/MCO.0000000000000107
Colman et al (2009). https://doi.org/10.1126/science.1173635
Civitarese et al (2007). https://doi.org/10.1371/journal.pmed.0040076
Cunningham et al (2021). https://doi.org/10.1016/j.tim.2021.01.003
del Castillo and Iriondo-DeHond (2018). https://doi.org/10.3390/nu10101358
Davari et al (2019). https://doi.org/10.3390/foods8030092
de Toro-Martín J, Arsenault BJ, Després JP, Vohl MC (2017) Precision nutrition: a review of personalized nutritional approaches for the prevention and management of metabolic syndrome. Nutrients 9(8):913. https://doi.org/10.3390/NU9080913
Deehan et al (2020). https://doi.org/10.1016/j.chom.2020.01.006
Depommier et al (2019). https://doi.org/10.1038/s41591-019-0495-2
Diaz et al (2019). https://doi.org/10.1093/advances/nmy063
Dimitrijevic R, Ivanovic N, Mathiesen G, Petrusic V, Zivkovic I, Djordjevic B, Dimitrijevic LJ (2014) J Dairy Res 81(2):202–207. https://doi.org/10.1017/S0022029914000028
Di Salle A et al (2016) Polyphenols nanoencapsulation for therapeutic applications. J Biomol Res Ther 5(2). https://doi.org/10.4172/2167-7956.1000139
Ferguson (2015). https://doi.org/10.1161/HCG.0000000000000030
Fontana et al (2004). https://doi.org/10.1073/pnas.0308291101
Frame et al (2020). https://doi.org/10.1093/nutrit/nuz106
Fu et al (2021). https://doi.org/10.1016/j.ijbiomac.2021.05.135
Gibson et al (2017). https://doi.org/10.1079/NRR200479
Gomez et al (2016). https://doi.org/10.1016/j.chom.2016.09.001
Gough SM, Casella A, Ortega KJ, Hackam AS (2021) Neuroprotection by the ketogenic diet: evidence and controversies. Front Nutr 8:782657. https://doi.org/10.3389/FNUT.2021.782657
Green et al (2020). https://doi.org/10.3390/ijms21082890
Gruendler R, Hippe B, Sendula Jengic V, Peterlin B, Haslberger AG (2020) Nutraceutical approaches of autophagy and neuroinflammation in Alzheimer's disease: a systematic review. Molecules (Basel, Switzerland) 25(24). https://doi.org/10.3390/molecules25246018
Guarente L (2011) Sirtuins, aging, and metabolism. Cold Spring Harb Symp Quant Biol 76:81–90. https://doi.org/10.1101/sqb.2011.76.010629
Guarner et al (2017). https://doi.org/10.3389/fmicb.2020.01662
Gutiérrez et al (2020). https://doi.org/10.1080/10408398.2022.2099807
Healey et al (2018). https://doi.org/10.1017/S0007114517003440
Heilbronn et al (2006). https://doi.org/10.1001/jama.295.13.1539
Hickson, LTJ, Langhi Prata LGP, Bobart SA, Evans TK, Giorgadze N, Hashmi SK, Herrmann SM, Jensen MD, Jia Q, Jordan KL, Kellogg TA, Khosla S, Koerber DM, Lagnado AB, Lawson DK, LeBrasseur NK, Lerman LO, McDonald KM, McKenzie TJ, Passos JF, Pignolo RJ, Pirtskhalava T, IMSaadiq, Schaefer KK, Textor SC, Victorelli SG, Volkman TL, Xue A, Wentworth MA, Wissler Gerdes EO, Zhu Y, Tchkonia T, Kirkland JL (2019) Senolytics decrease senescent cells in humans: Preliminary report from a clinical trial of Dasatinib plus Quercetin in individuals with diabetic kidney disease. EBioMedicine 47:446–456. https://doi.org/10.1016/j.ebiom.2019.08.069
Hill et al (2014). https://doi.org/10.1038/nrgastro.2014.66
Hofer et al (2008). https://doi.org/10.1089/rej.2008.0712
Holscher (2017). https://doi.org/10.1080/19490976.2017.1290756
Holscher et al (2015). https://doi.org/10.1080/19490976.2017.1290756
Ivanovic N, Đorđević B (2018) Probiotics and dietary intervention in the prevention of metabolic syndrome. Zadužbina Andrejević, Belgrade, p 204(Biblioteka Educatio). ISBN 978-86-525-0336-0

Ivanovic N, Minic R, Dimitrijevic LJ, Radojevic Skodric S, Zivkovic I, Djordjevic B (2015) Lactobacillus rhamnosus LA68 and Lactobacillus plantarum WCFS1 differently influence metabolic and immunological parameters in high fat diet-induced hypercholesterolemia and hepatic steatosis. Food Funct 6(2):558–565. https://doi.org/10.1039/C4FO00843J

Jiang Y, Li K, Li X, Xu L, Yang Z (2021) Sodium butyrate ameliorates the impairment of synaptic plasticity by inhibiting the neuroinflammation in 5XFAD mice. Chem Biol Interact 341:109452. https://doi.org/10.1016/J.CBI.2021.109452

Kalra EK (2003) No Title. AAPS PharmSci 5:25

Kerimi A, Kraut NU, da Encarnacao JA, Williamson G (2020) The gut microbiome drives inter- and intra-individual differences in metabolism of bioactive small molecules. Sci Rep 10(1). https://doi.org/10.1038/S41598-020-76558-5

Khan SU, Lone AN, Khan MS, Virani SS, Blumenthal RS, Nasir K, Miller M, Michos ED, Ballantyne CM, Boden WE, Bhatt DL (2021) Effect of omega-3 fatty acids on cardiovascular outcomes: a systematic review and meta-analysis. EClinicalMedicine 38:100997. https://doi.org/10.1016/J.ECLINM.2021.100997/ATTACHMENT/079022C4-42B6-4816-A327-3B5BF0CF0017/MMC1.DOCX

Kim et al (2010). https://doi.org/10.1016/j.ccr.2009.11.023

Korčok D, Tršić-Milanović D, Ivanović Dj N, Đorđević B (2018) Development of probiotic formulation for the treatment of iron deficiency anemia. Chem Pharm Bull 66(4):347–352. https://doi.org/10.1248/cpb.c17-00634

Kumari et al (2020). https://doi.org/10.1016/B978-0-12-820528-0.00012-0

Lefevre et al (2009). https://doi.org/10.1172/JCI36714

Leo et al (2019). https://doi.org/10.1016/B978-0-12-816397-9.00014-5

Leyrolle et al (2021). https://doi.org/10.1016/j.bbi.2021.01.014

Li W, Qin L, Feng R, Hu G, Sun H, He Y, Zhang R (2019) Emerging senolytic agents derived from natural products. Mech Ageing Dev 181:1–6. https://doi.org/10.1016/j.mad.2019.05.001

Liao J, Xiong Q, Yin Y, Ling Z, Chen S (2022) The effects of fish oil on cardiovascular diseases: systematical evaluation and recent advance. Front Cardiovasc Med 1959. https://doi.org/10.3389/FCVM.2021.802306

Lilja S, Stoll C, Krammer U, Hippe B, Duszka K, Debebe T, Höfinger I, König J, Pointner A, Haslberger A (2021) Five days periodic fasting elevates levels of longevity related christensenella and sirtuin expression in humans. Int J Mol Sci 22(5):1–15. https://doi.org/10.3390/ijms22052331

Liu Y, Tran DQ, Rhoads JM (2018) Probiotics in disease prevention and treatment. J Clin Pharmacol 58(Suppl 10):S164. https://doi.org/10.1002/JCPH.1121

Ma et al (2020). https://doi.org/10.1016/j.foodres.2020.109496

Madeo F, Carmona-Gutierrez D, Hofer SJ, Kroemer G (2019) Caloric restriction mimetics against age-associated disease: targets, mechanisms, and therapeutic potential. Cell Metab 29(3):592–610. https://doi.org/10.1016/j.cmet.2019.01.018

Madeo F, Eisenberg T, Pietrocola F, Kroemer G (n.d.) Review summary. Spermidine in health and disease. https://doi.org/10.1126/science.aan2788

Martín et al (2020). https://doi.org/10.1002/mnfr.201900952

Martínez et al (2019). https://doi.org/10.1093/advances/nmy078

Mazhar et al (2020). https://doi.org/10.2217/fmb-2019-0143

Merksamer PI, Liu Y, He W, Hirschey MD, Chen D, Verdin E (2013) The sirtuins, oxidative stress and aging: an emerging link. Aging 5(3):144–150. https://doi.org/10.18632/aging.100544

Meyer et al (2006). https://doi.org/10.2337/dc05-2565

Mills et al (2019). https://doi.org/10.3390/nu12061776

Morris BJ (2010) Calorie restriction mimetics and aging. In: Calorie restriction, aging and longevity. Springer, Netherlands, pp 141–175. https://doi.org/10.1007/978-90-481-8556-6_9

Mukherjee PK (2019) Quality control and evaluation of herbal drugs: evaluating natural products and traditional medicine. In: Quality control and evaluation of herbal drugs: evaluating natural products and traditional medicine. Elsevier. https://doi.org/10.1016/C2016-0-04232-8

Nording ML, Yang J, Georgi K, Hegedus Karbowski C, German JB, Weiss RH, Hogg RJ, Trygg J, Hammock BD, Zivkovic AM (2013) Individual variation in lipidomic profiles of healthy subjects in response to omega-3 fatty acids. PLoS ONE 8(10). https://doi.org/10.1371/JOURNAL.PONE.0076575

Ojima et al (2022). https://doi.org/10.3389/fmicb.2020.01349

Ortiz AM, Simpson J, Langner CA, Baker PJ, Aguilar C, Brooks K, Flynn JK, Vinton CL, Rahmberg AR, Hickman HD, Brenchley JM (2022) Butyrate administration is not sufficient to improve immune reconstitution in antiretroviral-treated SIV-infected macaques. Sci Rep 12(1):1–14. https://doi.org/10.1038/s41598-022-11122-x

Piwpong (2018). https://doi.org/10.1016/B978-0-08-100596-5.22721-2. https://doi.org/10.1016/j.freeradbiomed.2012.10.525

Piqué N, Berlanga M, Miñana-Galbis D (2019) Health benefits of heat-killed (tyndallized) probiotics: an overview. Int J Mol Sci 20(10). https://doi.org/10.3390/IJMS20102534

Poblocka M, Bassey AL, Smith VM, Falcicchio M, Manso AS, Althubiti M, Sheng XB, Kyle A, Barber R, Frigerio M, Macip S (2021) Targeted clearance of senescent cells using an antibody-drug conjugate against a specific membrane marker. Sci Rep 11(1):1–10. https://doi.org/10.1038/s41598-021-99852-2

Porwal et al (2020). https://doi.org/10.1016/j.biopha.2020.110448

Pushpass et al (2021). https://doi.org/10.1017/S0029665121003165

Reider et al (2020). https://doi.org/10.3390/nu12051257

Remely M, Stefanska B, Lovrecic L, Magnet U, Haslberger AG (2015) Nutriepigenomics: the role of nutrition in epigenetic control of human diseases. Curr Opin Clin Nutr Metab Care 18(4):328–333. https://doi.org/10.1097/MCO.0000000000000180

Rodriguez et al (2020). https://doi.org/10.22175/mmb.11687

Sánchez et al (2017). https://doi.org/10.1002/mnfr.201600240

Sanders et al (2019). https://doi.org/10.1038/s41575-019-0173-3

Santos et al (2020). https://doi.org/10.1039/C9FO02478F

Satoh A, Stein L, Imai S (2011) The role of mammalian sirtuins in the regulation of metabolism, aging, and longevity. Handb Exp Pharmacol 206:125–162. https://doi.org/10.1007/978-3-642-21631-2_7

Shepherd et al (2018). https://doi.org/10.1016/j.tim.2021.01.003

Shimizu T (2002) Newly established regulation in Japan: foods with health claims. Asia Pac J Clin Nutr 11(2):S94–S96. https://doi.org/10.1046/j.1440-6047.2002.00007.x

Shimizu T (2003) Health claims on functional foods: the Japanese regulations and an international comparison. Nutr Res Rev 16(2):241–252. https://doi.org/10.1079/nrr200363

Shokryazdan et al (2022). https://doi.org/10.1007/s00430-016-0481-y

Sierra et al (2019). https://doi.org/10.1093/advances/nmy078

Singh TP, Natraj BH (2021) Next-generation probiotics: a promising approach towards designing personalized medicine. Crit Rev Microbiol 47(4):479–498. https://doi.org/10.1080/1040841X.2021.1902940

Slavin (2018). https://doi.org/10.1093/cdn/nzy005

Sniffen et al (2018). https://doi.org/10.1371/journal.pone.0209205

Spacova I, Dodiya HB, Happel AU, Strain C, Vandenheuvel D, Wang X, Reid G (2020) Future of probiotics and prebiotics and the implications for early career researchers. Front Microbiol 11:1400. https://doi.org/10.3389/FMICB.2020.01400/BIBTEX

Spadaro O, Youm Y, Shchukina I, Ryu S, Sidorov S, Ravussin A, Nguyen K, Aladyeva E, Predeus AN, Smith SR, Ravussin E, Galban C, Artyomov MN, Dixit VD (2022) Caloric restriction in humans reveals immunometabolic regulators of health span. Science 375(6581):671–677. https://doi.org/10.1126/SCIENCE.ABG7292

Stefanska B, Karlic H, Varga F, Fabianowska-Majewska K, Haslberger AG (2012) Epigenetic mechanisms in anti-cancer actions of bioactive food components—the implications in cancer prevention. Br J Pharmacol 167(2):279–297. https://doi.org/10.1111/j.1476-5381.2012.02002.x

Szajewska (2014). https://doi.org/10.1111/apa.14270

Tuncil et al (2020). https://doi.org/10.1128/mSphere.00180-20
Vandeputte (2020). https://doi.org/10.1093/nutrit/nuaa098
Veiga et al (2020). https://doi.org/10.1038/s41564-020-0721-1
Walker et al (2011). https://doi.org/10.1038/ismej.2012.4
Wang J, Hodes GE, Zhang H, Zhang S, Zhao W, Golden SA, Bi W, Menard C, Kana V, Leboeuf M, Xie M, Bregman D, Pfau ML, Flanigan ME, Esteban-Fernández A, Yemul S, Sharma A, Ho L, Dixon R, Pasinetti GM (2018) Epigenetic modulation of inflammation and synaptic plasticity promotes resilience against stress in mice. Nat Commun 9(1). https://doi.org/10.1038/s41467-017-02794-5
Wegh CAM, Geerlings SY, Knol J, Roeselers G, Belzer C (2019) Postbiotics and their potential applications in early life nutrition and beyond. Int J Mol Sci 20(19):4673. https://doi.org/10.3390/ijms20194673. PMID: 31547172; PMCID: PMC680192
Weindruch et al (2001). https://doi.org/10.1093/gerona/56.suppl_1.20
Wu et al (2017). https://doi.org/10.1016/j.anaerobe.2017.08.016
Zalar B, Haslberger A, Peterlin B (2018) The role of microbiota in depression—a brief review. Psychiatr Danubina 30(2):136–141. https://doi.org/10.24869/spsih.2018.136
Zec M (2022) Emerging role of microbiota in precision nutrition approaches. Compr Gut Microbiota 220–229. https://doi.org/10.1016/B978-0-12-819265-8.00046-2
Zeisel SH (1999) No Title. Science 285:185–186
Zhang W, Qu J, Liu GH, Belmonte JCI (2020) The ageing epigenome and its rejuvenation. Nat Rev Mol Cell Biol 21(3):137–150. https://doi.org/10.1038/s41580-019-0204-5
Zhang X, Ritonja JA, Zhou N, Chen BE, Li X (2022) Omega-3 Polyunsaturated Fatty Acids Intake and Blood Pressure: A Dose-Response Meta-Analysis of Randomized Controlled Trials. Journal of the American Heart Association 11(11) e025071 10.1161/JAHA.121.025071
Zmora et al (2018). https://doi.org/10.1016/j.cell.2018.08.041
Zmora (2020). https://doi.org/10.1038/s41577-019-0198-4

Precision Nutrition from a Practical Clinical View, Case Study

11

Ursula Jacob

Contents

Abstract

The translation of scientific progress in the multiple areas of molecular biology into clinical praxis is certainly challenging. Already the integration of standard clinical data with results from analytical concepts using genetic, epigenetic, microbiota and metabolic markers needs a profound understanding of the medical relevance of these new markers. Improved ambitious and adaptive bioinformatic platforms for integration and appropriate storage of data are essential. There is also an urgent need for improved and harmonised concepts to explain the benefit of the new analytical possibilities, their relevance and the consequences of the results to patients. Whereas these new requirements are a logistic and financial burden for doctors, the advent of personalised medicine including personalized diets addresses our expectation of how to serve and treat patients the best possible.

U. Jacob (✉)
Clinic for Prevention and Regeneration, Munich, Germany
e-mail: u.jacob@doc-jacob.com
URL: https://doc-jacob.de

Clinic for Prevention and Regeneration, Dubai, United Arab Emirates

A. G. Haslberger (ed.), *Advances in Precision Nutrition, Personalization and Healthy Aging*, https://doi.org/10.1007/978-3-031-10153-3_11

11.1 Introduction

Running a clinic with a focus on individual counseling, personalized diagnostics and preventive therapy often reminds on the old nice saying: *We are what we eat*. And although we learned so much about scientific aspects of genetics, epigenetics, microbiota and metabolism this still holds true in many cases.

Understanding some of our principal health problems in the area of nutrition we need to reconsider important changes of the last few centuries. Before the recent times of globalized food trading most people sticked to local foods. The time of an exponential population growth resulted in the need of an increased food production. In the following green revolution, food yields increased, but monocultures, enhanced needs of pesticides, artificially breed organism, elite lines of staple crops with modified ingredients developed. Often this resulted in a loss of food diversity, vitamins, minerals and local marketing and distribution of foods (Figs. 11.1 and 11.2).

By the improved possibilities to travel worldwide and to get all kind of exotic fruits, corns or vegetables our genetic and biochemical "fabrics" in our cells as well as our immune system came into contact with new nutritional or microbial, especially viral challenges. Over the years it came to epigenetically triggered changes of our inner cell metabolism and in the end to changes in our digestive processes, building up different structures of microbiota and metabolism.

If we want to understand precision nutrition, we have to bear this in our thoughts.

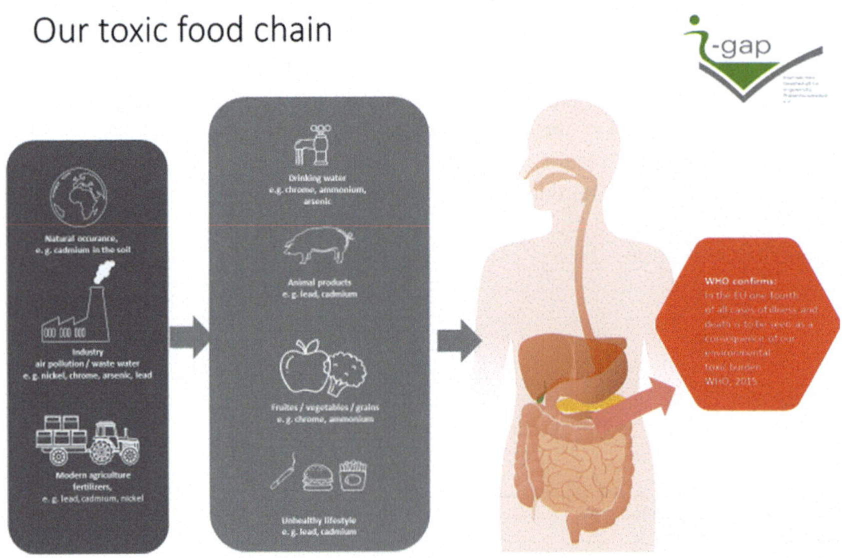

Fig. 11.1 Hazard from the food chain. International Society of Applied Preventive Medicine i-g-a-p (https://www.i-gap.org/index_en.php)

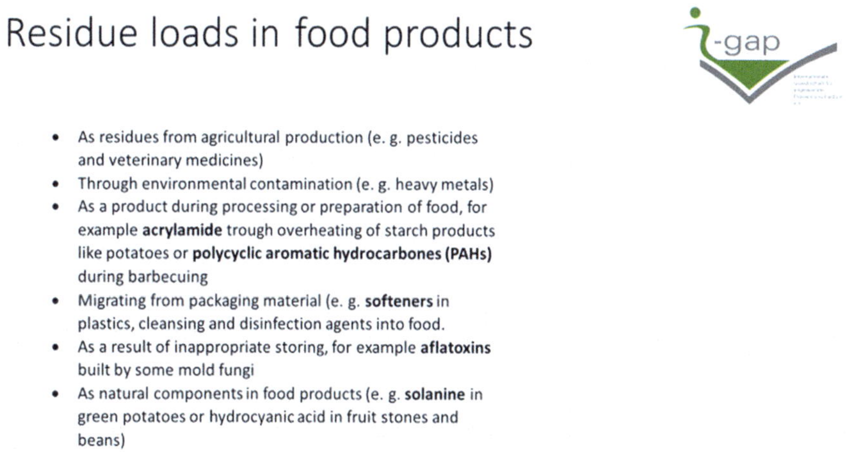

Fig. 11.2 Residual loads in foods. International Society of Applied Preventive Medicine i-g-a-p (https://www.i-gap.org/index_en.php)

This means: It is not enough to design diets for specific clients but control of food quality as well as compliance and adherence to advice needs a careful control.

We must consider that everyone has a different genetical disposition, different metabolic pathways as well as different bacterial mucosal flora in the digestive system and this not only in the big intestinal/bowel part. We have to analyze personal aspects from our mouth to the end of our bowels:

Alone, if we look into the mouth we have different parts, which are involved in the digestion process:

- Teeth, which often had or still have potentially toxic fillings
- The tongue, which has millions of nerves for taste or misguided sensitivity
- A big mucosal layer, which covers all the areas of the mouth, helping to harbor appropriate or problematic bacteria. Personal differences of mucosal sites and their consequences are right now an rapidly developing scientific area.
- Lots of glands to produce saliva with digestive properties.

If these co-workers get messed up through foreign materials like amalgam, gold, titan implants, etc. all of this perfect collaboration gets irritated and as a consequence of that, the first digestion steps of our food are already concerned. Also personal lifestyle and nutritional behaviors can change the milieu of our digestive tract, e.g., from alkaline to acidic which can trigger the problematic growth of bacteria which start processes of wrong food fermentation (Fig. 11.3).

When we come to the stomach, we again have bacteria there and different enzymes, which we need for further digestive processes.

If this part is getting "wrong information" from the mouth (like over-fermentation), then bacteria like helicobacter can become more active and change

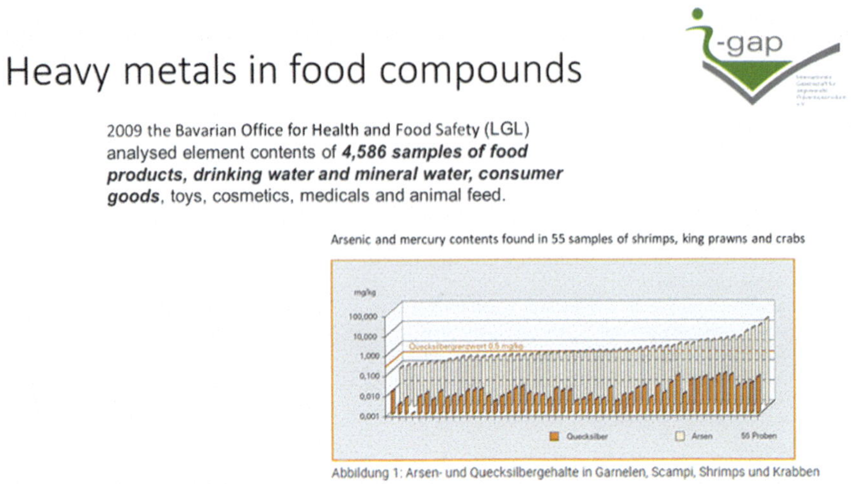

Fig. 11.3 Heavy meals in food components. International Society of Applied Preventive Medicine i-g-a-p (https://www.i-gap.org/index_en.php)

the alkaline milieu to acidic and it can end up in inflammation of the mucosal gastric layer (like gastritis or in the worst-case ulcers). This means that again digestive processes will change (Figs. 11.4 and 11.5).

The upper intestines are regulating not only digestive processes, but as well act as an important controller for building neurotransmitters, which our brain needs

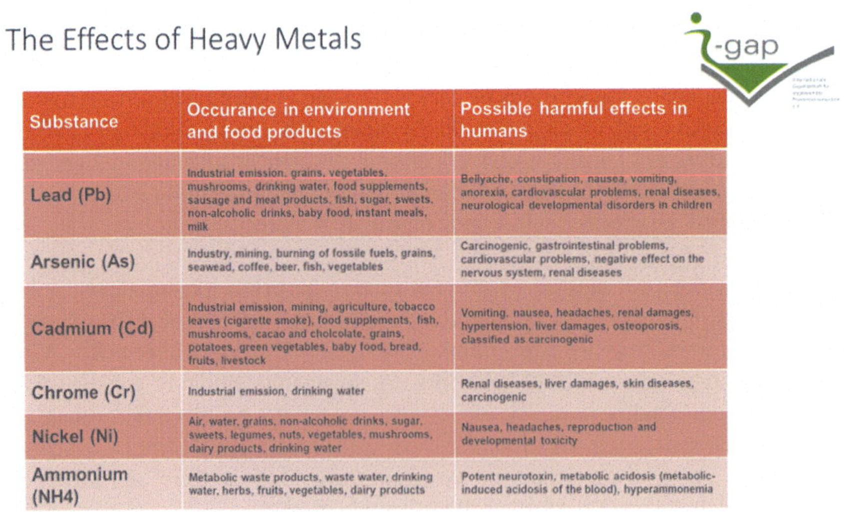

Substance	Occurance in environment and food products	Possible harmful effects in humans
Lead (Pb)	Industrial emission, grains, vegetables, mushrooms, drinking water, food supplements, sausage and meat products, fish, sugar, sweets, non-alcoholic drinks, baby food, instant meals, milk	Bellyache, constipation, nausea, vomiting, anorexia, cardiovascular problems, renal diseases, neurological developmental disorders in children
Arsenic (As)	Industry, mining, burning of fossile fuels, grains, seawead, coffee, beer, fish, vegetables	Carcinogenic, gastrointestinal problems, cardiovascular problems, negative effect on the nervous system, renal diseases
Cadmium (Cd)	Industrial emission, mining, agriculture, tobacco leaves (cigarette smoke), food supplements, fish, mushrooms, cacao and cholcolate, grains, potatoes, green vegetables, baby food, bread, fruits, livestock	Vomiting, nausea, headaches, renal damages, hypertension, liver damages, osteoporosis, classified as carcinogenic
Chrome (Cr)	Industrial emission, drinking water	Renal diseases, liver damages, skin diseases, carcinogenic
Nickel (Ni)	Air, water, grains, non-alcoholic drinks, sugar, sweets, legumes, nuts, vegetables, mushrooms, dairy products, drinking water	Nausea, headaches, reproduction and developmental toxicity
Ammonium (NH4)	Metabolic waste products, waste water, drinking water, herbs, fruits, vegetables, dairy products	Potent neurotoxin, metabolic acidosis (metabolic-induced acidosis of the blood), hyperammonemia

Fig. 11.4 Effects of heavy metals. International Society of Applied Preventive Medicine i-g-a-p (https://www.i-gap.org/index_en.php)

Arsenic levels in food products are significantly too high

aerzteblatt.de

-gap

Berlin – metal pollution through metalloid arsenic is too high in Europe.

Mainly rice, but also other food products, show too high arsenic levels, so the European Food Safety Authority (EFSA) and the European Society for Toxicology are warning .

Arsenic is a metalloid frequently found in our environment. It occurs naturally in minerals and ores, but it can also accumulate through the burning of fossil energy sources. There are geographical differences. Over the soil inorganic arsenic can also get into drinking water and food products.

According to the EFSA, young children in Europe take up each day an average of 0.61 to 2.09 micrograms of arsenic per kilogram bodyweight and are in the relevant range for an increased risk for cancer. This assessment has now been confirmed by the consultative commission of the European Society for Toxicology. Measures to reduce the arsenic burden are therefore "urgently necessary", says the expert association.

Based on epidemiologic studies we can calculate what daily intake of inorganic arsenic will lead to an increased risk for cancer. These calculations show an increased risk for humans by one percent for lung, skin and bladder cancer in humans in case of an arsenic intake of 0.3 – 8 micrograms per kilogram bodyweight. And moreover, an increased arsenic load over long periods will lead to chronic effects like skin lesions, development toxicity, neurotoxicity, cardiovascular diseases and diabetes.

Fig. 11.5 Arsenic levels in food products, Aerzteblatt

to function. This is called "gut-brain-axis". If this gets under stress because the food is not digested well enough (if we have gastritis, ulcers, etc.) the buildup of neurotransmitters is changed as well.

The bowels at the end have again different mucosal layers harboring microbiota which we need to finish the digestive process. But if the other parts of the GI tract are disturbed we can develop what is so called irritable bowls syndrome (IBS).

This can lead to "secondary food allergies or intolerances" and to a deficit in processing minerals, vitamins, etc.

All of the digestive processes, the bacterial flora, is influenced by what we eat, but as well at the same time through inner and outer factors.

This all we must bear in our mind if we want to understand, why nowadays precision nutrition is so important in prevention and intervention.

11.2 Chances of Diagnostics and Markers

With the new diagnostic concepts including genetics, epigenetics, biochemical markers and microbiota we can better understand what influences our personal digestive biochemical system. We need to know from the genetic, epigenetic and microbiota side which kind of metabolic type we are and how we can modulate our metabolism with personalized nutrition and lifestyle adequate exercise.

Our diagnostic system comprises following markers:

Lifestyle and nutrition Lifestyle and nutrition is not only analyzed by anamnesis but also by food frequency questionnaires and follow up recording.

Clinical data, including micronutrients.

Genetics, Selected panels of SNPS are tested according to clinical situation.

Epigenetics: Epigenetic markers including CpG methylation of selected genes, such as general aging and metabolic aspects as well as miRNAs regulating genes of relevance such as inflammation, immune responses, aging, metabolic, stress and fitness and even cosmetic aspects.

Telomere length as a marker of aging is analyzed as lifestyle factors, such as obesity, an unhealthy diet, smoking, alcohol, stress (also environmental stress) have negative effects on the telomere length. The telomere length can be modulated through dietary and lifestyle interventions including specific supplements or through a therapy of underlying diseases.

Inflammation and immune markers. TNF-alpha and IL-6, e.g., have a central regularity role in inflammation and immune response since it influences immune cells. A high occurrence of TNF-alpha in adipose tissue leads to a deterioration of insulin signaling and can contribute to the development of insulin resistance and weight gain (Fig. 11.6).

A	B
ZYTOKINE/ chron. Inflammation:	
	Norm:
CD86 auf APC (Monozyten)	20 - 100
CD80 auf APC (B Zellen)	50 - 125
B-ZELLEN (CD19)	120 - 630
CIC-IgA	< 4,6
CIC-IgG	< 12,8
CIC-IgM	< 3,6
CIC-C3	< 2,50
CRP high sensitiv	< 1 mg/l
alpha TNF (S)	< 9
Interleukin 6 (S)	< 4
Interleukin 8 (S)	< 62
Nagalase-Aktivität	< 200
TRION:	
positivity of enriched B-cells	<25%
gp350 cells:	0
PCR: >35 negativ, >28 positiv	
EBV	Normal
HHV	Normal
INTERLEUKINE:	
IL-1-r	Normal
IL-3-r	Normal
IL-6-r	Normal
IL-7-r	Normal
IL-9-r	Normal
IL-12-r	Normal

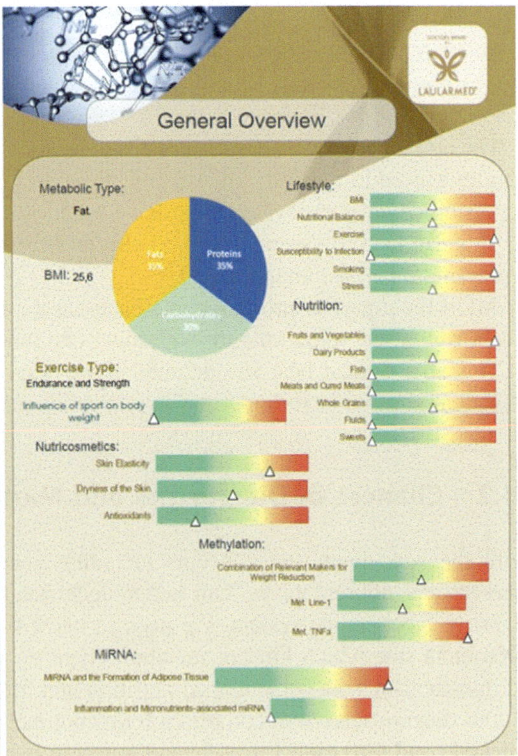

Fig. 11.6 Examples of clinical analysis

11.3 Case Study

The following case study wants to give an example how marker-based personalized medical- and nutritional advice can work.

Case 1: Male 51 years old

Existing risk factors:

Genetical risk factors: slow metabolism, slow detoxifying function.

Chronic heavy metal load (arsenic/mercury).

Chronic viral load (different herpes viruses).

Biochemical, microbial and epigenetic markers indicated significant metabolic, inflammatory and stress relevant burden as well as an active EBV virus disease.

Anamnesis:

Since 1960	CFS (chronic fatigue syndrome) IBS (irritable bowel syndrome in direction of diarrhea) Eczema, mononucleosis
1970	Chronic respiratory infections
1976	Massive food poisoning
Since 2015	Because of chronic infections the patient frequent antibiotics, therapy, worsening of IBS Worsening of chronic activity of different herpes viruses. Chronic inflammation
11/2018	Still IBS, despite sportive activities, and a good diet no improvement of weight and muscle buildup, he has increasing problems with muscle pain after exercising, no loss of abdominal fat, still massive fatigue

Gradual improvement until 2021. At this time an personalized therapy program was adapted according to indicated marker-based results:

- Start with different targeted antiviral therapies, anti-inflammatories
- Detox protocol with chelation therapies as well as herbal detox
- Tailormade supplements containing antioxidative, anti-inflammatory, immune supportive supplements especially white tea extract, resveratrol, omegas 3 fatty acid supplements,
- Lifestyle: The patient changed his diet following his genetic based metabolic type (balanced type). The patient followed his genetic (SNP-analyzed) exercise type.

The following specific improvements in clinical markers could be seen until 2021:

- Weight loss
- Formation of body, improvement of muscles, no pain any longer
- Improvement of IBS, CFS

- Less risk factors for chronic inflammation
- Improved EBV disease (Fig. 11.7).

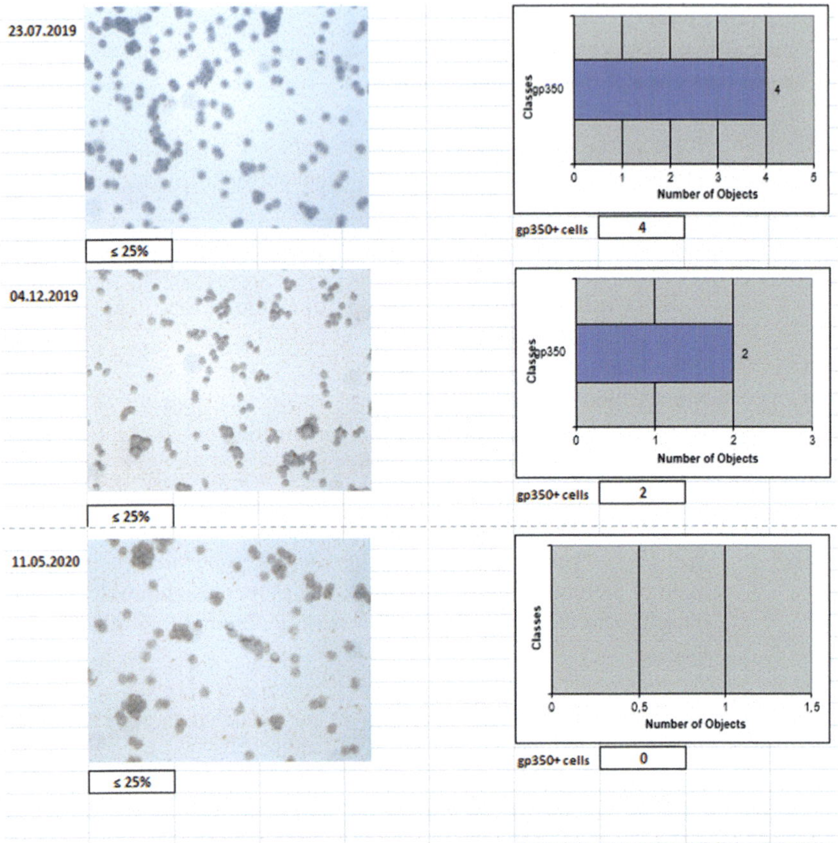

Fig. 11.7 Epstein Barr virus activity and occupancy: The specific B-cell testing for chronic occupation of B-cells by Epstein Barr virus and activity of Epstein Barr virus from 2016 to 2021. Reduction of chronic occupation of B-cells with Epstein Barr virus and change of activity of correlates with an improvement of epigenetic damages as well as improvement of symptoms and life-quality

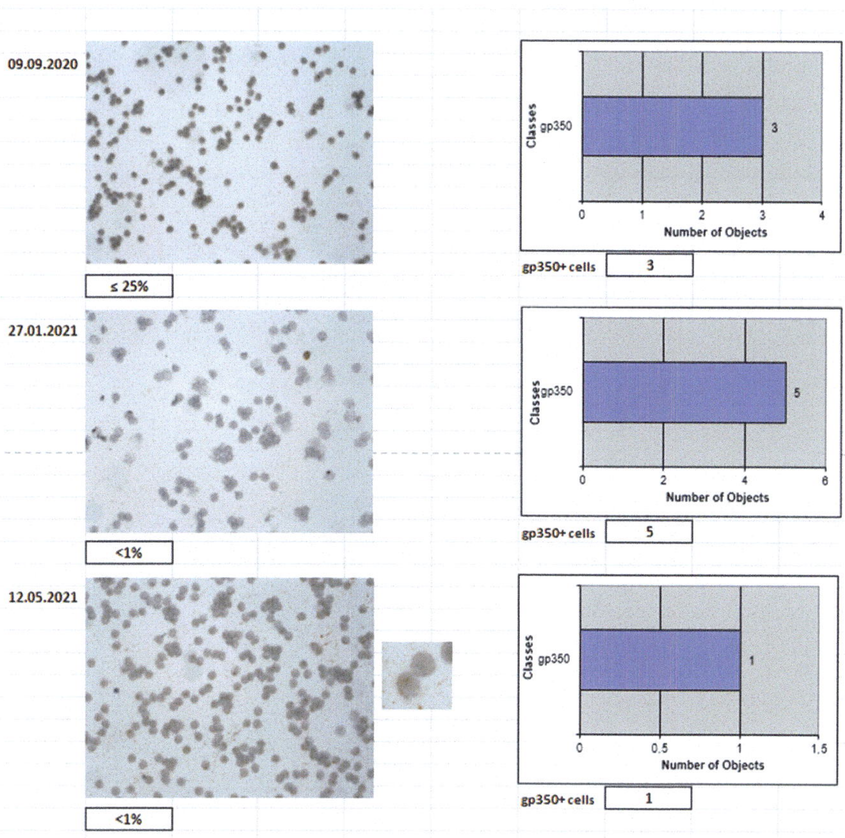

Fig. 11.7 (continued)

11.4 Conclusion

This case study suggests that the use of marker-based—precision medicine and precision nutrition can optimize health and quality of life.

Epigenetic markers early indicate ongoing pathogenic developments but also improvements by intervention.

Especially with aging, when our cells develop toward senescence and lifestyle is rather inflexible and hard to modulate it is difficult to influence nutrition and metabolism. Especially in this time of life it makes sense to use analytical marker-based personalized intervention and marker-based monitoring. We observed a widely improved compliance with this concept.

Translational Aspects in Precision Nutrition, Personalization, Biomarkers and Healthy Aging

12

Berit Hippe and Martin Schiller

Contents

Abstract

Scientific progress enables the analysis of biomarkers which can enable a personal disease prevention and improved healthcare. Whereas these opportunities can improve personal health management significantly, translation of these opportunities into practice faces social, economic and ethical aspects which need to be addressed. The article discusses some of these aspects. The health

B. Hippe (✉)
Department of Nutritional Sciences and HealthBioCare, University of Vienna, Vienna, Austria
e-mail: bh@healthbiocare.at

M. Schiller
Vienna, Austria
e-mail: m.schiller@medmedia.at

A. G. Haslberger (ed.), *Advances in Precision Nutrition, Personalization and Healthy Aging*, https://doi.org/10.1007/978-3-031-10153-3_12

system has changed dramatically over the past 100 years. Driven by the growing understanding of body functions and new technologies to measure them, a multitude of technological, clinical and biomolecular changes and related data resulted (Towse and Garrison in J Cancer Policy 11, 2017). Based on this enormous knowledge and the collected data, it is both technically feasible and economically sensible to strongly revise the "one fits all" strategy. Considering the ineffectiveness, side effects, and high cost of some standard treatments, it is now also unethical and uneconomical to offer non-personalized services (Doble and Lorgelly in Per Med 12, 2015). Not only the personalization of treatments is extremely important, but also the prevention of diseases is even more important. One of the most common preventive measures is a healthy diet that is adapted to the needs of each individual (Kirk et al. in Comput Biol Med 133, 2021). Many people deal with the topic of maintaining health in contrast to the previous strategy of disease treatment. Even if this topic was not sufficiently advertised during the corona pandemic, there has been an upturn in direct-to-consumer (B2C) and business to distributors (B2D) analysis tests. The demand for contactless test systems has risen sharply during this time. Since fitness studios and sports courses were closed, new ways of self-optimization had to be found. The possibilities range from epigenetic analyzes to determine the biological age, physical sport- or stress level, to the determination of genetic diet or sport types, to the determination of the hormone status from urine to complex microbiota analyzes out of feces samples. Also, the new hype of DIY (do it yourself) especially concerning nutrition, pushed this economic branch of individual nutrition and health advice.

12.1 Boom in Medical Self-tests—Since Corona Pandemic

Due to the national test strategy, PCR tests and quick tests for at home, have greater acceptance in the population. For that there are currently many new products or companies on the market that offer self-tests.

The business model of blood tests or other preventive tests, like feces or urine, could now be more promising than it was a few years ago. Because the public is now much more open to tests than before the corona pandemic. Michael Neumaier, German specialist in laboratory medicine, thinks that the term: "test, test, test" has also caught on. In the past, you only tested if you went to the doctor when you were sick. Now you do antigen tests or PCR tests routinely even if you are healthy. The willingness to test oneself has increased compared to before 2020 (Nützel 2021).

Since the outbreak of the corona pandemic, many people have experienced that even technically well-developed tests can deliver incorrect results. The specificity is one of the measures that indicate how high the probability is that a test will produce a correct result. That indicates the number of times a test will confirm the same result being tested for. A specificity of 99% sounds great to consumers, but it must be interpreted correctly. It means that one out of every hundred people has a false positive test. So, if thousand people are tested, 10 people will be false positive, if tested 10,000 healthy people, already 100 false cases will be analyzed. A false positive test means a person could be told to have any bad result or even a sickness. Also, a false good result can lead to the failure to intervene, and increase the risk.

Doctors also used the opportunity to outsource some of their analyzes and treatments. For example, patients did not have to sit in the doctor's office for long to be tested for fructose intolerance. Doctors used the option of a B2D methane breath gas analysis. Sending their patients the kit to measure any food intolerances (fructose, lactose, histamine, ...) home, the patients sends the applied kit to a standardized laboratory and the doctor gets the result from the laboratory.

Caused by the dynamic situation in the pandemic, it has become more difficult for companies to always guarantee all services. In many consulting and analysis companies, it was possible to switch to e-commerce.

12.2 From One Fits-All Recommendations to Personalized Tests

When the food pyramid was introduced in the 1970s, it was a compromise between easy-to-understand recommendations and economic feasibility. At that time, Luise Light was commissioned in the USA to revise the post-war guidelines. She should create a more contemporary model for feeding the people. Dr. Light and her team worked very scientifically, correlating individual nutrients with diseases, bioavailability, biochemical relationships, and more. Political contexts such as food availability or ecological resources were also considered. The new guideline should hardly have any negative effects on consumers and the food industry (Minger 2014). When the pyramid was published, there were many deviations from the submitted project with serious, long-term consequences for the health status. Even then, Dr Light valued the influence of diet on quality of life, for many healthy years of life, very highly.

Why have people different health outcomes despite eating similar diets and practicing similar lifestyles?

Current nutrient intake guidelines are based on population estimates of the nutrient intake required to prevent malnourishment according to sex, age, and other physiological states such as pregnancy or lactation (Ardini 2019). Nevertheless, the "one fits all" approach was unable to reduce the risk of diet-related diseases and provided very different results in nutrition studies.

When in Oct. 2004, the International Human Genome Sequencing Consortium publishes its scientific description of the finished human genome sequence the expectations for improved, personalized nutritional recommendations for disease reduction and prevention were very high (International Human Genome Sequencing Consortium 2021). The subsequent mapping of human DNA sequencing heralded in the "era of big science" (Mathers 2017).

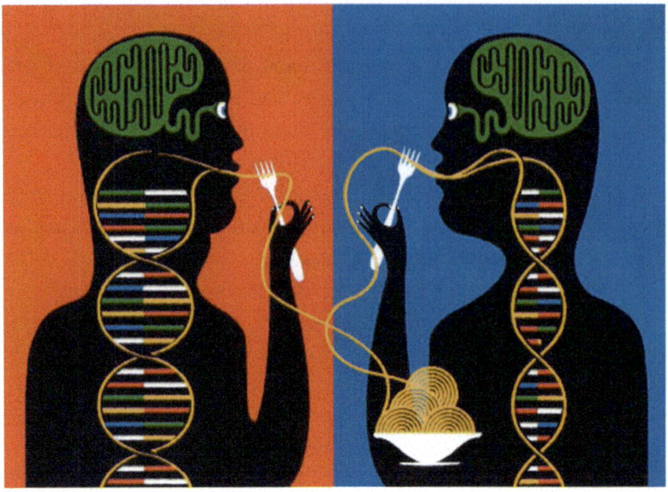

Fig. 12.1 The genetic makeup of two people is on average almost 99.7% the same. The small difference, however, can mean that food, such as a serving of pasta, is digested very differently. One person uses the same portion far better than another—and gains weight (Eberle 2021) © Karsten Petrat

The interaction between genetic and genomic information with environmental data such as diet could now be explored. Due to the research of nutrigenomics (how nutrition affects gene function) and nutrigenetics (how genetic variation affect nutritional response, food intake, and eating behavior) a better insight into the variability in biological response to nutrients was discovered (Fig. 12.1).

12.3 Nutrigenetics

The human genome contains approximately 3 billion nucleotides, forming 25,000 genes (Zahn 2021). Most of the genes that have been identified do not directly cause complex disease but enhance the susceptibility and predisposition. A wind range of biochemical and molecular pathways are involved in digestion pathways. Several research studies have demonstrated that single nucleotide polymorphisms can partly explain why foods can have different effects on different people or different risks for diet-related complex disease traits (Robino et al. 2019). The genetic makeup of two people is on average almost 99.7% the same. The differences in genes are called SNPs, Single Nucleotide Polymorphisms. These are variations at a single position in a DNA sequence among individuals. Recall that the DNA sequence is formed from a chain of four nucleotide bases: A, C, G, and T. If more than 1% of a population does not carry the same nucleotide at a specific position in the DNA sequence, then this variation can be classified as a SNP. If a SNP occurs within a gene, then the gene is described as having more than one allele. In these cases, SNPs may lead to variations in the amino acid sequence. SNPs, however,

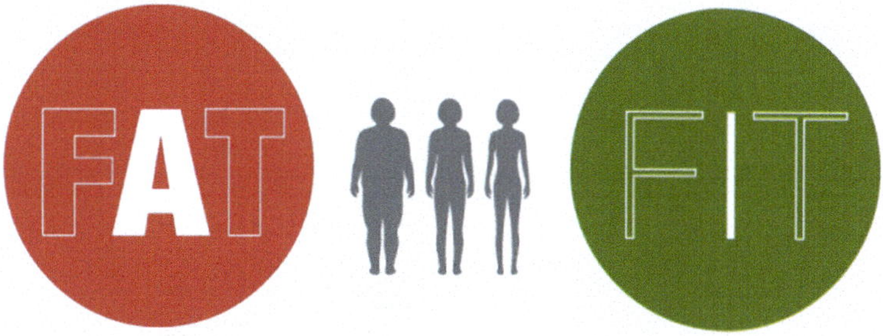

Fig. 12.2 A SNP, single nucleotide polymorphism, can change the interpretation of a gene by changing only one nucleotide (Konstantinidou 2020)

are not just associated with genes; they can also occur in noncoding regions of DNA (Karki et al. 2015). A mutation in the sequence of a functional gene can make that the body still can read, but the interpretation will change (Fig. 12.2).

12.4 Impact of Genetic Lifestyle Tests

Genetic testing is being increasingly used in a growing number of healthcare settings and in direct-to-consumer testing for a range of common complex disorders and nutrition advice. There is an expectation that communicating DNA-based disease risk estimates, will motivate changes in key health behaviors, including smoking, diet, or physical activity. A Cochrane meta-study was conducted in 2010 to examine whether knowledge of a genetic predisposition leads to behavioral changes. This study was expanded in 2016 by 7 publications. The results of this updated systematic review with meta-analysis using Cochrane methods suggest that communicating DNA-based disease risk estimates has little or no impact on risk-reducing health behavior. Existing evidence did not support expectations that such interventions could play a major role in motivating behavior change to improve population health (Hollands et al. 2016). Despite the clarity of the study situation, the authors of the Cochrane Review point to the high or unclear risk of bias in the analyzed studies and the low quality of the evidence obtained. More than 100 gene variants often play a role in regulating the metabolism. This shows the need to view genetic data as a complex interaction. Nor should consumers be left alone with the interpretation of such highly complex findings. Behavioral optimization can only succeed with long-term care from well-trained therapists (doctors, nutritionists).

A study, published in nature—European journal of human genetics that participants who stated they intended to modify their behavior after genetic testing results, effectively did so over time. This result held both for participants who received a positive or negative test result. In general, a healthier diet was the most

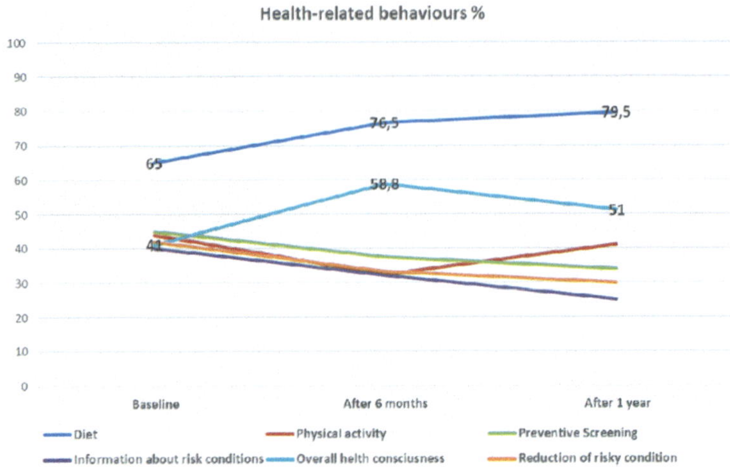

Fig. 12.3 Changes in health-related behaviors over time. Shown in the figure are the percentages of subjects who decided to change specific aspects of their lifestyle at T1 (6 months) and at T2 (1 year) (McBride et al. 2010)

frequently observed long-term behavioral change. As regards psychological variables, a risk-taking attitude and risk tolerance did not seem to affect the decision to change the lifestyle. They found an overall reduction in anxiety and worry over health over time, but also a reduction in the motivation for health promotion and prevention, health esteem, and positive expectations for their health in the future (McBride et al. 2010) (Figs. 12.3 and 12.4).

12.5 Compliance of Lifestyle Change

In principle, this does not mean all markers of genetic analyses are useless. Testing a genetic risk may not change the behavior, but in situations when a lifestyle adaption already has taken place, the compliance will increase with the knowledge of strengths and weaknesses. Especially the knowledge of the potential bioavailability of nutrients has an impact on the success of lifestyle change. Knowing the screws that must be tighten, is less elaborate than to fix the whole system. The more effective the change is, the longer it can be implemented. Goal setting is also important. What would you like to change? Which markers are important to know in order not to be overwhelmed by a huge amount of information? So, lifestyle analyzes can achieve a great deal with a specific question and the science-based personalized intervention.

Using the advances in DNA analysis and the resulting insights into the complexity of personalization, or of the affected subgroups and the possible positive effects on the development of effective therapeutics is one further step in personalized, marker-based interventions and treatments.

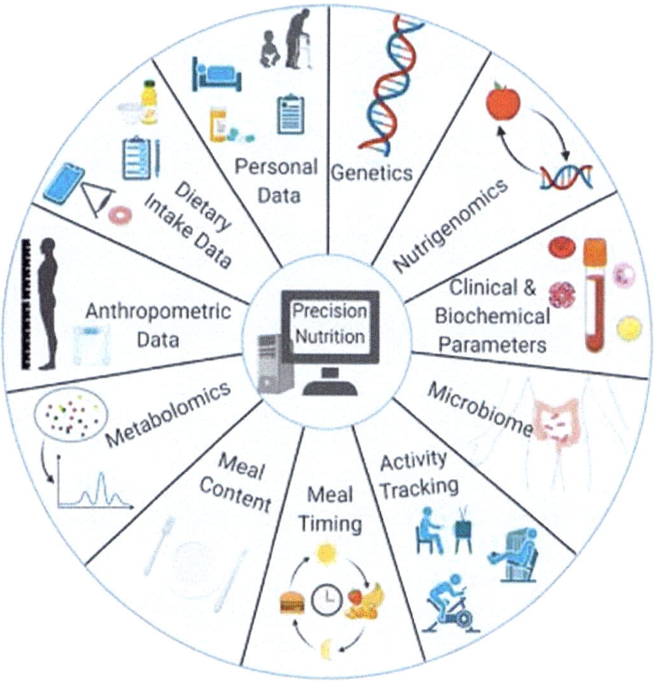

Fig. 12.4 Precision nutrition is a complex combination of science-based biomarkers

12.6 Nutrigeneomics

The interaction between genes and diets is fundamental in human evolution. Studying the interactions between nutrition and genetic provides a better insight into the variability in biological response to nutrients, but still, this explains around 40–70% variability (Friedman 2004). Nutrigenomics, also known as nutritional genomics, is broadly defined as the relationship between nutrients, diet, and gene expression (Chadwick 2004). Nutrigenomics is the foundation of "personalized nutrition" approaches tailored to individuals (Ordovas et al. 2018). Many diseases are diet-related or are significantly influenced by diet. So gluten can change in people with a gastrointestinal autoimmune disorder the expression patterns and gene networks, leading to celiac disease (Banaganapalli et al. 2020).

12.7 Personalized Epigenetic Testing

Now it is known that genetics can only answer a fraction of nutrition-related questions. Epigenetics is a stream of genetics, initiated by external or ecological aspects, which turn genes on and off and affect cellular ability to read genes without being affected by changes in genotype. Epigenetics results into changes in phenotype of an organism rather than genotype, wherein underlying DNA or RNA sequence remains unchanged. Epigenetic alterations are essential for development as they are dynamic and change with respect to environmental stimuli. However, these changes can be stable and could pass from one generation to another. Biology and genetic expressions of most organisms are affected by epigenetics, which makes it one of the most useful tools in personalized nutrition intervention analysis.

Over the last 20 years, however, an increase in research studies has demonstrated that epigenetics can ultimately affect nutrition and decrease the risk of developing a variety of health issues, including oxidative stress or inflammation (Tiffon 2018). The mechanisms by which environmental influences affect gene regulations are: Methylation and Demethylation, Histonmodifications and miRNA expression (Miller and Grant 2013).

Epigenetic Clocks

Since many companies focus on healthy aging, methylation analysis of aging relevant genes is the most common used type of personalized analysis which is done in routine. The change in DNA-methylation patterns are associated to a specific biological age (Unnikrishnan et al. 2019). Depending on the question of the analysis, different genes, and different numbers of CpGs are examined. Sometimes combined with further biomarkers as Telomere length, or miRNAs (Table 12.1).

In epigenetic analyses, single markers should represent entire metabolic systems. Since so many factors must be considered, it is often not easy to distinguish cause and effect: do the changes to epigenetic clocks contribute to aging or are they a consequence of aging? Or are they helpful ways in which our bodies compensate as a response to aging and disease, in which case reversing their pattern

Table 12.1 List of epigenetic clocks developed for humans (Unnikrishnan et al. 2019)

Epigenetic clock	Platform used	Tissues used in training	CpG sites identified	Accuracy in predicting age
Horvath Clock (2013)	27 k and 450 K arrays	Blood, brain, breast, buccal cells, colon, heart, liver, lung, placenta, saliva, CD4 cells, immortalized B cells, adipose, kidney, muscle, uterine tissue	353	$r^2 = 0.96$
Hannun Clock (2013)	27 k and 450 K arrays	Blood	71	$r^2 = 0.91$
Weidner Clock (2014)	27 k and 450 K arrays	Blood	3	$r^2 = 0.98$
PhenoAge Clock (2018)	27 k, 450 K and EPIC arrays	Blood	513	$r^2 = 0.92$

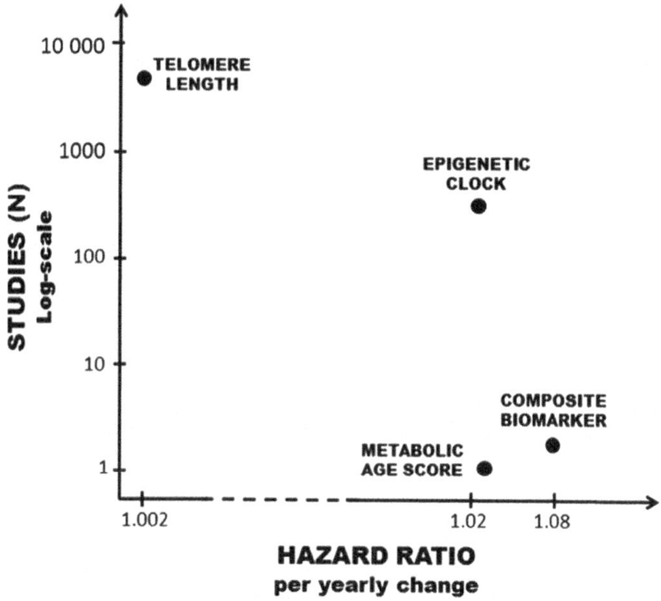

Fig. 12.5 Number of studies versus mortality hazards for the biological age predictors

would not be advantageous? Different epigenetic clocks are linked to different age-related conditions (Unnikrishnan et al. 2019). Even scientific accepted epigenetic clocks are not always associated with things like alcohol use, smoking, diabetes, and hypertension (Jylhävä et al. 2017) (Fig. 12.5).

Overview of the four biological age predictors telomere length, epigenetic clock, Metabolic Age Score, and composite biomarker which have all been used in survival models. The hazard ratio per yearly change in biological age (de-) acceleration for each predictor is presented on the x-axis. The y-axis presents an approximation of the number of studies on a log-scale where the predictor has been used (Jylhävä et al. 2017).

miRNA-Based Lifestyle Tests

The selection of biomarkers in nutrigenomics needs to reflect subtle changes in homoeostasis representing the relation between nutrition and health, or nutrition and disease. Noncoding RNAs, such as circulating microRNAs (miRNAs), represent a new class of integrative biomarkers to reflect complex metabolic processes (Rome 2015).

MicroRNAs (miRNAs) are small non-protein-coding RNA molecules that regulate gene expression. Diet and lifestyle factors have been published in several studies to be involved in the regulation of miRNA expression, and vice versa (Slattery et al. 2016). MiRNAs derived from capillary blood, whole blood and also from feces, provide additional support for the influence of nutritional factors, oxidative balance, stress response or sport intervention (Slattery et al. 2016; Francavilla et al. 2021; Ulrike et al. 2021).

Until now, the most relevant body fluids for miRNA quantification in response to nutrition are plasma or serum, but also capillary blood could be used to quantify the physiological impacts of diet or lifestyle. In addition, a number of recent studies also indicate that miRNAs could permit to monitor the impact of diet on gut microbiota (Rome 2015).

miRNA Tests and Stress Response

Through many studies on metabolism and stress, the connection between epigenetics and the most important disease processes are becoming clearer. The identification of the most important molecular biomarkers lead to new therapy strategies. MicroRNAs (miRNAs) play a central role in the regulation of cellular processes, including physiological and psychological stress response pathways (Olejniczak et al. 2018). miRNAs are small, noncoding RNAs that have important regulatory roles in gene expression. They are excellent biomarkers as they are present in all organs and are very stable. Several miRNAs are involved in the regulation of stress and stress responses. Changes of stress-associated miRNA levels in the blood reflect the changes in the brain contrary (Du et al. 2019). Analysis of the amount of different miRNAs involved can used to assess stress, but also for diagnosis, prevention and treatment of stress-related illnesses (Fig. 12.6).

Fig. 12.6 Example of miRNA lifestyle analysis test

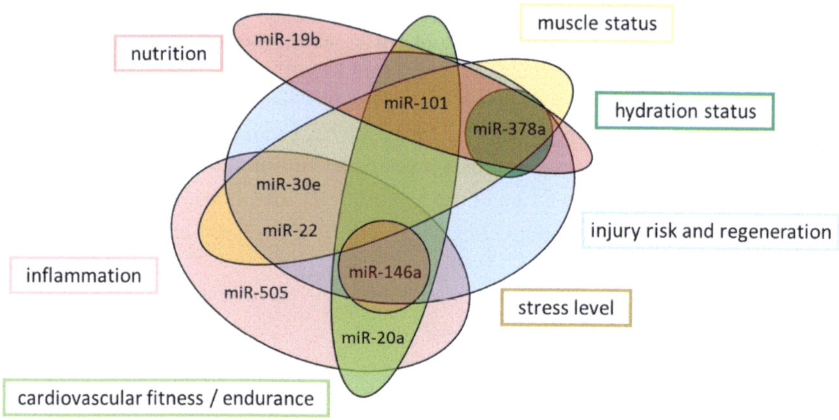

Fig. 12.7 The properties of the individual miRNAs and their importance and classification as sports-relevant biomarkers (Krammer et al. 2021)

Sport Relevant miRNA and Nutrition Intervention Response

In a 2-year study with 160 participants, HealthBioCare tested over 460 different metabolic factors and circulating miRNAs for their use as biomarkers to investigate systemic and cellular changes. miRNAs reflect the complex metabolic processes that take place in the body during a training cycle. The analysis of blood-borne microRNAs, additional with genetic factors, enables HealthBioCare to create recommendations on the intention and frequency of endurance and strength sports, nutrition and lifestyle factors (Ulrike et al. 2021). Many secondary plant ingredients, so called Phytoceuticals are known for its epigenetically effects. Several studies supporting the strategy for targeted intervention with Phytoceuticals to modify the expression patterns of relevant miRNAs (Haslberger et al. 2020; Gruendler et al. 2020) (Fig. 12.7).

12.8 Personalized Microbiota Analysis

With nearly $160 million in government funding in the USA, the NIH Common Fund's Nutrition for Precision Health research program, expected to launch 2022, seeks to enroll 1 million people to study the interactions among diet, the microbiome, genes, metabolism and other factors (Nutrition for Precision Health, powered by the All of Us Research Program 2022).

In the intestine, host-derived factors are genetically hardwired and difficult to modulate. However, the intestinal microbiome is more plastic and can be readily modulated by dietary factors. It is accepted that the microbiome can potentially impact physiology by participating in digestion, the absorption of nutrients, shaping of the mucosal immune response, energy homeostasis, and the synthesis or modulation of several potential bioactive and epigenetic active metabolites. Thus, diet-induced microbiota alterations may be harnessed to induce changes in host physiology, stress management, sleep quality, energy metabolism, blood lipid levels, or disease development and progression (Kogut 2022).

Gut microbiota analysis companies offer individualized diet regimens based on analyzing gut bacteria to answer to health problems ranging from irritable bowel syndrome to obesity. Venture capitalists invested $1 billion into these startups from 2015 to 2020, according to Crunchbase (Hall 2020). But not all companies have the scientific background which is needed to interpret these complex data.

12.9 Legal Responsibility

Who owns the data?

Scale development and validation are critical and most important for personalized tests. For this, consumers must sign that their data can be used for study purposes.

It is an ethical question who owns the results of the analyses. On the one hand, the consumer pays for his/her test and the analysis of his/her results, on the other hand, the analyzing companies use the results at best to validate their data. Worst of all, as you can see from the example of 23 and me to develop and patent new markers and sell them back to consumers. Some companies offer people to opt-in or opt-out of having their DNA used by law enforcement, further studies, or validation purposes. Recent studies of actual and potential users have demonstrated that individuals' responses to the use of these tests for these purposes are complex, with privacy, disruptive consequences, potential for misuse, and secondary use by law enforcement cited as potential concerns (Hazel et al. 2021). Users should consider where genetic testing data ends up and define personal boundaries what can be used with its own data. The companies are responsible for the sensitive handling of personal data.

12.10 Validation of Study Results and Generation of Limit Values

Validation of Personalized Tests

Scale development is not, however, an obvious or a straightforward endeavor. There are many steps to scale development, there is significant jargon within these techniques, the work can be costly and time consuming, and complex statistical analysis is often required (Boateng et al. 2018) (Fig. 12.8).

Legal Aspects

Preventive analysis kits do not need to be approved by the FDA or EMA. Some providers use alternative seals of approval. Since these are preventive tests, they are not subject to the regulations of the EU Medical Device Regulation (MRD) or the in Vitro Diagnostic Medical Devices Regulation (IVD). The responsibility lies with the provider. But there are also epigenetic IVD Tests for Personalized

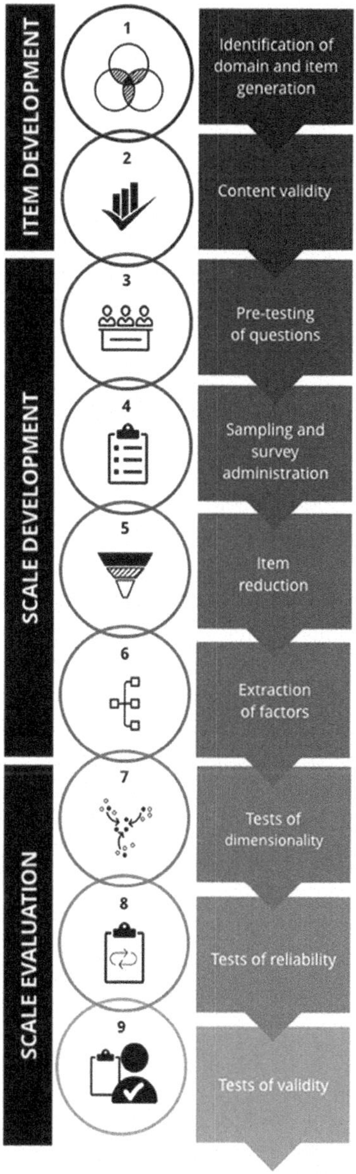

Fig. 12.8 An overview of the three phases and nine steps of scale development and validation (Boateng et al. 2018)

Precision Medicine in Cancer (Beltrán-García et al. 2019). A quality factor lies in the reproducibility of the tests since epigenetic tests can be repeated regularly to check the influence of the intervention.

Data Processing and Artificial Intelligence
The era of general dietary recommendations is over. It is becoming increasingly clear that blanket recommendations are not suitable for everyone. An individually appropriate diet is particularly good for health and takes into account the individual, genetically determined metabolic needs as well as the needs resulting from different environmental influences (epigenetic regulated metabolism). Since the publication of the study by Zeevi et al. (2015) at the latest, it has been clear that recommendations for a healthy diet can no longer be general recommendations (Zeevi 2015). "The modern view of food and medicine has led to a significant shift in nutritional research and practice toward precision nutrition" (Rodgers and Collins 2020).

Due the complex interplay among genetics, microbiome, antibiotic and probiotic use, metabolism, food environment, and physical activity, as well as economic, social, and other behavioral characteristics the data which could be assessed is enourm.

Data from any of these feature elements can be integrated into machine learning models and used to generate nutritional advice on a personalized basis (Kirk et al. 2021). Large data processing companies are now playing a major role in the healthcare system.

Bioinformatic tools are basic for the interpretation of changes in genes related and gene regulation to specific nutrients or dietary patterns. Grouping biomarkers based on functional similarity can help to enhance the biological interpretation (Figs. 12.9 and 12.10).

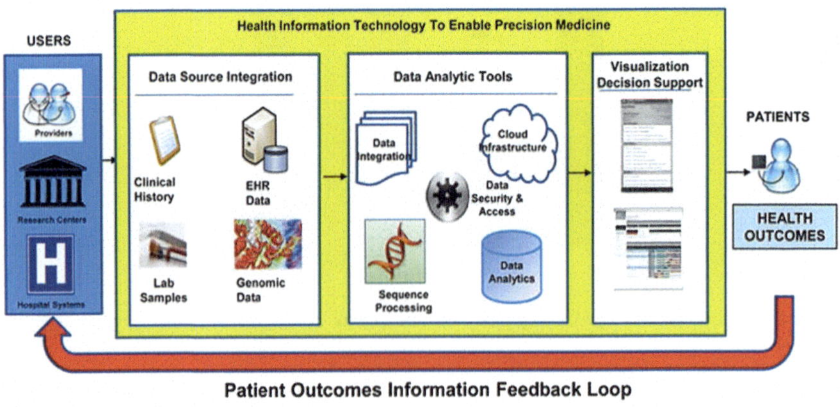

Fig. 12.9 Patient outcome information feedback loop

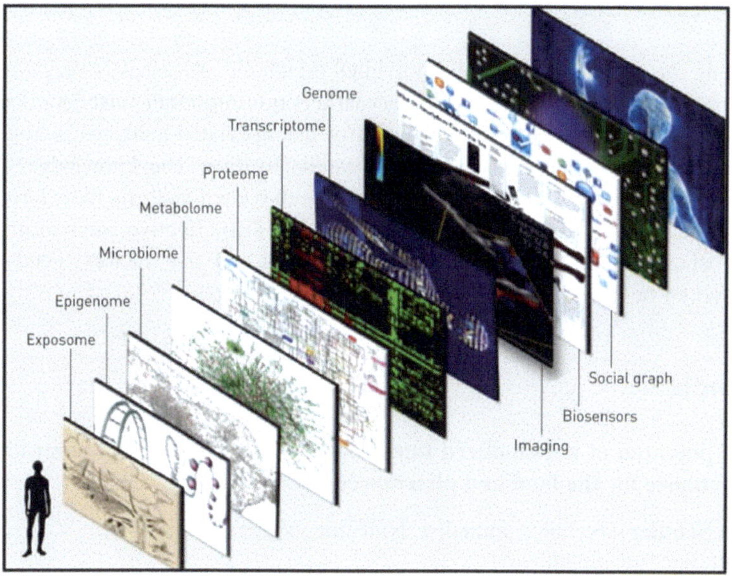

Fig. 12.10 Omic overviev—complexity of analyzed data

Do Personalized Tests Reflect the Lifestyle?

The tests do not reflect lifestyle but how well the body can handle the lifestyle. Consumers must be well-trained to correctly interpret the test results. The risk of negative test results, in particular, is difficult for laypeople to assess. But also positive results can have adverse effects on the consumer behavior. Psychology has shown that many people tend to interpret test results differently than they should. A positive screening test does not mean the life the person lived has been healthy. It does not validate the way of life. Before testing the consumer needs to consider what screening test is needed, can the recommendations be implemented and what will be the outcome (Nützel 2021).

What Are the Consequences of the Tests for the Consumers?

The consequences for the consumers are not always clear. What do you do with the knowledge of your biological age without the appropriate intervention?

A tailored health advice is still hard for consumers to implement in daily life without the support of trained doctors or nutritionists. It has been estimated that about 12% of American adults can understand the majority of health information, and only 66% of smokers shown a test result for their genetic susceptibility to lung cancer can accurately interpret the result (McBride et al. 2010).

The type of evaluation must be adapted to the level of education of the consumers, without leaving them alone in interpreting the risk assessment. The benefits of dietary changes versus supplements or trends like eating intuitively versus marker-based diets, need to be weighed.

12.11 Conclusion

Reflecting the entire body with all individual metabolic processes is very complex and expensive. Therefore, functional biomarkers are important to reflect complex systems. The responsibility of companies for health-related analyzes is huge. All analysis needs to be based on scientific, validated data. The knowledge gained through rigorous research into personalized, health related analysis tools have concrete applications in treatment and prevention. But also effective ways to improve health behaviors and interpretation of test results and risk factors, need to be supported by health experts, as doctors and nutritionist.

Appendix

Future potential of personalized nutrition from the consumer's point of view and as chance for the food and pharmaceutical industry

Martin Schiller Freelance Jounalist, Nutrition, Vienna
m.schiller@medmedia.at

Interest in personalization of nutrition has increased in the last years—not only among scientists but also among consumers, nutritional consulters and in the food and pharmaceutical industry.

This trend has 3 dimensions:

- Individual composition of food, e.g., a modularized system for creating people's own muesli (Buxel 2019) or individually designed wraps and burgers in restaurants.
- Digitally supported food shopping: Using applications/artificial intelligence enables consumers a choice specific to their needs and preferences. Only a few seconds are needed to personalize and order the product. To mention a few examples: individual variation of fruits in yoghurts, the desired shape of cookies creating individual labels of jams as a gift, etc. All those concepts address the growing interest in individualization, above all in the younger population. Additionally, preferences regarding packaging design, style of food preparation and volumes can be covered (Buxel 2019). Using artificial intelligence also means that, for example, people only get suggestions which consider people's health condition (allergies and intolerances, blood levels, nutrition deficiency, etc.)
- Personalized concepts für daily nutrition: A survey has shown that consumers have a high interest using digital products which enable to generate individual personalized nutritional recommendations (Markant Magazin 2020). Also, using counseling services that provide dietary suggestions based on individual's nutritional have shown an increased popularity. The potential of those concepts and the challenges in addressing consumers will now be discussed.

Personalized Concepts

A person who is interested in personalized nutritional concepts has the target to maximize the potential of maintaining health through daily nutrition and improve the and simplify everyday life. Looking at current consumers of personalized nutrition, the majority can be termed as, early adaptors'. This kind of consumer is strongly interested in wellness, ready to invest in personal health and actively seeking new personalized solutions to further optimize health. There are also some consumers who joined the personalization approach due to specific medical needs (metabolic disorders, weight problems, etc.). The majority of consumer uptake is in Europe, North America and Japan (Spitzer 2020).

Looking at the demographics, two main parameters can be identified: an aging society in the industrialized countries and the digitalized, health-focused consumer. While the number of elderly people is increasing in the first world countries, a so called, silver generation 'is staying in health for longer periods in the aging process. This is a completely new phase in life, compared to former generations. Those consumers have a higher quality of life and are increasingly interested in products offering support for healthy aging. Nevertheless there is also an increased incidence of lifestyle diseases and numbers go up for age-related chronic diseases. Many of those diseases are strongly associated with risk factors such as lack of physical activity, unhealthy diet, obesity and smoking. There is a growing movement toward prevention instead of the, wait an treat' scenario. Personalized nutrition solutions will support the preventative aspect of health and by using digital technologies the development will be further promoted. Self-Care aspects become more and more popular among consumers. A growing spectrum of health information and advice via digital devices is used and the freedom to diagnose and intervene in the field of problem solving and prevention are high on the agenda of these proactive and highly engaged consumers (Spitzer 2020).

What do people expect from personalized nutrition and which kind of personalization is expected? The highly engaged consumer expects to be able to access personal information to adapt the health behavior and to purchase decisions. Access to this information shall be available everytime and wherever they are. A personalized digital healthy life outside the healthcare system is also one of the future scenarios when looking at the demands of modern engaged consumers (Spitzer 2020).

Challenges

Consumer acceptance is crucial for the success of personalized nutrition services (Reinders et al. 2020). Although consumer attitudes toward genetic tests aiming to reveal the risks of a predisposition to various illnesses have been examined by several research studies worldwide, the aspect of acceptance is only examined by a low number of papers (Szakály et al. 2021).

In a study published in 2020 was explored whether ambivalent feelings and contextual factors could help to further explain consumers' usage intentions of those services. An online survey was conducted with a final sample of 797 participants. According to the results, weighing personalization benefits and privacy risks is positively related to the intention to use personalized nutrition advice. There is a more positive intention when more benefits than risks are perceived. Ambivalent feelings are related to a lower intention to use personalized advice. The more the eating context is perceived as a barrier to use personalized nutrition advice, the more ambivalent feelings are perceived. Predicting the consumers intentions means that there is need to address affective concerns and consider an individual's eating context (Reinders et al. 2020).

Several studies have confirmed consumer's fear of high costs of personalized diets. This could hinder the acceptance of new technologies. Moreover, social aspects and norms must be considered. The acceptance of personalized nutrition also depends of the preferences of family/family members, friends and opinion leaders (Ghosh 2014; Stewart-Knox et al. 2016; Ronteltap et al. 2009).

Consumer acceptance is also an important factor regarding nutrigenomics-based level of personalization. A study from Hungary had the aim to create more information on the consumers' point of view, using a survey involving 1000 persons. Results showed that 23.5% of respondents accept genetic test-based personalized nutrition. A gender gap was also found out: The technology was rejected more often by men compared to women. Persons over 70 years rejected genetically-based personalized nutrition significantly more than expected (49.1%). In the age group 40–49 the rejection was in a smaller proportion than expected (21.9%). Results also indicate that it is perceived cost/benefit that is most related to genetically-based personalized dietary preferences, followed by perceived risk and subjective norms. Perceived uncertainty and perceived behavioral control, however, have only a weak relationship with genetic-based personalized dietary preferences. The results also showed that individual psychological processes have a greater influence on the development of preferences than any socio-demographic factor. For product development this means that psychological characteristics should be given more emphasis among the segmentation criteria (Szakály et al. 2021).

Communication Tasks

First surveys show that often genetic analyses are associated with personalized nutrition. It is not in the thought of many consumers that creating individual profiles can be based on far more tests, which is a challenge for all stakeholders working in nutritional communication, nutritional counseling, and nutritional education. A problem for the more interested consumer is that with increasing individualization less studies exist, because no big trials examine the effects of individual nutrition for every test person. This is also a matter of communication: Interested consumers will not find results of big trials for evidence of a certain concept, they must be persuaded by the quality of testing and of product development.

Another important aspect to bring personalized nutrition to the proactive consumer is the strong integration of those concepts in personal nutritional counseling and in health promotion projects. This offers support of the client beyond the counseling unit and is also a good opportunity to clear out myths and misunderstandings of personalized nutrition.

Individualization also has high importance regarding the low health literacy in the general population even after decades of nutritional journalism. It seems that general recommendations have not fulfilled the hopes of scientists and institutions regarding a healthy lifestyle and many people doubt that those recommendations address their life-situation and their needs. Self-tests can help to find out which kind of nutrition is individually best suited to fulfill health targets, achieve a balanced gut microbiota, lose weight, build muscle mass, etc. without strictly looking on general recommendations. Those tests already exist, but for consumers it is not always clear which tests deliver valid results and are based on serious scientific work. A simple search on Google offers a broad variety of self-testing options, leaving consumers in irritation, which tests they can trust. So it is a big challenge for scientists and industry to address people directly and explain the methods of tests and why it is possible to create personalized concepts based on the results. Otherwise, producers of unscientific tests will take over the field. The need and demand among consumers is here, now it is time to address them more intensively.

Chances for Industry

Industry is asked to address the growing need and demand for self-care. In the upcoming years a growing market for products and services regarding personalized nutrition is expected. This refers to personalized food as well as to personal recommendations based on various scientific tests.

In the area of food production the focus is on personalization of taste. A survey showed that 40% of the respondents already bought a product for which individual taste preferences were considered in the manufacturing process. Trend products could be mainly yoghurts, read-to-eat foods, meat and meat products, pizza and cheese. An important reason to buy is also the personalization of ingredients (favored by 28% of the respondents) and the personalization of the product volume. Besides the individual manufacturing for the consumer modularized systems for salads, bread, nuts, muesli, sweets and confectionary, fruits vegetables, yoghurt, pasta products, cheese and tea could be considered. The market development will also bear more products based on nutrigenomic and metabolic testing.

Personalized nutrition will establish as business model in the next years. Therefore investment in artificial intelligence is needed, because large analysis capacities are a main criterion für competing in a growing market und enable to be part of the value-chain. At the same time these developments will need transparency as consumers want to know which technologies are used and what will happen to their data. For this reason, questions of ethical and moral will be companions of the trend of personalized nutrition.

References

Ardini W (2019) Personalized nutrition: how to make it possible? World Nutr J 3

Banaganapalli B et al (2020) Exploring celiac disease candidate pathways by global gene expression profiling and gene network cluster analysis. Sci Rep 10:16290

Beltrán-García J, Osca-Verdegal R, Mena-Mollá S, García-Giménez JL (2019) Epigenetic IVD tests for personalized precision medicine in cancer. Front Genet 10

Boateng GO, Neilands TB, Frongillo EA, Melgar-Quiñonez HR, Young SL (2018) Best practices for developing and validating scales for health, social, and behavioral research: a primer. Front Public Heal 6

Buxel H (2019) My Food—Personalisierung und Ernährung. Auf: https://www.dlg.org/de/lebensmittel/themen/publikationen/studien/dlg-my-food/personalisierung-und-ernaehrung-personalisierte-lebensmittel [Auszüge aus der Studie „MyFood" der Deutschen Landwirtschaftsgesellschaft (DLG)]

Chadwick R (2004) Nutrigenomics, individualism and public health. Proc Nutr Soc 63:161–166

Doble B, Lorgelly P (2015) Clinical players and healthcare payers: aligning perspectives on the cost–effectiveness of next-generation sequencing in oncology. Per Med 12

Du J et al (2019) The critical role of microRNAs in stress response: therapeutic prospect and limitation. Pharmacol Res 142:294–302

Eberle U (2021) Nutrigenetik: Unser Erbgut, unser Essen. https://webcache.googleusercontent.com/search?q=cache:LOj4ClpwGecJ:www.geo.de/magazine/geo-wissen-ernaehrung/47-rtkl-gesunde-ernaehrung-nutrigenetik-unser-erbgut-unser-essen+&cd=1&hl=de&ct=clnk&gl=at&client=firefox-b-d

Francavilla A et al (2021) Faecal miRNA profiles associated with age, sex, BMI, and lifestyle habits in healthy individuals. Sci Rep 11:20645

Friedman JM (2004) Modern science versus the stigma of obesity. Nat Med 10:563–569

Ghosh D (2014) The drivers and consumer attitudes in the personalisation of health and nutrition. Agro Food Ind Hi Tech 25:48–49

Gruendler R, Hippe B, Sendula Jengic V, Peterlin B, Haslberger AG (2020) Nutraceutical approaches of autophagy and neuroinflammation in Alzheimer's disease: a systematic review. Molecules 25:6018

Hall C (2020) Startups focused on gut microbiome for health clues raised $1B in VC funding. https://news.crunchbase.com/news/startups-focused-on-gut-microbiome-for-health-clues-raised-1b-in-vc-funding/

Haslberger A, Jacob U, Hippe B, Karlic H (2020) Mechanisms of selected functional foods against viral infections with a view on COVID-19: mini review. Funct Foods Heal Dis 10:195

Hazel JW et al (2021) Direct-to-consumer genetic testing: prospective users' attitudes toward information about ancestry and biological relationships. PLoS ONE 16:e0260340

Hollands GJ et al (2016) The impact of communicating genetic risks of disease on risk-reducing health behaviour: systematic review with meta-analysis. BMJ i1102. https://doi.org/10.1136/bmj.i1102

International Human Genome Sequencing Consortium (2021) Human genome project timeline of events. https://www.genome.gov/human-genome-project/Timeline-of-Events

Jylhävä J, Pedersen NL, Hägg S (2017) Biological age predictors. Ebiomedicine 21:29–36

Karki R, Pandya D, Elston RC, Ferlini C (2015) Defining "mutation" and "polymorphism" in the era of personal genomics. BMC Med Genomics 8

Kirk D, Catal C, Tekinerdogan B (2021) Precision nutrition: a systematic literature review. Comput Biol Med 133

Kogut MH (2022) Role of diet-microbiota interactions in precision nutrition of the chicken: facts, gaps, and new concepts. Poult Sci 101:101673

Konstantinidou V (2020) Nutrigenetic testing and weight losss. https://www.dnanutricoach.com/why-nutrigenetic-testing-is-not-about-weight-loss/

Markant Magazin (2020) Individuell statt Standard. Auf: https://www.markant-magazin.com/markant/individuell-statt-standard

Mathers JC (2017) Nutrigenomics in the modern era. Proc Nutr Soc 76:265–275

McBride CM, Koehly LM, Sanderson SC, Kaphingst KA (2010) The behavioral response to personalized genetic information: will genetic risk profiles motivate individuals and families to choose more healthful behaviors? Annu Rev Public Health 31:89–103

Miller JL, Grant PA (2013) The role of DNA methylation and histone modifications in transcriptional regulation in humans. 289–317. https://doi.org/10.1007/978-94-007-4525-4_13
Minger D (2014) Death by food pyramid: how shoddy science, sketchy politics and shady special interests have ruined our health, vol 1. In: Primal nutrition
Nutrition for precision health, powered by the All of Us research program (2022)
Nützel N (2021) Boom der medizinischen Selbsttests. Deutschlandfunk
Olejniczak M, Kotowska-Zimmer A, Krzyzosiak W (2018) Stress-induced changes in miRNA biogenesis and functioning. Cell Mol Life Sci 75:177–191
Ordovas JM, Ferguson LR, Tai ES, Mathers JC (2018) Personalised nutrition and health. BMJ bmj.k2173. https://doi.org/10.1136/bmj.k2173
Reinders MJ et al (2020) Consumer acceptance of personalised nutrition: the role of ambivalent feelings and eating context. PLoS ONE 15(4):e0231342. https://doi.org/10.1371/journal.pone.0231342
Robino, Concas, Catamo, Gasparini (2019) A brief review of genetic approaches to the study of food preferences: current knowledge and future directions. Nutrients 11
Rodgers GP, Collins FS (2020) Precision nutrition—the answer to "What to Eat to Stay Healthy". JAMA 324
Rome S (2015) Use of miRNAs in biofluids as biomarkers in dietary and lifestyle intervention studies. Genes Nutr 10:33
Ronteltap A, van Trijp JCM, Renes RJ (2009) Consumer acceptance of nutrigenomics-based personalised nutrition. Brit J Nutr. https://doi.org/10.1017/S0007114508992552
Slattery M, Herrick J, Mullany L, Stevens J, Wolff R (2016) Diet and lifestyle factors associated with miRNA expression in colorectal tissue. Pharmgenomics Pers Med 10:1–16
Spitzer V (2020) Who are today's personalised nutrition consumers? Auf: https://www.vitafoodsinsights.com/personalised-nutrition/who-are-todays-personalised-nutrition-consumers
Stewart-Knox BJ, Markovina J, Rankin A et al (2016) Making personalised nutrition the easy choice: creating policies to break down the barriers and reap the benefits. Food Policy. https://doi.org/10.1016/j.foodpol.2016.08.001
Szakály Z, Kovács B, Szakály M et al (2021) Consumer acceptance of genetic-based personalized nutrition in Hungary. Genes Nutr 16(1):3. https://doi.org/10.1186/s12263-021-00683-7
Tiffon C (2018) The impact of nutrition and environmental epigenetics on human health and disease. Int J Mol Sci 19:3425
Towse A, Garrison L (2017) Value assessment in precision cancer medicine. J Cancer Policy 11
Ulrike DB, Krammer UD, Tschida S, Berner J, Lilja S, Switzeny OJ, Hippe B, Rust P, Haslberger AG (2021) miRNA-based "Fitness Score" to assess the individual response to diet, metabolism and exercise. Preprint
Unnikrishnan A et al (2019) The role of DNA methylation in epigenetics of aging. Pharmacol Ther 195:172–185
Zahn LM (2021) The human genome. Science 80:373
Zeevi D et al (2015) Personalized nutrition by prediction of glycemic responses. Cell 163

The manufacturer's authorised representative in the EU is Springer Nature Customer Service Centre GmbH, Europaplatz 3, 69115 Heidelberg, Germany. If you have any concerns regarding our products, please contact ProductSafety@springernature.com

Printed and bound by CPI Group (UK) Ltd, Croydon, CR0 4YY
30/09/2026
02225304-0011